Technology & Civilization

Dr. Vera Pavri

York University

Kendall Hunt
publishing company

Cover image © Shutterstock, Inc.

Kendall Hunt
publishing company

www.kendallhunt.com
Send all inquiries to:
4050 Westmark Drive
Dubuque, IA 52004-1840

Copyright © 2010 by Kendall Hunt Publishing Company

ISBN 978-0-7575-8272-1

Printed in Canada
10 9 8 7 6 5 4

Table of Contents

Introduction

Have you ever fallen asleep reading an article given to you in class? Have you ever found yourself reading the same line over and over again and still do not understand what you have read? This is a common problem most students have when tackling articles and book chapters. One of the reasons for this is because students try to read materials assigned to them much in the same way they try to read books or magazines they choose for themselves for pleasure. THIS DOES NOT WORK. In order to keep your mind focused and active on an article that you would not necessarily choose to read yourself, you must be aware of why you have been assigned that article and what approach you are going to take when you finally decide to try and read it.

When reading any scholarly article, the important thing to remember is PURPOSE. Why are you reading the article in the first place? Is it to gain new knowledge or supplement ideas that have already been introduced to you in lecture? In this class, unless stated otherwise, all readings are designed to supplement knowledge—that is, to further ingrain information that has already been presented in class. As such, a good approach to reading in this class is to tackle articles following a lecture. After all, how will you know what to look for in a reading if you don't know what are the most important concepts that have been covered in lecture?

If you attend the lecture and are comfortable with the material that has been introduced to you, you are probably going to be reading to supplement knowledge. Here, you are actively looking for material that is familiar to you and which can be used to solidify ideas and arguments in your mind. This way, you are doing more than just memorizing facts, which will soon be forgotten—you are actually learning (and hopefully retaining) important information. This will allow you to practice the art of skimming texts, which takes both time and practice to perfect.

In contrast, if you missed the lecture or really didn't understand what was being said, then you will be reading to gain new knowledge. This may take you longer to get through an article. This is because you are being introduced to new concepts and ideas that are unfamiliar to you, and this simply requires more time and effort.

During this year, I will be providing you the opportunity to learn ways to better your critical reading skills. Our first tutorial seminar will provide you with fundamental information of what to look for when reading any academic work, such as key words or important sentences. Following this tutorial, other examples will also be provided in lecture about critical reading skills, so that hopefully by the end of term, you won't be falling asleep as frequently as you have in the past!

As for the readings themselves, here are a few things to consider. Remember that all of our discussions in this class will focus on one or more of the course theories and themes that were presented to you in the first lecture. Articles such as Sergio Sismondo's "Two Questions Concerning Technology," Sungook Hong's "Historiographical Layers in the Relationship between Science and Technology," and Melvin Kranzberg's "Overview: Technology and History: Kranzberg's Laws" are especially important when trying to understand theories and themes in our course such as technological determinism, necessity is the mother of invention, technology as applied science, and the relationship between science and technology. Similarly, Nelly Oudshoorn and Trevor Pinch's "How Users and Non-users Matter," Claude Fischer's "'Touch Someone': The Telephone Industry Discovers Sociability," Paul Ceruzzi's "The Personal Computer, 1972–1977" and "Shaping the Early Development of Television" by Jan van den Ende et al are excellent accounts of looking at how users shape new technology.

Lynne Osman Elkin's article "Rosalind Franklin and the Double Helix" provides us with a vivid illustration of gender and science issues, while Carolyn Merchant's piece "The Scientific Revolution and *The Death of Nature*" takes this one step further by examining the relationship between science, technology, gender and religion. Articles such as Nataraja Sarma's "Diffusion of Astronomy in the Ancient

World," James Frederick Edwards' "Building the Great Pyramid: Probable Construction Methods Employed in Giza," Justin Yifu Lin's "The Needham Puzzle: Why the Industrial Revolution did Not Originate in China," Pamela Long's "Technology in the Medieval West" and Nathan Rosenberg's "Why in America" offer an interesting study of science and technology from both a historical and a cross-culture perspective. Michael Adas' article "Contested Hegemony: The Great War and the Afro-Asian Assault on the Civilizing Mission," and Robert Proctor's article "Nazi Science and Nazi Medical Ethics: Some Myths and Misconceptions" are excellent pieces that tackle the issue of science, technology, and race, while the role of politics in science and technology decisions is made clear in articles such as Barton Bernstein's "The Atomic Bombings Reconsidered" and Michael Gordin et al.'s "Ideologically Correct Science." Finally, articles such as David Hounshell's "Mass Production," Thomas Hughes' "The System Must Be First," Lawrence Lessig's "Code is Law," the US Department of Energy's "Genomics and its Impact on Society," and Peter Bowler and Iwan Rhys Morus' "Ecology and Environmentalism" all discuss the theme of management and control—whether it be machines, natural resources, or the human body.

Good luck and happy reading!

Dr. Vera Pavri

Two Questions Concerning Technology

Sergio Sismondo

Is Technology Applied Science?

The idea that technology is applied science is now centuries old. In the early seventeenth century, Francis Bacon and René Descartes both argued for the value of scientific research by claiming that it would produce useful technology. In the twentieth century this view was championed most importantly by Vannevar Bush, one of the architects of the science policy pursued by the United States after the Second World War: "Basic research . . . creates the fund from which the practical applications of knowledge must be drawn. New products and new processes do not appear full-grown. They are founded on new principles and new conceptions, which in turn are painstakingly developed by research in the purest realms of science. . . . Today, it is truer than ever that basic research is the pacemaker of technological progress."

The view that technology is applied science has been challenged from many directions. In particular, accounts of artifacts and technologies show that scientific knowledge plays relatively little direct role in the development even of many state-of-the-art technologies. Historians and other theorists of technology have argued that there are technological knowledge traditions that are independent of scientific knowledge traditions, and that to understand the artifacts one needs to understand those knowledge traditions. At the same time, however, some people working in science and technology studies (S&TS) have argued that science and technology are not sufficiently well defined and distinct for there to be any determinate relationship between them.

Because of its large investment in basic research, in the mid-1960s the US Department of Defense conducted audits to discover how valuable that research was. Project Hindsight was a study of the key events leading to the development of 20 weapons systems. It classified 91 percent of the key events as technological,

8.7 percent as applied science, and 0.3 percent as basic science. Project Hindsight thus suggested that the direct influence of science on technology was very small, even within an institution that invested heavily in science, and which was at some key forefronts of technological development. A subsequent study, TRACES, challenged that picture by looking at prominent civilian technologies and following their origins further back in the historical record.

Among historians of technology it is widely accepted that "science owes more to the steam engine than the steam engine owes to science." For example, work on the history of aircraft suggests that aeronautical engineering is relatively divorced from science: engineers consult scientific results when they see a need to, but there is no sense in which their work is driven by science or in which it is the application of science (Vincenti 1990). Engineers develop their own mathematics, their own experimental results, and their own techniques. Or, similarly, the innovative electrical engineer Charles Steinmetz did not either apply physical theory or derive his own theoretical claims from it (Kline 1992), instead developing theoretical knowledge in purely engineering contexts. This is despite the fact that he considered his work to be applied science; it was applied science in that it fit Steinmetz's understanding of the scientific method.

Edwin Layton (1971, 1974) argues that the reason why the model of technology as applied science is so pervasive, and yet individual technologies cannot be seen to depend on basic science, is that technological knowledge is downplayed. Engineers and inventors participate in knowledge traditions, which shape the work that they do. Science, then, does not have a monopoly on technical knowledge. In the nineteenth century, for example, American engineers developed their own theoretical works on the strength of materials, drawing on but modifying earlier scientific research. When engineers needed results that bore on their practical problems, they looked to engineering research, not pure science.

The idea that there are important traditions of technological knowledge has resonated with and influenced the thinking of other historians and philosophers of technology. Rachel Laudan's problem-solving model of technology (1984) assumes that the development of technologies is a research process, driven by interesting problems. The sources of problems Laucan identifies are: actual and potential functional failure of current technologies, extrapolation from past technological successes, imbalances between related technologies, and, much more rarely, external needs demanding a technical solution. It is notable that all but the last of these problem sources stem from within technological knowledge traditions. Edward Constant's (1964) Kuhnian model of normal and revolutionary technology adopts a very similar picture of technology as research, and a similar picture of engineering and other technological knowledge forming its own traditions.

As we will see in later chapters, for a group of people to have its own tradition of knowledge suggests that that knowledge will be tied to the group's social networks and material circumstances. There is a measure of practical incommensurability between knowledge traditions, seen in difficulties of translation. In addition, some knowledge within a tradition is tacit, not fully formalizable, and requires socialization for it to be passed from person to person.

So far, we have seen arguments for the autonomy of technological practice. A separate set of arguments challenges the picture of technology as applied science by insisting on the lack of distinctness of science and technology. For example, in a study of the idea of large technological systems, the historian of technology Thomas Hughes claims that "persons committed emotionally and intellectually to problem solving associated with system creation and development rarely take note of disciplinary boundaries, unless bureaucracy has taken command" (Hughes 1987: 64). "Scientists" invent, and "inventors" do scientific research—whatever is necessary to push forward their systemic program.

Actor-network theory's term *technoscience* suggests a pragmatic characterization of scientific knowledge. For the pragmatist, scientific knowledge is about what natural objects can be made to do. Thus Knorr Cetina's book *The Manufacture of Knowledge* (1981) argues that laboratory science is about what can be constructed, not about what exists independently. Peter Dear (1995) argues that the key to the scientific revolution was a new orientation to experimental inquiry. Even though it produced unnatural objects, experimentation became acceptable because it showed what could be constructed of the natural world. Patrick Carroll-Burke (2001) argues that this practical orientation has remained intact, that science is defined by its use of a number of different types of "epistemic engines," devices for making natural objects more knowable. For the purposes of this chapter, the pragmatic orientation is relevant in that it draws attention to the ways in which science depends upon technology, both materially and conceptually.

The term *technoscience* also draws attention to the increasing interdependence of science and technology. We might see it as odd that historians are insisting on the autonomy of technological traditions and cultures precisely when there is a new spate of science-based technologies and technologically-oriented science—biotechnologies, new materials science, and nanotechnology all cross obvious lines.

Latour's networks and Hughes's technological systems bundle many different resources together. Thomas Edison freely mixed economic calculations, the properties of materials, and sociological concerns in his designs (Hughes 1985). Technologists need scientific and technical knowledge, but they also need material, financial, social, and rhetorical resources. Even ideology can be an input, in the sense that it might shape decisions and the conditions of success and failure (e.g. Kaiserfeld 1996). For network builders nothing can be reduced to only one dimension. Technology requires heterogeneous engineering of a dramatic diversity of elements (Law; 1937; Bucciarelli 1994). A better picture of technology, then, is one that incorporates many different inputs, rather than being particularly dependent upon a single stream. It is possible that no one input is even essential, and could not be worked around, given enough hard work, ingenuity and other resources.

To sum up, scientific knowledge is one resource on which engineers and inventors can draw, and perhaps on which they are drawing increasingly. But there is no reason to see it as a dominant resource. Rather, the development of technology is a complex process that integrates many different things: different lands of knowledge—including its own knowledge traditions—and different kinds of material resources. At the same time, it is clear that science draws on technology for

Whatever the relations are between science and technology, they are shifting. For example, although corporate research has always been supposed to lead to profits, many companies created research units the goals of which were shielded from business concerns so that they could pursue open-ended questions. Since the 1980s, restructuring of many high-technology companies especially in the United States has led to the shrinking of those laboratories, and to their changing relationships with the rest of the company. One researcher comments that "I have jumped from theoretical physics to what customers would like the most" (Varmi 2000: 405). In part this is a change in the notion of the purposes of science, but it may also represent a change in people's sense of science's technological potential—as the idea of the "knowledge economy" suggests.

Going alongside this is a sense that universities, and academic research, are also changing. Universities, and university researchers, are increasingly patenting their results, and entering into partnerships with corporations to fund research and develop products. There have been a number of different formulations of the changing structures of research. Much discussed is the idea that there is a new "mode" of knowledge production (Gibbons et al. 1994; Ziman 1994). Instead of the discipline-bound problem-oriented research of mode 1, mode 2 involves transdisciplinary work, possibly involving actors from a variety of types of organizations, in which the application is clearly in view. Some of the same authors as put forward the mode 2 concept have developed an alternative that sees an increase in "contextualization" of science, a process in which non-scientists become involved in shaping the direction and content of specific pieces of research (Nowotny, Scott, and Gibbons 2001). There might be an increase in the importance of research that is simultaneously basic and applicable (Stokes 1997). Another alternative formulation of the change is in the more organizational terms of a "triple helix" of university-industry-government in interaction (Etzkowitz and Leydesdorff 1997). Governments are demanding of universities that they be relevant in trade for support, universities are becoming entrepreneurial, and industry is buying research from universities. The change might also be put in more negative terms, that there is a crisis in key categories that support the idea of pure scientific research: the justification for and bounds of academic freedom, the public domain, and disinterestedness have all become unclear, disrupting the ethos of pure science (McSherry 2001).

Critics of each of these formulations charge that the changes described are not nearly as abrupt as they are portrayed to be, but these criticisms do not take away from the feeling that dramatic changes are under way for scientific research, and that these changes are connected to the potential application of that research. Whether or not there is less pure science, there is certainly more applied science. Science and technology are seen as increasingly connected.

its instruments, and perhaps also for some of its models of knowledge, just as some engineers may draw on science for their models of engineering knowledge. There are multiple relations of science and technology, rather than a single monolithic relation.

Does Technology Drive History?

A few of Karl Marx and Friedrich Engels's memorable comments on the influence of technology on economics and society can stand in for the position of the technological determinist, though they are certainly not everything that Marx and Engels had to say about the determinants of social structures. Looking at large-scale structures, Marx famously said: "The hand-mill gives you society with the feudal lord, the steam-mill, society with the industrial capitalist." Engels, talking about smaller-scale structures, claimed that "the automatic machinery of a big factory is much more despotic than the small capitalists who employ workers ever have been."

Technological determinism is the position that material forces, and especially the properties of available technologies, determine social events. The reasoning behind it is usually economistic: the available material resources, and the technologies for manipulating those resources, form the environment in which rational economic choices are made. In addition, technological determinism emphasizes "real-world constraints" and "technical logics" that shape technological trajectories (Vincenti 1995). Therefore, social variables ultimately depend upon material ones.

There are a number of different technological determinisms (see Bimber 1994), but the central idea is that technological changes force social adaptations, and through this narrowly constrain the trajectories of human history. Robert Heilbroner, supporting Marx, says that

> the hand-mill (if we may take this as referring to late medieval technology in general) required a work force composed of skilled or semiskilled craftsmen, who were free to practice their occupations at home or in a small atelier, at times and seasons that varied considerably. By way of contrast, the steam-mill—that is, the technology of the nineteenth century—required a work force composed of semi-skilled or unskilled operatives who could work only at the factory site and only at [a] strict time schedule. (Heilbroner 1994 [1967]: 60)

Because economic actors make rational choices, class structure is determined by the dominant technologies. This sort of reasoning applies to the largest scales, but also to much more local decisions. Thus technology shapes economic choices, and through those shapes history.

Some technologies appear particularly compatible with some types of political and social arrangements. In a well-known essay, Langdon Winner (1986a) asks: "do artifacts have politics?" He concludes that they do. Following Engels, he argues that some complex technological decisions will lend themselves to more hierarchical organization than others, in the name of efficiency—the complexity of modern industrial production does not lend itself to consensus decision-making. In addition, Winner argues, some technologies, such as nuclear power, are dangerous enough that they may bring their own demands for policing, and other forms of state power. And finally, individual artifacts may be constructed to achieve political goals. In a much-cited example, Winner describes how New York's "Master Builder" of roads, Robert Moses, designed overpasses on Long Island's parkways that were low enough to discourage buses, thus reserving Long Island's beaches for the car-owning classes. This example has been shown to be less straightforward than

Winner originally portrayed it (Joerges 1999), but the point can be made with mundane examples. Speed bumps perform the political purpose of reducing and slowing traffic on a street, and thus increasing the property values of a family-oriented neighborhood; this technology becomes a form of long-distance control (Latour 1993a). To take a different type of example, David Noble argues that in the history of industrial automation choices were made to disempower key groups (Noble 1984). Numerical control automation, the dominant form, was developed to eliminate machinist skill altogether from the factory floor, and therefore to eliminate the power of key unions. While also intended to reduce factories' dependence on skilled labor, record-playback automation, a technology not developed nearly as much, would have required the maintenance of machinist skill, to reproduce it in machine form (for some related issues, see Wood 1982).

Even for non-determinists, the effects of technologies can form an important site of study. As we saw in chapter 1, a key part of the pre-S&TS constellation of ideas on science and technology was the study of the positive and negative effects of technologies, and the attempt to think systematically about these effects. That type of work continues, and is part of the more practical and applied portion of S&TS. At the same time, some researchers have been challenging a seemingly unchallengeable assumption, the assumption that technologies have any systematic effects at all! In fact, they challenge something slightly deeper, the idea that technologies have essential features. If technologies have no essential features, then they should not have systematic effects, and if they do not have any systematic effects then they cannot determine the structure of the social world.

No technology—and in fact no object—has only one potential use. Even something as apparently purposeful as a watch can be simultaneously constructed to tell time, to be attractive, to make profits, to refer to a well-known style of clock, to make a statement about its wearer, etc. Even the apparently simple goal of telling time might be seen as a multitude of different goals: within a day one might use a watch to keep on schedule, to find out how long a bicycle ride took, to regulate the cooking of a pastry, to notice when the sun set, and so on. Given this diversity, there is no essence to a watch. And if the watch has no essence, then we can only say that it has systematic effects within a particular human environment. Change that environment and one changes what the watch does.

Trevor Pinch and Wiebe Bijker (1987), in their work on "Social Construction of Technology" (SCOT), develop this point into a framework for thinking about the development of technologies. In their central example, the development of the safety bicycle, the basic design of most twentieth-century bicycles, there is an appearance of inevitability about the outcome. The standard modern bicycle is stable, safe, efficient, and fast, and therefore we might see its predecessors as important, but ultimately doomed, steps toward the safety bicycle. On Pinch and Bijker's analysis, though, the safety bicycle did not triumph because of an intrinsically superior design. Some users felt that other early bicycle variants represented superior designs, at least superior to the early versions of the safety bicycle with which they competed. For many young male riders, the safety bicycle sacrificed style for a claim to stability, even though new riders did not, of course, find it very stable. Young male riders were one *relevant social group* that was not appeased by the new design. There is *interpretive flexibility* both in the understanding of technologies and in their design. We should see trajectories of technologies as the result of rhetorical operations, defining the users of artifacts, their uses, and the problems that particular designs solve.

On a SCOT analysis, the success of an artifact depends in large part upon the strength and size of the group that takes it up and promotes it. Its definition depends upon the associations that different actors make. Interpretive flexibility is thus a necessary feature of artifacts, because what an artifact does and how well it performs are the results of a competition of different groups' claims. Thus the good design of the safety bicycle cannot be the mover behind its success; good design is instead the result of its success.

Keith Grint and Steve Woolgar (1997) have developed this argument further. They take the metaphor of "technology as text," and try to show that technologies are interpretable as flexibly as are texts. They show, for example, that the Luddites of early nineteenth-century Britain adopted a variety of interpretations of the factory machines that they did and did not smash. Although some saw the factory machines as upsetting their preferred modes of work, others saw the problem in the masters of the factories. And,

they argue, resistance to the new technologies diminished when new left-wing political theories articulated the machines as saviors of the working classes. The machines, then, did not have any single set of effects.

A study done by Woolgar shows some of the work needed to give technologies determinate meanings. Woolgar acted as participant-observer in a computer firm that was in the process of developing a computer, with bundled software, for educational markets. At a point well into the development process, the firm needed to test prototypes of their package, to see how easy it was for unskilled users to figure out how to perform some standard tasks. On the one hand these tests could be seen as revealing what needed to be done to the computers in order for them to be more user-friendly. On the other hand they could be seen as revealing what needed to be done to the users—how they needed to be defined, educated, controlled—to make them more computer-friendly: successful technologies require *configuring the user*. The computer, then, does what it does only in the context of an appropriate set of users.

But surely some features of technologies defy interpretation? We might, for example, ask with Rob Kling (1992): "What's so social about being shot?" Everything, say Grint and Woolgar. In a tour de force of anti-essentialist argumentation, Grint and Woolgar argue that a gun being shot is not nearly as simple a thing as it might seem. It is clear that the act of shooting a gun is intensely meaningful—some guns are, for example, more manly than others. But more than that, even injuries by gunshot can take on different meanings. When female Israeli soldiers were shot in 1948, "men who might have found the wounding of a male colleague comparatively tolerable were shocked by the injury of a woman, and the mission tended to get forgotten in a general scramble to ensure that she received medical aid" (R. Holmes, quoted in Grint and Woolgar 1997: 159). Even death is not so certain. Leaving aside common uncertainties about causes of death and the timing of death, there are cross-cultural differences about death and what happens following it. No matter how unmalleable a technology might look, there are always situations, some of them highly theoretical, in which the technology can take on unusual uses or interpretations.

To accept that technologies do not have essences is to pull the rug out from under technological determinism. If they do nothing outside of the social and material contexts in which they are developed and used, technologies cannot be the real drivers of history. Rather these contexts are in the drivers' seats. This recognition is potentially useful for political analyses, as particular technologies can be used to affect social relations (Hård 1993). This has been explored most in the context of labor relations (e.g. Noble 1984) and in the context of gender relations (e.g. Cockburn 1985; Cowan 1983; Kirkup and Smith Keller 1992).

There should, then, be no debate about technological determinism. However, in practice nobody holds a determinism that is strict enough to be completely overturned by these arguments. Even the strictest of determinists, like Heilbroner, admit that social forces play a variety of important roles in producing and shaping technology's effects. Such a "soft" determinism is an interpretive stance, according to Heilbroner, that directs us to look first to technological change to understand economic change. Choices are made in relation to material resources and opportunities. To the extent that we can see social choices as economic choices, technology will play a key role.

The anti-essentialist strain that has developed within S&TS is a counterbalancing interpretive stance. Anti-essentialists show us that even soft determinism must be understood within a social framework, in that the properties of technologies can be determinative of social events only once the social world has established what the properties of technologies are. It thus directs us to look to the social world to understand technological change and its effects. This is perhaps most valuable for its constant reminder that things could be different.

Just paying attention to anti-essentialist lessons would remove the interest in studying technology. We study technology because artifacts appear to do things, or at least are made to do things. Therefore we need a nondeterministic theory of technology that can accommodate the obduracy of artifacts. We need something like Wiebe Bijker's theory of *sociotechnology*, a theory of the thorough intertwining of the social and technical (Bijker 1995). This theory draws heavily on work on technology as heterogeneous engineering, and on Bijker's work with Trevor Pinch on SCOT. A key concept in Bijker's theory is that of the *technological frame*, the set of practices and the material and social infrastructure built up around an artifact or collection of similar artifacts (Bijker 1995: 123). As the frame is developed, it guides future actions. A technological frame, then, may reflect engineers' understandings of the key problems of the artifact, and the directions in which solutions should be sought. It may also reflect understandings of the potential users of the artifact, and users' understanding of its functions. If a strong technological frame has developed, it will cramp interpretive flexibility. The concept is therefore useful in helping to understand how technologies can appear deterministic, while only appearing so in particular contexts.

Even if technologies do not have essential forms—properties that they have independent of their interpretations, or functions that they perform independently of what they are made to do— essences can return, in muted form, as dispositions. For Francis Bacon, with whom we started this chapter, the reason that scientific research should translate into technological benefits was that scientific research investigated what substances could be made to do. The form of a substance, for Bacon, is its response to circumstances. The forms of substances are those underlying natures that express the potentialities of those substances. This is why Bacon says that "knowledge and human power are synonymous."

One might make the same argument about the properties of technological artifacts. Artifacts do nothing by themselves, though they can be said to have effects in particular social contexts. To the extent that we can specify the relevant features of their social contexts, we might say that technological artifacts have Baconian forms. Particular pieces of technology can be said to have definitive properties, though they change depending upon context. Material reductionism, then, only makes sense in a given social context, just as social reductionism only makes sense in a given material context.

Does technology drive history, then? History could be almost nothing without it. As Bijker puts it, "purely social relations are to be found only in the imaginations of sociologists or among baboons." But, equally, technology could be almost nothing without history. Bijker continues, "and purely technical relations are to be found only in the wilder reaches of science fiction" (Bijker 1995).

Box 1.2 Were electric automobiles doomed to fail?

David Kirsch's history of the electric vehicle illustrates both the difficulty and power of deterministic thinking (Kirsch 2000). The standard history of the internal combustion automobile portrays the electric vehicle as doomed to failure. Compared with the gasoline-powered vehicle, the electric vehicle suffered from lack of power and lack of range, and therefore could never be the all-purpose vehicle that consumers wanted. Kirsch argues, however, that "technological superiority was ultimately located in the hearts and minds of engineers, consumers, and drivers, not programmed inexorably into the chemical bonds of refined petroleum" (Kirsch 2000: 4).

Until about 1915 electric cars and trucks could compete with gasoline-powered cars and trucks in a number of market niches. In man ways electric cars and trucks were the more natural successors to horse-drawn carriages. Gasoline-powered trucks were faster than electric trucks, but for the owners of delivery companies speed was more likely to damage goods, to damage the vehicles themselves, and anyway was effectively limited in cities. Because they were easy to restart, electric trucks were better suited to making deliveries than early gasoline-powered trucks; this was especially true given the horse-paced rhythm of existing delivery service, which demanded interaction between driver and customer. Electric taxis were fashionable, comfortable, and quiet, and for a time were successful in a number of American cities, so much so that in 1900 the Electric Vehicle Company was thy largest manufacturer of automobiles in the United States.

As innovators, electric taxi services were burdened with early equipment, they sometimes suffered poor management, and they were hit by expensive strikes. They also failed to participate in an integrated urban transit system that linked rail and road, to create a niche that they could dominate and in which they could then innovate. Meanwhile, Henry Ford's grand experiment in producing low-cost vehicles on assembly lines helped to spell the end of the electric vehicle. The First World War created a huge demand for gasoline-powered vehicles, which were better suited to war conditions than were electric ones. Increasing suburbanization of US cities meant that electric cars and trucks were restricted to a smaller and smaller segment of the market. Of course, that suburbanization was helped along by the successes of gasoline, and thus we might argue that the demands of consumers not only shaped, but were shaped by the technologies.

In 1900, then, the fate of the electric vehicle was not sealed. Does this failure of technological determinism mean that electric cars could be rehabilitated? According to Kirsch, that is unlikely. In 1900, both gasoline and electric cars fit the material and social contexts they faced, albeit very imperfectly. In the year 2000, material and social contexts have been shaped around the internal combustion engine. Given how unlikely it is that electric cars could compete directly with gasoline cars in these new contexts, their future does not look bright. Thus while a global technological determinism fails, within some contexts a soft determinism serves a useful heuristic purpose.

Overview: Technology and History
"Kranzberg's Laws"

Melvin Kranzberg

A few months ago I received a note from a longtime collaborator in building the Society for the History of Technology, Eugene S. Ferguson, in which he wrote, "Each of us has only one message to convey." Ferguson was being typically modest in referring to an article of his in a French journal[1] emphasizing the hands-on, design component of technical development, and he claimed that he had been making exactly the same point in his many other writings. True, but he has also given us many other messages over the years.

However, Ferguson's statement of "only one message" might indeed be true in my case. For I have been conveying basically the same message for over thirty years, namely, the significance in human affairs of the history of technology and the value of the contextual approach in understanding technical developments.

Because I have repeated that same message so often, utilizing various examples or stressing certain elements to accord with the interests of the different audiences I was attempting to reach, my thoughts have jelled into what have been called "Kranzberg's Laws." These are not laws in the sense of commandments but rather a series of truisms deriving from a longtime immersion in the study of the development of technology and its interactions with sociocultural change.

* * *

We historians tend to think of historical change in terms of cause and effect and of means and ends. Although it is not always easy to find causative elements and to distinguish ends from means in the interactions between technology and society, that has not kept scholars from trying to do so.

Indeed one of the intellectual clichés of our time, whose scholarly statement is embodied in the writings of Jacques Ellul and Langdon Winner, is that technology is pursued for its own sake and without regard to human need.[2] Technology, it is said, has become autonomous and has outrun human control; in a startling reversal, the machines have become the masters of man. Such arguments frequently result in the philosophical doctrine of technological determinism, namely, that technology is the prime factor in shaping our life-styles, values, institutions, and other elements of our society.

Not all scholars accept this version of technological omnipotence. Lynn White, jr., has said that a technical device "merely opens a door, it does not compel one to enter."[3] In this view, technology might be regarded as simply a means that humans are free to employ or not, as they see fit—and White recognizes that many nontechnical factors might affect that decision. Nevertheless, several questions do arise. True, one is not compelled to enter White's open door, but an open door is an invitation. Besides, who decides which doors to open—and, once one has entered the door, are not one's future directions

---

DR. KRANZBERG, now deceased, was one of the pioneers of the history of technology in the United States. He was the founding editor of *Technology & Culture,* recipient of the Society for the History of Technology's Leonardo da Vinci Medal (1967), and the Society's president (1983–84). He was Callaway Professor of the History of Technology at the Georgia Institute of Technology at the time this paper was presented as his delayed presidential address on October 19, 1985, at the Henry Ford Museum in Dearborn, Michigan.

[1] Eugene S. Ferguson, "La Fondation des machines modernes: des dessins," *Culture technique* 14 (June 1985): 182–207. *Culture technique* is the publication of the Centre de Recherche sur la Culture Technique, located in Paris under the direction of Jocelyn de Noblet. The June 1983 edition of *Culture technique,* dedicated to *Technology and Culture,* contained French translations of a number of articles from the SHOT journal.

[2] Jacques Ellul, *The Technological Society* (New York, 1964), and Langdon Winner, *Autonomous Technology: Technics Out-of-Control as a Theme in Political History* (Cambridge, Mass., 1977).

[3] Lynn White, jr., *Medieval Technology and Social Change* (Oxford, 1962), p. 28.

guided by the contours of the corridor or chamber into which one has stepped? Equally important, once one has crossed the threshold, can one turn back?

Frankly, we historians do not know the answer to this question of technological determinism. Ours is a new discipline; we are still working on the problem, and we might never reach agreement on an answer—which means that it will provide employment for historians of technology for decades to come. Yet there are several things that we do know, and that I summarize under the label of Kranzberg's First Law.

Kranzberg's First Law reads as follows: Technology is neither good nor bad; nor is it neutral.

By that I mean that technology's interaction with the social ecology is such that technical developments frequently have environmental, social, and human consequences that go far beyond the immediate purposes of the technical devices and practices themselves, and the same technology can have quite different results when introduced into different contexts or under different circumstances.

Many of our technology-related problems arise because of the unforeseen consequences when apparently benign technologies are employed on a massive scale. Hence many technical applications that seemed a boon to mankind when first introduced became threats when their use became widespread. For example, DDT was employed to raise agricultural productivity and to eliminate disease-carrying pests. Then we discovered that DDT not only did that but also threatened ecological systems, including the food chain of birds, fishes, and eventually man. So the Western industrialized nations banned DDT. They could afford to do so, because their high technological level enabled them to use alternative means of pest control to achieve the same results at a slightly higher cost.

But India continued to employ DDT, despite the possibility of environmental damage, because it was not economically feasible to change to less persistent insecticides—and because, to India, the use of DDT in agriculture was secondary to its role in disease prevention. According to the World Health Organization, the use of DDT in the 1950s and 1960s in India cut the incidence of malaria in that country from 100 million cases a year to only 15,000, and the death toll from 750,000 to 1,500 a year. Is it surprising that the Indians viewed DDT differently from us, welcoming it rather than banning it? The point is that the same technology can answer questions differently, depending on the context into which it is introduced and the problem it is designed to solve.

Thus while some American scholars point to the dehumanizing character of work in a modern factory,[4] D. S. Naipaul, the great Indian author, assesses it differently from the standpoint of his culture, saying, "Indian poverty is more dehumanizing than any machine."[5] Hence in judging the efficacy of technological development, we historians must take cognizance of varying social contexts.

It is also imperative that we compare short-range and long-range impacts. In the 19th century, Romantic writers and social critics condemned industrial technology for the harsh conditions under which the mill workers and coal miners labored. Yet, according to Fernand Braudel, conditions on the medieval manor were even worse.[6] Certain economic historians have pointed out that, although the conditions of the early factory workers left much to be desired, in the long run the worker's living standards improved as industrialization brought forth a torrent of goods that were made available to an ever-wider public.[7] Of course, those long-run benefits were small comfort to those who suffered in the short run; yet it is the duty of the historian to show the differences between the immediate and long-range implications of technological developments.

Although our technological advances have yielded manifold benefits in increasing food supply, in providing a deluge of material goods, and in prolonging human life, people do not always appreciate technology's contributions to their lives and comfort. Nicholas Rescher, citing statistical data on the way people perceive their conditions, explains their dissatisfaction on the paradoxical ground that technical progress inflates their expectations faster than it can actually meet them.[8]

Of course, the public's perception of technological advantages can change over time. A century ago, smoke from industrial smokestacks was regarded as a sign of a region's prosperity; only later was it recognized that the smoke was despoiling the environment. There were "technological fixes," of course. Thus, one of the aims of the Clean Air Act of 1972

[4]E.g., Christopher Lasch, *The Minimal Self: Psychic Survival in Troubled Times* (New York, 1984).
[5]Quoted in Dennis H. Wrong, "The Case against Modernity," *New York Times Book Review,* October 28, 1984, p. 7.
[6]Fernand Braudel, *The Structures of Everyday Life,* vol. 1 of *Civilization and Capitalism, 15th–18th Century* (New York, 1981).
[7]E.g., T. S. Ashton, *The Industrial Revolution, 1760–1830* (Oxford, 1948), and David S. Landes, *The Unbound Prometheus: Technological Change and Industrial Development in Western Europe from 1750 to the Present* (Cambridge, 1969).
[8]Nicholas Rescher, *Unpopular Essays on Technological Progress* (Pittsburgh, 1980).

was to prevent the harmful particulates emitted by smokestacks from falling on nearby communities. One way to do away with this problem was to build the smokestacks hundreds of feet high; then a few years later we discovered that the sulfur dioxide and other oxides, when sent high into the air, combined with water vapor to shower the earth with acid rain that has polluted lakes and caused forests to die hundreds of miles away.

Unforeseen "dis-benefits" can thus arise from presumably beneficent technologies. For example, although advances in medical technology and water and sewage treatment have freed millions of people from disease and plague and have lowered infant mortality, these have also brought the possibility of overcrowding the earth and producing, from other causes, human suffering on a vast scale. Similarly, nuclear technology offers the prospect of unlimited energy resources, but it has also brought the possibility of worldwide destruction.

That is why I think that my first law—Technology is neither good nor bad; nor is it neutral—should constantly remind us that it is the historian's duty to compare short-term versus long-term results, the utopian hopes versus the spotted actuality, the what-might-have-been against what actually happened, and the trade-offs among various "goods" and possible "bads." All of this can be done only by seeing how technology interacts in different ways with different values and institutions, indeed, with the entire sociocultural milieu.[9]

* * *

Whereas my first law stresses the interactions between technology and society, my second law starts with internalist elements in technology and then stretches to include many nontechnical factors. Kranzberg's Second Law can be simply stated: Invention is the mother of necessity.

Every technical innovation seems to require additional technical advances in order to make it fully effective. If one invents a lathe that can cut metal faster than existing machines, this necessitates improvements in the lubricating system to keep the mechanism running efficiently, improved grinding materials to stand up under the enhanced speed, and new means of taking away quickly the waste material from the item being turned.

Many major innovations have required further inventions to make them completely effective. Thus,

Alexander Graham Bell's telephone spawned a variety of technical improvements, ranging from Edison's carbon-granule microphone to central-switching mechanisms. A variation on this same theme is described in Hugh Aitken's book on the origins of radio, in which he indicates the various innovative steps whereby the spark technology that produced radio waves was tuned into harmony (syntonized) with the receiver.[10] In more recent times, the design of a more powerful rocket, giving greater thrust, necessitates innovation in chemical engineering to produce the thrust, in materials to withstand the blast, in electronic control mechanisms, and the like.

A good case of invention mothering necessity can be seen in the landmark textile inventions of the 18th century. Kay's "flying shuttle" wove so quickly that it upset the usual ratio of four spinners to one weaver; either there had to be many more spinners or else spinning had to be similarly quickened by application of machinery. Thereupon Hargreaves, Cartwright, and Crompton improved the spinning process; then Cartwright set about further mechanizing the weaving operation in order to take full advantage of the now-abundant yarn produced by the new spinning machines.

Thomas P. Hughes would refer to the phenomenon that I have just described as a "reverse salient";[11] but I prefer to call it a "technological imbalance," a situation in which an improvement in one machine upsets the previous balance and necessitates an effort to right the balance by means of a new innovation. No matter what one calls it, Hughes and I are talking about the same thing. Indeed, Hughes has gone further in discussing technological systems, for he shows how, as a system grows, it generates new properties and new problems, which in turn necessitate further changes.

The automobile is a prime example of how a successful technology requires auxiliary technologies to make it fully effective, for it brought whole new industries into being and turned existing industries in new directions by its need for rubber tires, petroleum products, and new tools and materials. Furthermore, large-scale use of the auto demanded a host of auxiliary technological activities—roads and highways, garages and parking lots, traffic signals, and parking meters.

While it might be said that each of these other developments occurred in response to a specific

[9]The "New Directions" program session at the 1985 SHOT annual meeting indicated that historians of technology are continuing to broaden their concerns and are indeed investigazing new areas of the sociocultural context in relation to technological developments.

[10]Hugh G. J. Aitken, *Syntony and Spark: The Origins of Radio* (New York, 1976).

[11]Thomas P. Hughes, "Inventors: The Problems They Choose, the Ideas They Have, and the Inventions They Make," in *Technological Innovation: A Critical Review of Current Knowledge,* ed. Patrick Kelly and Melvin Kranzberg (San Francisco, 1978), pp. 166–82.

need, I claim that it was the original invention that mothered that necessity. If we look into the internal history of any mechanical device, we find that the basic invention required other innovative changes to make it fully effective and that the completed mechanism in turn necessitated changes in auxiliary and supporting technological systems, which, taken all together, brought many changes in economic and sociocultural patterns.

* * *

What I have just said is virtually a statement of my Third Law: Technology comes in packages, big and small.

The fact is that today's complex mechanisms usually involve several processes and components. Radar, for example, is a very complicated system, requiring specialized materials, power sources, and intricate devices to send out waves of the proper frequency, detect them when they bounce off an object, and then interpret them and place the results on a screen.

That might explain why so many different people have laid claim to inventing radar. Each is perfectly right in pointing out that he provided an element essential to the final product, but that final product is composed of many separate elements brought together in a system that could not function without every single one of the components. Thus radar is the product of a packaging process, bringing together elements of different technologies into a single device.

In his fascinating account of the development of mass production, David A. Hounshell tells how many different experiments and techniques were employed in bringing Ford's assembly line into being.[12] Although many of the component elements were already in existence, Ford put these together into a comprehensive system—but not without having to develop additional technical capabilities, such as conveyor lines, to make the assembly process more effective.

My third law has been extended even further by Thomas P. Hughes's 1985 Dexter Prize—winning book *Networks of Power*. What I call "packages" Hughes more precisely and accurately calls "systems," which he defines as coherent structures composed of interacting, interconnected components.[13] When one component changes, other parts of the system must undergo transformations so that the system might continue to function. Hence the parts of a system

cannot be viewed in isolation but must be studied in terms of their interrelations with the other parts.

Although Hughes concentrates on electric power systems, what he provides is a paradigm that is applicable to other systems—transportation, water supply, communications, and the like. And because entire systems interact with other systems, a system cannot be studied in isolation any more than can its component parts; hence one must also look at the interaction of these systems with the entire social, political, economic, and cultural environment. Hughes's book thus provides excellent case studies proving the validity of the first three of Kranzberg's Laws, and also of my fourth dictum.

* * *

Unfortunately, Kranzberg's Fourth Law cannot be stated so pithily as the first three. It reads as follows: Although technology might be a prime element in many public issues, nontechnical factors take precedence in technology-policy decisions.

Engineers claim that their solutions to technical problems are not based on mushy social considerations; instead, they boast that their decisions depend on the hard and measurable facts of technical efficiency, which they define in terms of input-output factors such as cost of resources, power, and labor. However, as Edward Constant has shown in studying the Kuhnian paradigm's applicability to technological developments, many complicated sociocultural factors, especially human elements, are involved, even in what might seem to be "purely technical" decisions.[14]

Besides, engineers do not always agree with one another; different fields of engineering might have different solutions to the same problem, and even within the same field they might disagree on what weight to assign to different trade-off factors. Indeed, as Stuart W. Leslie demonstrated in his Usher Prize article on "Charles F. Kettering and the Copper-cooled Engine,"[15] the most efficient device does not always win out even in what we might regard as a narrowly technical decision within a single industrial corporation. Although Kettering regarded his copper-cooled engine as a technical success, it never went into production. Why not? True, it had some technical "bugs," but these could not be successfully ironed out because of divisions between the research engineers and the production people—and because

---

[12]David A. Hounshell, *From the American System to Mass Production 1800–1932: The Development of Manufacturing Technology in the United States* (Baltimore, 1984), chap. 6.

[13]Thomas P. Hughes, *Networks of Power: Electrification in Western Society, 1880–1930* (Baltimore, 1983), p. ix.

[14]Edward W. Constant, *The Origins of the Turbojet Revolution* (Baltimore, 1980). This book was awarded the Dexter Prize by SHOT in 1982.

[15]Stuart W. Leslie, "Charles F. Kettering and the Copper-cooled Engine," *Technology and Culture* 20 (October 1979): 752–76.

of the overall decision that the copper-cooled engine could not meet the corporate demand for immediate profit. So technical worth, or at least potential technical capability and efficiency, was not the decisive element in halting the copper-cooled engine.

In *Networks of Power* Hughes likewise demonstrates how nontechnical factors affected the efficient growth of electrical networks by comparing developments in Chicago, Berlin, and London. Private enterprise in Chicago, in the person of Samuel Insull, followed the path of the most efficient technology in seeking economies of scale. In Berlin and London, however, municipal governments were more concerned about their own authority than about technical efficiency, and political infighting meant that they lagged behind in developing the most economical power networks.

Technologically "sweet" solutions do not always triumph over political and social forces.[16] The debate a dozen years ago over the supersonic transport (SST) provides an example. Although the SST offered potential advantages, its development to the point where its feasibility and desirability could be properly determined was never allowed to take place. Economic factors might have underlain the decision to cut R&D funds for the SST, but the public decision seems also to have been based on a fear of the environmental hazards posed by the supersonic aircraft in commercial aviation.

Environmental concerns have indeed assumed a major place in public decisions regarding technical initiatives. These concerns are not groundless, for we have seen how certain technologies, employed without awareness of potential environmental effects, have boomeranged to present hazardous problems, despite their early beneficial effects. Many engineers believe that hysterical fear about technological development has so gripped our nation that people overlook the benefits provided by technology and concentrate on the dangers presented either by ill-conceived technological applications or by human error or oversight in technical operations. But who can blame the public, with Love Canal and Bhopal crowding the headlines?[17]

American politics has now become the battleground of special-interest groups, and few of these groups are willing to make the trade-offs required in many engineering decisions. In the case of potential environmental hazards, Daniel A, Koshland has stated that we can satisfy one or the other of the different groups, but only at a cost of something undesirable to the others.[18]

Especially politicized has been the question of nuclear power. The nuclear industry itself has been partly to blame for technological deficiencies, but the presumption of risk by the public, especially following the Three Mile Island and Chernobyl accidents, has affected the future of what was once regarded as a safe and inexhaustible source of power. The public fears possible catastrophic consequences from nuclear generators.

Yet the historical fact is that no one has been killed by commercial nuclear power accidents in this country. Contrast this with the 50,000 Americans killed each year by automobiles. But although antinuclear protestors picket nuclear power plants under construction, we never see any demonstrators bearing signs saying "Ban the Buick"!

Partly this is due to the public's perception of risk, rather than to the actual risks themselves.[19] People seek a zero-risk society. But as Aaron Wildavsky has so aptly put it, "No risk is the highest risk of all."[20] For it would not only petrify our technology but also stultify developmental growth in society along any lines.

Nevertheless, the fact that political considerations take precedence over purely technical considerations should not alarm us. In a democracy, that is as it should be. To deal with questions involving the interactions between technology and the ecology, both natural and social, we have devised new social instruments, such as "technology assessment," to evaluate the possible consequences of the applications of technologies before they are applied.

Of course, political considerations often continue to take precedence over the commonsensible results of comprehensive and impartial technological assessments. But at least there is the recognition that technological developments frequently have social,

---

[16]Eugene B. Skolnikoff states, "Technology alters the physical reality, but is not the key determinant of the political changes that ensue," in *The International Imperatives of Technology: Technological Development and the International Political System* (Berkeley, Calif.: University of California Institute of International Studies, n.d.), p. 2.

[17]Speaking of the Bhopal tragedy, President John S. Morris of Union College has said: "Methyl isocyanate makes it possible to grow good crops and feed millions of people, but it also involves risks. And analyzing risks is not a simple matter" (*New York Times*, April 14, 1985).

[18]Daniel A. Koshland, "The Undesirability Principle," *Science* 229 (July 5, 1985): 9.

[19]See Dorothy Nelkin, ed., *Controversy: The Politics of Ethical Decisions* (Santa Monica, Calif., 1984).

[20]Aaron Wildavsky, "No Risk Is the Highest Risk of All," *American Scientist* 67 (1979): 32–37.

human, and environmental implications that go far beyond the intention of the original technology itself.

* * *

The fact that historians of technology must be aware of outside forces and factors affecting technology—from the human personality of the inventor to the larger social, economic, political, and cultural milieu—has led me to Kranzberg's Fifth Law: All history is relevant, but the history of technology is the most relevant.

In her presidential address to the Organization of American Historians several years ago, Gerda Lerner pointed out how history satisfies a variety of human needs, serving as a cultural tradition that gives us personal identity in the continuum of the past and future of the human enterprise.[21] Other apologists for the profession point out that history is one of the fundamental liberal arts and is essential as a key to an understanding of the future.

No one would quarrel with such worthy sentiments, but, to repeat questions raised by Eugene D. Genovese, "If so, how can we explain the dangerous decline in the teaching of history in our schools; the cynical taunt, 'What is history good for anyway?' "[22] Although historians might write loftily of the importance of historical understanding by civilized people and citizens, many of today's students simply do not see the relevance of history to the present or to their future. I suggest that this is because most history, as it is currently taught, ignores the technological element.

Two centuries ago the great German philosopher Immanuel Kant stated that the two great questions in life are (1) What can I know? and (2) What ought I do?

To answer Kant's first question, we can learn the history of the past. I look on history as a series of questions that we ask of the past in order to find out how our present world came into being. We call ours a "technological age." How did it get to be that way? That indeed is the major question that the history of technology attempts to answer. Our students know that they live in a technological age, but any history that ignores the technological factor in societal development does little to enable them to comprehend how their world came into being.

True, economic and business historians have perforce taken cognizance of those technological elements that had a mighty effect on their subject matter. Similarly, social historians of the *Annales* school have stressed how technology set the patterns of daily life for the vast majority of people throughout history, and Brooke Hindle, in a fine historicgraphical article, has indicated how some of our fellow historians have begun to see how technology impinges on their special fields of study.[23] But for the most part, social, political, and intellectual historians have been oblivious to the technological parameters of their own subjects.

Perhaps most guilty of neglecting technology are those concerned with the history of the arts and with the entire panoply of humanistic concerns. Indeed, in many cases they are disdainful of technology, regarding it as somehow opposed to the humanities. This might be because they regard technology solely in terms of mechanical devices and do not even begin to comprehend the complex nature of technological developments and their direct influences on the arts, to say nothing of their indirect influence on mankind's humanistic endeavors.

Yet anyone familiar with Cyril Stanley Smith's writings would be aware of the importance of the aesthetic impulse in technical accomplishments and of how these in turn amplified the materials and techniques available for artistic expression.[24] And any historian of art or of the Renaissance should perceive that such artistic masters as Leonardo and Michelangelo were also great engineers. That relationship continues today, as David Billington has shown in stressing the relationship of structural design and art.[25]

Today's technological age provides new technical capabilities to enlarge the horizons and means of expression for artists in every field. Advances in musical instruments have given larger scope to the imagination of composers and to musical interpretation by performers. The advent of photography, the phonograph, radio, movies, and television have not

[21]Gerda Lerner, "The Necessity of History and the Professional Historian," *Journal of American History* 69 (June 1982): 7–20.
[22]Eugene D. Genovese, "To Celebrate a Life—Biography as History," *Humanities* 1 (January–February 1980): 6. An analysis of today's low state of the history profession is to be found in Richard O. Curry and Lawrence D. Goodheart, "Encounters with Clio: The Evolution of Modern American Historical Writing," *OAH Newsletter* 12 (May 1984): 28–32.

[23]Brooke Hindle, "'The Exhilaration of Early American Technology': A New Look," in *The History of American Technology: Exhilaration or Discontent?* ed. David A. Hounshell (Wilmington, Del., 1984).
[24]See especially Cyril Stanley Smith's Usher Prize article, "Art, Technology, and Science: Notes on Their Historical Interaction," *Technology and Culture* 11 (October 1970): 493–549.
[25]See David Billington's Dexter Prize–winning book, *Robert Maillart's Bridges: The Art of Engineering* (Princeton, N.J., 1979), and "Bridges and the New Art of Structural Engineering," *American Scientist* 72 (January–February 1984): 22–31.

only given artists, composers, and dramatists new tools with which to exercise their vision and talents but have also enlarged the audience for music, drama, and the whole panoply of the arts. They also extend our audio and visual memory, enabling us to see, hear, and preserve the great works of the past and present.

In the field of learning and education, there is little point in belaboring the impact of writing tools, paper, the printing press, and, nowadays, radio and TV. But there is also an indirect influence of technology on education, one that makes it more possible than ever before in human history for larger numbers of people in the industrialized nations to take advantage of formal schooling.

Let me give a brief example drawn from American history. Thomas Jefferson was very proud of the educational system that he devised for the state of Virginia. But in his educational scheme, only a very small percentage could ever hope to ascend to the heights of a university education.

This is not because Jefferson was an elitist. Far from it! But the fact is that the agrarian technology of his time was not productive enough to allow large numbers of youth to participate in the educational process. From a very early age, children worked in the fields alongside their parents or, if they were town dwellers, were apprenticed to craftsmen. Only when great increases in agricultural and industrial productivity were made possible by revolutionary developments in technology did society acquire sufficient wealth to keep children out of the work force and enable them to attend school. As the 19th century progressed, first elementary education was made compulsory, then secondary education, and by the mid-20th century, America had grown so wealthy that it could afford a college education for all its citizens. True, some students drop out of high school before completing it, and not everyone going to college takes full advantage of the educational opportunities. But the fact is that the majority of Americans today have the equivalent education of the small segment of the upper-class elite in pre-industrial society. In brief, technology has been a significant factor, not only in the pattern of our daily lives and in our workaday world, but also in democratizing education and the intellectual realm of the arts and humanities.

However, such vast generalizations might do little to convince the public of the wisdom of Stanley N. Katz's vision of scholars participating "in public discourse in order to recover the traditional role of the humanist as a public figure."[26] But the relevance of the history of technology to today's world can be spelled out in very specific terms. For example, because we live in a "global village," made so by technological developments, we are conscious of the need to transfer technological expertise to our less fortunate brethren in the less developed nations. And the history of technology has a great deal to say about the conditions, complexities, and problems of technology transfer.

Likewise, we are faced with public decisions regarding global strategy, environmental concerns, educational directions, and the ratio of resources to the world's burgeoning population. Technological history can cast light on many parameters of these very specific problems confronting us now and in the future—and that is why I say that the history of technology is more relevant than other histories.

One proof of this is that the outside world, especially the political community, is becoming increasingly cognizant of the contributions that historians of technology can make to public concerns. Whereas several decades ago historians were rarely called on to provide information to Congress on matters other than historical archives, memorials, and national celebrations, nowadays it is almost commonplace for historians of technology to testify before congressional committees dealing with scientific and technological expenditures, aerospace developments, transportation, water supplies, and other problems having a technological component. Congressmen obviously think that the information provided by historians of technology is relevant to coping with the problems of today and tomorrow.

Leaders in all fields are increasingly turning to historians of technology for expertise regarding the nature of the sociotechnical problems facing them. Let me give a few more specific examples. SHOT is an affiliate of the American Association for the Advancement of Science (AAAS), and there was a time when historians of technology appeared only on the program sessions of Section L of the AAAS, the History and Philosophy of Science. But historians of technology also have important things to say to a public larger than that composed of their historical colleagues. Hence it was a source of great personal pride to me—almost paternal pride—when, at the 1985 AAAS meeting, Carroll Pursell appeared on a program session with a congressman and a former

---

[26]Stanley N. Katz, "The Scholar and the Public," *Humanities* 6 (June 1985): 14–15.

assistant secretary of commerce; the program dealt with certain social and economic problems affecting the United States today, and Pursell's historical account of the technological parameters was truly germane to the thrust of the discussion. Similarly, at a recent conference, at my own Georgia Tech, on the problems expected to affect the workplace in the future, David Hounshell provided a meaningful technological historical context for a discussion that involved top labor leaders, political figures, and corporate executives. (I took family pride in that too!)

I regard this entrance of historians of technology into the public arena as empirical evidence of the true relevance of the history of technology to the worlds of today and tomorrow. To reiterate, all history is relevant, but the history of technology is most relevant. The rest of the world realizes that, and SHOT is working to make our historical colleagues from other fields recognize it too.

* * *

This brings me to my final law, Kranzberg's Sixth Law: Technology is a very human activity—and so is the history of technology.

Anthropologists and archaeologists studying primate evolution tell us of the importance of purposive toolmaking in the formation of *Homo sapiens.* The physical development of our species is apparently inextricably bound up with cultural developments, so that technology is classed as one of the earliest and most basic of human cultural characteristics, one helping to develop language and abstract thinking. Or, to put it another way, man could not have become *Homo sapiens,* "man the thinker," had he not at the same time been *Homo faber,* "man the maker."

Man is a constituent element of the technical process. Machines are made and used by human beings. Behind every machine, I see a face—indeed, many faces: the engineer, the worker, the businessman or businesswoman, and, sometimes, the general and admiral. Furthermore, the function of the technology is its use by human beings—and sometimes, alas, its abuse and misuse.

To those who identify technology simply with the machines themselves, I use the computer as a metaphor to show the importance of the interaction of human and social factors with the technical elements—for computers require both the mechanical element, the "hardware," and the human element, the "software"; without the software, the machine is simply an inert device, but without the hardware, the software is meaningless. We need both, the human and the purely technical components,

in order to make the computer a usable and useful piece of technology.

Those of you who were at our Silver Anniversary meeting in 1983 will recall that I told an anecdote, which I sometimes use to quiet my most voluble anti-technological humanistic colleagues. A lady came up to the great violinist Fritz Kreisler after a concert and gushed, "Maestro, your violin makes such beautiful music." Kreisler held his violin up to his ear and said, "I don't hear any music coming out of it."

You see, the instrument, the hardware, the violin itself, was of no use without the human element. But then again, without the instrument, Kreisler would not have been able to make music. The history of technology is the story of man and tool—hand and mind—working together. If the hardware is faulty or if the software is deficient, the sounds that emerge will be discordant; but when man and machine work together, they can make some beautiful music.

People sometimes speak of the "technological imperative," meaning that technology rules our lives. Indeed, they can point to many technical elements, such as the clock, that determine the character and pace of our daily existence. Likewise, the automobile determines where and how we Americans live, work, think, play, and pray.

But this does not necessarily mean that the "technological imperative," usually based on efficiency or economy, necessarily directs all our thoughts and actions. We can point to many technical devices that would make life simpler or easier for us but which our social values and human sensibilities simply reject. Thus, for example, Ruth Schwartz Cowan has shown in her Dexter Prize–winning book, *More Work for Mother*, how communal kitchens would be feasible and save the mother from much drudgery of food preparation. But our adherence to the concept of the home has made that technical solution unworkable; instead we have turned to other technologies to ease the housework and cooking chores, albeit requiring more time and attention from mother.[27]

In other words, technological capabilities do not necessarily determine our actions. Indeed, how else can we explain why we have spent billions of dollars on nuclear power plants that we have had to abandon before they were completed? Obviously, other human factors proved more powerful than the combined technical and economic pressures.

---

[27]Ruth S. Cowan, *More Work for Mother: The Ironies of Household Technology from the Open Hearth to the Microwave* (New York, 1983), chap. 5.

Building the Great Pyramid: Probable Construction Methods Employed at Giza

James Frederick Edwards

Every year, droves of visitors travel to Egypt to gaze upon the oldest survivor of the seven wonders of the ancient world. Most of them ask the same question: "How was it built?"

The largest in a group of three, the Great Pyramid was begun by King Khufu during the Fourth Dynasty of the Old Kingdom in Egypt, which commenced with the reign of King Sneferu, approximately 2613 B.C.E., and ended with the death of King Shepseskaf circa 2500 B.C.E.[1] Its original outer casing stone and some other blocks have been removed, but at 147.5 meters high and 230 meters square at the base, its volume, when first built, would have exceeded 2,600,000 cubic meters. It has been estimated that 2,300,000 separate blocks of stone, the majority weighing between 2 and 3 tonnes, were used in its construction.[2] There has always been much speculation about how it was constructed, and Egyptologists and historians are divided about the building techniques employed. The majority favor the idea that gigantic ramps were used to lift the building blocks to their locations within the structure, while others claim that levering systems were employed.

The principal theory is that a massive ramp was built against one full face of the pyramid, and was lengthened as construction proceeded.[3] Various gradients have been proposed for such a ramp, although a slope of 1 in 10 is considered the most practical. Such a ramp would have been about 1 ½ kilometers long and have required more than three times the volume of material used in the completed pyramid. Apart from the mammoth task of building it, maintaining the ramp during construction of the pyramid would have been a colossal undertaking in its own right. It is difficult to guess where such a ramp might have been located. A 1-in-10 gradient could not have been achieved from the adjacent quarry area, and the local topography and other building works in the vicinity would have placed severe restrictions upon its location.

Another proposal is that there was a spiral ramp or combination of ramps around the structure of the pyramid. Numerous virtually insurmountable problems would have been associated with physically supporting and constructing such ramps.[4] Furthermore, they would have afforded only a relatively narrow hauling surface, a problem exacerbated by the simultaneous use of the ramp by both the ascending and

Dr. Edwards is a chartered consultant engineer and physicist. He is head of rehabilitation engineering services at South Manchester University Hospitals NHS Trust, Manchester, England, where he is currently working on a number of research and development projects in the field of medical physics. He is a coauthor of *Properties of Materials* (1986), *Statics* (1989), and *Motion and Energy* (1993), and has also written numerous other technical and nontechnical books. He thanks Chris Scarre of the McDonald Institute for Archaeological Research, Cambridge, for pointing him in the direction of *Technology and Culture,* and Kath Mannion for putting his handwritten notes into a typed format. He also thanks the *Technology and Culture* referees and editors for their helpful suggestions.

[1]Of the numerous books written about Egyptian pyramids, a good general source is Mark Lehner, *The Complete Pyramids* (London, 1997). For Egyptian construction methods, see Dieter Arnold, *Building in Egypt: Pharaonic Stone Masonry* (New York and Oxford, 1991).

[2]Lehner, 202. Most of the building stone used for the core blocks was quarried close by the pyramid. The stone for the outer casing blocks, which was a harder, more homogeneous white limestone, was brought from Tura, across the Nile Valley from Giza. The small number of large granite blocks used for the burial chamber, and for plugging up passages, were brought from Aswan, some 500 miles to the south.

[3]The various ramp theories are fully explained by Lehner, 215–17. See also Zahi Hawass, "The Pyramids," in *Ancient Egypt*, ed. David P. Silverman (London, 1997), 168–91; Arnold, 98–101.

[4]Peter Hodges, *How the Pyramids Were Built*, ed. Julian Keable (Dorset, 1989), 125–27. As a spiral ramp progressively increased in height its sides would need to be vertical in order for it not to encroach upon its own lower stages. For such a structure to be inherently stable it would need to be constructed from material meeting criteria approaching those for the pyramid itself (i.e., stone). It is also extremely doubtful that support for such a structure could be provided off the angled faces of the pyramid.

descending hauling teams. The hauling teams would also have encountered great difficulty negotiating the tight right-angled turns at each corner of the pyramid.

Although the foundations of a number of small, embankment-like structures have been discovered adjacent to the Great Pyramid, it seems likely that these were only used for elevating blocks at a very low level during the initial stages of the pyramid's construction.[5] Large ramps of any type would have generated an enormous amount of material, and there is no such volume of material at or near the construction site. There is, therefore, a dearth of conclusive archaeological evidence supporting the theory that such massive ramps were constructed in the vicinity of the Great Pyramid.

Alternatively, it has been proposed that every individual block of stone used in the construction of the Great Pyramid was elevated into position using levers and packing pieces.[6] Such a technique involves jacking up a block at one side with lever and fulcrum, inserting a wooden packing piece, jacking up the opposite side of the block in a similar manner, and then repeating the process until the desired elevation is achieved. It has been estimated that a vertical distance equivalent to the thickness of one course of block in the Great Pyramid could have been achieved in about 5 minutes by a team men operating two levers per side.[7] The exposed core blocks of the Great Pyramid suggest that it was constructed as a "coursed" pyramid, as it consists of level courses of stones. Although these courses vary slightly in thickness, each separate course appears to be a level array of squared stones. Once a block had been lifted one course, it would have had to be moved horizontally onto the bottom of the next course, whereupon the procedure could be repeated, until the working plateau had been reached. By the time the Great Pyramid had reached half its completed height, some fifteen hundred separate jacking actions, together with approximately eighty horizontal transfers, would have

had to be undertaken for a single block, all requiring a degree of precision in order to avoid any mishaps.

Both the ramp and lever methods would have been inefficient in their deployment of personnel, for in both cases the haulers and lifters would have had to ascend and descend the pyramid structure as part of each elevating cycle. Such approaches would also have been extremely time consuming; at the halfway point in the pyramid's construction the elevating cycle for one core block would have been forty minutes using a straight ramp and seven hours using levers.[8]

It has also been proposed that a type of shaduf—a counterbalanced sweep used in the ancient world to raise water—could have been used to lift the pyramids' building blocks.[9] Such an approach would have necessitated the construction of substantial wooden towers in order to withstand the forces involved. It is proposed that the pyramids' outer building blocks were initially left square and untrimmed and that the wooden towers were moved up the stepped sides of the pyramid as construction proceeded—an operation fraught with danger, as well as an extremely time-consuming and impractical one.

Because of the problems alluded to, it must be concluded that these ramp and lever theories present unsatisfactory resolutions relating to the methodology employed for the elevation of the building blocks.

Hauling Stone Blocks

It is possible to deduce, from contemporary and even ancient evidence, certain scientific parameters relating to the hauling of stone blocks.

During recent experiments relating to the hauling of stone blocks at Karnak Temple, it was found that three men could pull a sledge-mounted block weighing one tonne over a stone surface that had been lubricated with water to reduce the effects of friction.[10] From this evidence we can, by making some practical assumptions, determine the frictional effects encountered by the haulers.

[5]Lehner, 217, 221. Shallow, low-level ramps were probably used during the construction of the lower courses of the pyramid. The exposed lower courses of outer casing blocks on the adjacent Menkaure pyramid reveal a number of undressed stones that were probably initially covered by such ramps.

[6]Hodges, chap. 1. One must conclude from a perusal of this book that Hodges is constantly attempting to make his theories fit in with the writings of Herodotus, who visited the pyramids at Giza some two thousand years after their construction. Herodotus' writings are open to many interpretations, and because of the length of time that elapsed between the pyramids' construction and his visit, they can at best only be taken as conjecture. For the translated details, see *Herodotus: The Histories*, trans. Aubrey de Selincourt (Harmondsworth, 1954).

[7]Hodges, 83.

[8]At the halfway point in the pyramid's construction, assuming a hauling speed of 0.6 meters per second (1.36 miles per hour) the time taken to achieve a nonstop ascent of a 1-in-10 ramp would be about twenty-one minutes. Allowing time for descending and contingencies, the "hauling cycle" can be estimated at forty minutes per block. For the levering technique described by Hodges, the average time to elevate a single block at this same point in construction works out to about seven hours, although there would have been scope for elevating numerous blocks in unit time using this method.

[9]Richard Koslow, "How the Egyptians Built the Pyramids," www.egyptspyramids.com/html/article.html.

[10]Lehner (n. 1 above), 224.

Friction is a resistive force that prevents two objects from sliding freely against each other. The relationship between the force of friction and the pressure between the two surfaces—called the normal pressure—is given by the coefficient of friction, which is generally denoted by the Greek letter μ. There are different types of and values for the coefficient of friction, depending on the type of resistive force. In the case of hauling stone blocks, we are interested in the kinetic coefficient of friction, which concerns the force restricting the movement of an object sliding on a relatively smooth hard surface.[11] This is represented by the equation $\mu = F \div N$, where F is the force of friction and N is the normal pressure between two surfaces.

At this stage we have to make an assumption regarding the individual force exerted by each hauler. It has been estimated that an individual man is capable of exerting a pulling force equal to 150 pounds, or 68 kilograms.[12] This would appear to be a credible number, as it seems reasonable to think that an adult male would be capable of exerting a force approaching his own body weight, and 68 kilograms would be 90 percent of the body weight of a man weighing 75 kilograms (165 pounds). Substituting known and assumed values for the example of hauling carried out at Karnak yields this estimated kinetic coefficient of friction:

$$\mu = \frac{F}{N} = \frac{3 \times 68}{1 \times 1,000} = 0.204$$

We can now turn to an example of hauling known to have been carried out in ancient Egypt. In the Twelfth Dynasty tomb of the nobleman Djehuty-hotep at Deir el-Bersha, there is a wall painting showing a statue of the tomb owner being hauled on a sledge. The statue, which is known to have weighed about 58 tonnes, is being hauled by 172 men in four files of 43. A man is shown standing on the base of the statue pouring liquid from a jar onto the ground in front of the statue/sledge assembly. Three other men are carrying yokes of two fresh jars of liquid each, while other men walk behind the statue. Three more men are carrying what appears to be a large

lever.[13] We can use the estimated kinetic coefficient of friction determined for hauling the Karnak block, 0.204, to test whether the painting is accurate in terms of the number of haulers depicted in it. If the force of friction $F = 68 \times H$ (where H is the number of haulers) and normal pressure $N = 58$ tonnes, or 58,000 kilograms, then

$$0.204 = \frac{68 \times H}{58,000}$$

$$H = \frac{0.204 \times 58,000}{68} = \frac{11,832}{68} = 174$$

As the number of haulers depicted in the tomb painting is 172, the correlation between the two sets of data is remarkably close.

What conclusions can be drawn from these results? First, assuming that 68 kilos is a reasonable estimate for the equivalent force exerted by one hauler, then the estimated kinetic coefficient of friction for hauling both the Karnak blocks and the statue of Djehuty-hotep is 0.204. Second, while the estimated kinetic coefficient of friction would vary depending upon the exact amount of pulling force required to be exerted by the haulers, the important factor is that there is a direct correlation between the contemporary and ancient estimates, which implies that the amount of required pulling force exerted by each individual hauler was similar in both cases. Third, the calculations support the assumption that the wall painting in the tomb of Djehuty-hotep at Deir el-Bersha is accurate in terms of the number of haulers depicted in it. Fourth, the estimates and calculations provide strong evidence that the lubricating medium used for moving the ancient statue was water.[14]

Adequate ropes would have been required to haul the blocks of stone. A rope of about 8 centimeters

[11]For examples of kinetic coefficients of friction for various mixes of materials, see www.physlink.com/Education/AskExperts/ae139.cfm. There appear to be no definitive data available relating to the value of kinetic coefficient of friction between wood and lubricated stone. The kinetic coefficient of friction between wood and wood (dry, smooth, and unlubricated) is 0.2. This value would diminish if a lubricant were introduced between the sliding surfaces.

[12]Koslow, 2.

[13]Lehner, 203.

[14]Other practical experiments relating to the hauling of stone blocks have also been carried out; see Michael Barnes et al, *Secrets of Lost Empires* (London, 1996), 61–62. In 1995, a team of Egyptologists built, in Giza, a small pyramid using blocks similar in size to those used for the core and outer casing blocks of the Great Pyramid. When completed, this tiny pyramid was 6 meters high and 9 meters square at the base. The team moved 2-tonne blocks mounted on wooden sledges over a surface of *tafla* (a type of clay) and wood, lubricated with water, using two files of men hauling on 4-centimeter-diameter ropes. It was found that twelve men could move the blocks with ease up an inclined roadway. This team favored the idea that massive ramps had been constructed in order to lift the building blocks of the Great Pyramid, and they used ramps and lever techniques to construct their tiny version of a pyramid. In effect, they constructed the very last few blocks of the Great Pyramid, but at ground level—which, although a useful exercise in some respects, did not meaningfully relate to the massive scale of work carried out on the ancient monument.

diameter would have been a practical size for a team of haulers to handle. Such ropes are capable of hauling loads in excess of 4 tonnes and can be made from the doum palm, a tree indigenous to Upper Egypt.[15]

Lifting the Stone for the Great Pyramid

From an engineering point of view, a basic question to ask, when attempting to propose probable methods and techniques involved with lifting the building blocks is, "why build separate ramps when the pyramid has four inclined planes as an integral part of its structure?" Granted, these inclined planes are steep, lying at 52 degrees to the horizon. But we can examine the forces and methods required to lift blocks up them based on the aforementioned parameters coupled with some simple mathematics.

The force P to pull a body up an inclined plane (the force being parallel to the plane) is given by the equation $P = W(\mu\cos\alpha + \sin\alpha)$, where W is the weight of the body, μ is the kinetic coefficient of friction between the body and the plane, and α is the angle that the inclined plane makes with the horizon.[16] As the majority of the core blocks in the Great Pyramid weigh about 2 tonnes, let us use such a block as an example in order to calculate the force required to haul it up one side of the pyramid. Let us make these further assumptions: that these core blocks were laid layer by layer and that the outer casing blocks were put into position on the faces up which hauling was taking place as each layer of core blocks was laid; that these outer casing blocks, which were of a harder and more durable material than the core blocks, would have been dressed by the stonemasons on their angled outside surfaces in order to provide a reasonably smooth surface for the blocks to be hauled up on; that they would have been made oversize so that enough material would remain to allow final dressing of the stones once construction had been completed. It is probable that the oversized angled profiles of the outer surfaces of the casing stones were undercut to provide a series of horizontal ledges that would facilitate the erection of scaffolding from which the final dressing could be achieved. It is also highly likely that the stones at each extreme corner of each course of blocks were cut to their final shapes as building work progressed; this would have ensured the geometrical accuracy of the four angled corners and provided a guide for the final dressing process of each separate face.

The force required to keep a sledge-mounted, 2-tonne block being hauled on a single 8-centimeter rope moving up an angled face of the pyramid is around 2½ tonnes.[17] The maximum number of haulers required to sustain this force is about fifty.[18] A greater force would have been required in order to commence motion. This could have been provided by, say, four additional workers at the base of the pyramid. Such a team would be required anyway in order to deal with positioning the assembly and fastening the rope to it. Once the team had completed these tasks it would then, in conjunction with the hauling team of fifty men on the pyramid's plateau, prepare the assembly for its journey up the face of the pyramid and assist in providing its initial movement.

---

[15]*Machinery's Handbook*, 20th ed. (New York, 1978), 1122–26. The working load of an 8-centimeter fiber rope, when used at low speeds (up to 1.5 meters per second), is about 4.2 tonnes. The ultimate tensile strength of such a rope is about 29 tonnes, and it weighs approximately 4.3 kilograms per meter. The working load is calculated conservatively and provides a safety margin of almost 7 (29 ÷ 4.2). The ancient Egyptians would not have been aware of such criteria, and they probably subjected such ropes to a higher working load. Because of the high safety factor, some excess loading would have been acceptable as long as the rope was not subjected to a suddenly applied shock load. When hauling blocks of stone it is reasonable to assume that loads would be gradually applied and that motion would be within the upper limit of 1.5 meters per second.

[16]Ibid., 307.

[17]Consider the situation when the pyramid had reached half its completed height (about 74 meters above ground level). At this height each angled face would be 94 meters long, and the flat plateau onto which the blocks would be hauled would be about 115 meters square. Assuming that a single 8-centimeter rope was used to haul one block up one face of the pyramid and that the block was mounted on a wooden sledge, the total weight W would equal the weight of the block plus the weight of the sledge plus the effective weight of the rope. If we stipulate 2 tonnes for the block, 0.3 tonnes for the sledge, and 0.5 tonnes for a 120-meter-long rope, we arrive at a value for W of 2.8 tonnes. We can substitute the estimated kinetic coefficient of friction from the previous examples, 0.204, and we know that the angle of inclination is 52 degrees to the horizontal. Thus, substituting in the formula

$$P = W(\mu\cos\alpha + \sin\alpha)$$

we see that

$$P = 2.8(0.204 \times 0.616 + 0.788) = 2.8(0.914) = 2.56$$

It should be noted that if the value of the kinetic coefficient of friction were doubled the resulting value of P would only increase by 13.7 percent, to 2.91 tonnes, which would require five additional haulers to sustain (see n. 18).

[18]As with the example of transporting the statue of Djehuty-hotep, it is assumed that one man exerts a force equivalent to 68 kilograms. The number of men required to keep the block and sledge assembly moving up an angled face of the pyramid is $(2.56 \times 1000) \div 68 = 37.6$ men. However, this is for a hauling force which is parallel to the pyramid's face, whereas the haulers would actually be pulling in a horizontal direction across the pyramid's plateau. It is suggested that protective wooden battens were used at the point where the rope passed over the lip of the top outer casing block, and an additional hauling force would be required in order to overcome the frictional effects between the battens and the rope. This is difficult to estimate, although it may have accounted for an increase in force of about 20 percent, requiring a total hauling team of forty-five men. Allowing for contingencies, then, stipulate a maximum of fifty men.

In order to provide the necessary lubrication at the interface between the block and sledge assembly and the outer casing blocks, water, which was in abundant supply from the adjacent canal and harbor complex, was probably poured down the face of the pyramid up which hauling was taking place.[19] Alternatively, a person of small stature, and thus light weight, may have ridden up with the assembly, applying lubricant from a vessel, as in the example of moving the statue of Djehuty-hotep. (The addition of a "lubricator" would not have had a significant effect on the required hauling force).

At the halfway point (about 74 meters high), the hauling team would have had enough available space on the plateau to have hauled the block and sledge assembly up the face of the pyramid and onto the plateau in one continuous movement. The assembly would have "tipped" easily onto the plateau due to the generous angle between the face of the pyramid and the plateau, and wooden battens were probably used to prevent the hauling ropes from fraying.[20] (It is possible that the outer casing blocks might have been covered with a latticework of wood over which the block and sledge assemblies were hauled). Once the blocks had reached the plateau, far less energy would have been required to move them into position than had been needed to lift them to the plateau. Levering techniques would probably have been used to remove the blocks from the sledges and position them accurately in their final locations. Obviously, as the height of construction increased the additional weight of rope would have had an effect, although not a significant one, and it would have remained feasible to lift blocks of up to about 4 tonnes on a single rope. Thus, the vast majority of the blocks used in the construction of the pyramid (the core blocks and outer casing blocks) could have been lifted using the methods described. For the relatively few heavier blocks within the structure it would have been necessary to use multiple ropes, with a corresponding increase in manpower in order to facilitate lifting.

Constructing the Great Pyramid

We have shown that the building blocks used to construct the Great Pyramid could have been hauled up a face of the pyramid. Using this approach as a reasonable model, we can now consider the probable building processes.

It would be reasonable to assume that a logical methodology was applied to the construction sequences and that a systematic approach was employed involving organized teams. We can only guess at what that approach might have been. Each team would have been responsible for hauling the blocks onto the plateau and then moving them to their final positions. Suppose that each team was assigned an area to work within and be responsible for. If this area were about 5 meters wide, it would have allowed sufficient room for teams to have kept clear of each other when hauling. This means that a 5-meter-wide "slipway" would have been assigned up a hauling face of the pyramid and then carried across the flat surface of the plateau. Let us consider the situation at different stages of the pyramids' construction.

When the pyramid had reached about a quarter of its height, the plateau would have been approximately 173 meters square and 37 meters above ground level. At this point each angled face would have been 47 meters in length. Approximately 1,327,100 blocks of stone would have been laid, accounting for about 58 percent of the volume of the completed pyramid. There would have been thirty-five 5-meter-wide slipways at this height, and, due to the size of the plateau, it would have been possible to simultaneously haul blocks up two opposing faces. The teams would have commenced by laying blocks at the center of the plateau and then working outward toward each hauling face. This would have produced a capacity of seventy blocks per lift at this height. As construction continued from the center outward the hauling teams would have moved onto the top of the current course of blocks in order to have effected lifting. This two-sided approach could have continued up to a height of about 40 meters, at which stage the hauling teams would have begun to intrude upon each other's hauling space, assuming that lifting took place in one continuous movement. Because on average each core block is about a one-meter cube, each hauling team would have been responsible for laying a course of blocks about 5 meters wide. At a height of 40 meters, over 60 percent of the volume of the pyramid would have been completed and approximately 1,400,000 blocks of stone laid.

An assumption at this stage is that a section of one of the faces not being used for hauling would have been kept free of outer casing blocks to provide a "stairway" for the workforce to climb to and from the

---

[19]Lehner (n. 1 above), 204–5. It is known, from archaeological evidence, that there was a harbor and interlinked canal system adjacent to the construction site, which were fed by the waters of the Nile.
[20]Apart from manufacturing protective battens to prevent the ropes from fraying, it is probable that joiners would have constructed many diverse devices in order to aid both the hauling and building processes.

plateau. Interior passages and stairs leading up and down to various chambers would also have served as temporary ways for the workforce to reach the construction site.

Between the heights of 40 meters and 74 meters (half the height of the completed pyramid) the hauling of the stone blocks could have been carried out up one face in a single continuous movement. At the halfway point the plateau would have been approximately 115 meters square, and each angled face would have been 94 meters long. Approximately 2,012,500 blocks of stone would have been laid, accounting for almost 88 percent of the volume of the completed pyramid. There would have been twenty-three 5-meter-wide slipways at this height, and upon reaching the plateau the blocks would have been taken to the opposite side of the plateau and laid back toward the hauling face, with each hauling team laying block in rows five blocks wide. The time taken to haul one block up the face of the pyramid at this point in the construction process would have been less than 3 minutes, as compared to 40 minutes for the ramp theory and 7 hours for the lever theory at this same point.

As the building work progressed, the plateau would have become progressively smaller, reducing the working area. Between 74 meters and 80 meters two ropes might have been used in order to maximize the working space. Two files would take up only half the length of a single rope hauling team, and the lift could still have been achieved in one continuous movement. At the 80-meter point over 90 percent of the volume of the pyramid would have been completed and approximately 2,076,900 blocks laid. Technically, the final 10 percent (by volume) of the pyramid would have been the most difficult to construct. Above 80 meters, the task of hauling the blocks up to the higher levels would have been more laborious as the surface area of the plateau decreased. As progress continued, it would be likely that the blocks would have been hauled up in stages.

Once all the building blocks had been positioned, the outer casing blocks would have required dressing in order to achieve a smooth outer surface. This not inconsiderable task was probably effected using wooden scaffolding from which the stonemasons could carry out their work. Chippings, assumed to be from this dressing work, have been discovered at the base of the Great Pyramid.[21]

---

[21]Michael Jones and Angela Milward, "Survey of the Temple of Isis, Mistress of the Pyramid at Giza," *Journal of the Society for the Study of Egyptian Antiquities* 12 (1982): 139–51.

It is difficult to estimate how much time was actually spent constructing the Great Pyramid. There would have been periods, as with any project of this nature, when inclement weather, illness, and the like would have delayed the building program, and some years would have been better than others from this point of view. However, based on the techniques described here we can make an estimate of the construction time.

With respect to the first 40 meters of the pyramid, we have seen that seventy blocks per lift for the core and outer casing blocks could have been achieved, assuming that two faces of the pyramid were used simultaneously. As the width of the base is 230 meters, decreasing at the height of 40 meters to a plateau 169 meters wide, the average number of 5-meter-wide slipways would have been forty each side between ground level and 40 meters. Given an adequate supply of blocks, there would have been a capacity to move forty blocks up one face in unit time. The next question to ask is, "how long did it take, on average, to lift and position the blocks?" An estimation can be made for this by looking at the sequence for one block, assuming that it had already been delivered to the base of the pyramid on its wooden sledge: (1) Connect the hauling rope to the block/sledge assembly, 10 minutes; (2) Haul the block/sledge assembly up the side of the pyramid and onto the flat plateau, 45 seconds assuming an average speed of 0.6 meters per second—say one minute; (3) Move the block/sledge assembly across the plateau, 5 minutes (again, at 0.6 meters per second); (4) Unload the block from the sledge, 10 minutes; (5) Position the block, lower the sledge and rope down to the base of the pyramid, and disconnect the rope from the sledge, 30 minutes. Adding up the individual elements gives an overall time of approximately 56 minutes for lifting and positioning one block. Allowing for contingencies, let us round this up to one hour per block. Therefore, for the first 40 meters of the pyramid's height, the time taken to lift and put into position the estimated 1,400,000 blocks would have been 1,400,000 hours. This time is, of course, for each individual block being dealt with as a separate entity. However, if, on average, eighty slipways were operational between ground level and 40 meters, then eighty blocks could have been processed at any one time. Therefore, the time spent on this section of the pyramid per team would have been 1,400,000 ÷ 80, or 17,500 hours. Let us assume that, each year, 10 hours per day for 320 days of the year were spent on construction.

This would give an approximate time of completion for the first 40 meters of the pyramid of 17,500 ÷ (10 × 320) = 5.47 years. Allowing additional time for moving the larger, heavier, burial chamber blocks, it could be estimated that this element of construction took about six years—assuming, again, that lifting took place simultaneously up two opposing faces of the pyramid and that all the other work associated with building (quarrying, transport, and so on) also took place simultaneously. We can use a similar approach for the next 40 meters of construction, up to a height of 80 meters. The average number of slipways between these two heights would have been twenty-eight, with lifting taking place up one face only. Therefore, the time taken to lift and put into position the estimated 676,900 blocks in this section of the pyramid would have been 676,900 hours. (The time taken to haul the blocks up the face of the pyramid to a height of 80 meters would have been twice that for hauling to a height of 40 meters, but because of its reduced size less time would have been required to move the blocks across the plateau, therefore the same overall time of one hour has been applied.) The individual team time would therefore be 676,900 ÷ 28 = 24,175 hours. Applying the same building time criteria as before gives an approximate time of completion for the second 40 meters of the pyramid of 24,175 ÷ (10 × 320) = 7.55 years. Again, allowing some additional time for moving the larger blocks required for the burial chamber and passages, it could be estimated that this element of construction took about eight years. Thus, the estimated time needed to build to a height of 80 meters would have been fourteen years. From 80 meters to completion of the block laying would probably have taken, based on the previous assumptions, an additional six years, meaning that the entire structure would have been completed in twenty years.

Following completion of the block laying, there would then have been the task of dressing the outer casing blocks. This is very difficult to estimate but could well have taken a further two years to complete, assuming that work progressed on all four faces simultaneously. All this gives an estimated time for completion of the actual building work carried out on the Great Pyramid of twenty-two years. To this must be added time for preparing the site prior to building, setting out, building sledges, making ropes, and so on—say another year. The estimated completion time fits in with the generally held view that the Great Pyramid was completed during the reign of

King Khufu, which is thought to have lasted for a minimum of twenty-three years.[22]

In order to keep up with the construction work, a constant supply of cut stone blocks would have had to be available. During the first 40 meters of construction, which was the most intensive, an average of eighty blocks would have been required every sixty minutes. The core blocks would have been delivered to the building site direct from the adjacent quarry, while the blocks from Tura and Aswan would have been brought overland.[23]

During the most intensive stage of construction, that up to the 40-meter point, it can be estimated that an average workforce along the following lines would have been required. For haulers and setters, 40 teams of 50 men times 2 hauling on the plateau, 40 teams of 5 times 2 setting the blocks, 40 teams of 4 times 2 working at the base, and 80 "lubricators," for a total of 4,800. During the building of the 6-meter-high pyramid in 1995, 12 Egyptian stonemasons quarried 186 blocks of similar size to those used for the core of the Great Pyramid in 22 days using iron tools. This equates to 0.7 blocks per day per man, or, over an 8 hour working day, 0.0875 blocks per hour. Compensating for less effective copper tools, let's say 0.07 blocks per hour for the ancient stonemasons. The time taken for the hauling teams to complete this stage of the pyramid was estimated at 17,500 hours per team, which is the overall time. Therefore, in this time, and using the adjusted 1995 work rate as a guide, one stonemason would cut 1,225 blocks (17,500 × 0.07). The total number of blocks within this section is estimated at 1,400,000 blocks. Therefore, it is estimated that the number of stonemasons would have been 1,400,000 ÷ 1,225, or 1,143—say 1,200 men. As well as stonemasons, there would have been workers removing and transporting the blocks between the quarry and the pyramid. The sequence of transportation would have involved a round-trip estimated to have taken around 60 minutes per block. For a rate of 80 blocks per hour, and a team comprising 12 men, 960 workers would have been required (80 × 12), so say 1,000 men. Obviously, there would have been many other workers involved with the construction process: joiners making and repairing the wooden sledges and other devices; rope makers making and repairing the

[22]Lehner, 206.
[23]Eugen Strouhal, *Life in Ancient Egypt* (Cambridge, 1992), 173–82. In the case of the blocks from Aswan a fair proportion of the journey would probably have been via boat along the Nile.

necessary ropes required for hauling, water carriers, surveyors, supervisors, additional quarrymen, purveyors of food, and so on. It could be estimated that perhaps as many as 3,000 people were involved in these ancillary activities.

These estimations do not attempt to include the workforce involved in transporting the outer casing blocks from Tura or the granite blocks from Aswan. The figures do, however, attempt to convey some idea of the likely workforce at the construction site. Adding up the various elements gives an approximate total workforce in the immediate environs of the pyramid of about ten thousand people during the most intensive period of construction. It has been proposed that as many as twenty-five thousand workers were involved at the building site during the most intensive period of construction.[24] However, a significant proportion of this workforce would have been involved with building and maintaining massive construction ramps, which, using the methodology put forward here, would not have been required.

Conclusion

The method of construction for the Great Pyramid proposed here, using the angled faces of the structure itself as surfaces on which to transport the blocks used to construct the pyramid, provides a more logical and practical alternative methodology to the view

that massive, separately constructed ramps were used to move the stone blocks. Apart from eliminating the need to build separate ramps, such a methodology is considerably more energy efficient and far less time consuming, as it removes the need for hauling teams to go trudging up and down ramps all day long because the teams would have remained on the pyramids' level plateau—where they may indeed have lived during the more intensive periods of construction. The proposition that every individual block was elevated into position using levers and packing pieces is also an unsatisfactory solution; such a process would have been extremely awkward and risky due to the numerous maneuvers involved with each individual elevation. As with the separate ramp theory, the lifting teams would have ascended with each block lifted, which is inefficient. It is therefore suggested that levering techniques were only utilized for assisting the builders in a very localized fashion, such as loading and unloading sledges and positioning the building blocks in their final locations. The proposal that a form of shaduf was used to elevate the building blocks is also deemed an impractical solution.

It is estimated that the Great Pyramid took about twenty-three years to complete, and that during the most intensive period of work around ten thousand people were involved in its construction at the building site.

It can be concluded, from contemporary technical evidence relating to the hauling of large blocks of stone, that the wall painting in the tomb of Djehutyhotep at Deir el-Bersha is accurate in terms of the number of haulers depicted in it.

---

[24]Lehner (n. 1 above), 225.

Diffusion of Astronomy in the Ancient World

Nataraja Sarma

Astronomical techniques, calendars and devices were developed independently in many places around the world. However, there was much cross-cultural exchange of technology over the centuries. The cultures of Egypt, Greece, India and China influenced each others' astronomy and each cannot be treated in isolation.

Early nomads, herders and farmers who found that nature was unpredictable correlated their successes and disasters with the clouds, eclipses and other celestial events. Markings on Palaeolithic artifacts with obviously calendric or astronomical connections[1] indicate that these people performed magical rites at times governed by the appearance of the stars. They also found that the sun, moon and stars provided them with a reliable calendar that guided their actions. The rising and setting of the sun, the waxing and waning of the moon, and the positions of prominent stars were directly connected with their occupations, be it hunting, fishing or farming. More significantly, they soon realised that sharing observational data as well as their theories and beliefs with neighbours proved to be very productive. As time went on, when they found that proximity encouraged such exchanges, people from villages and towns all over the world gathered to interact in urban conglomerations.

Ancient Astronomy

Gradually, great centres of civilization were established in the ancient world. The best known, investigated in depth by archaeologists and historians, were located in the valleys of the Tigris and Euphrates rivers, along the Nile, in Greece and Rome, in India, and in China. Each of these centres developed its own schools of astrology and astronomy, and its own catalogue of stars and star groups. Isolated cultures also saw patterns in star groups. The Greeks named the constellation Orion because it reminded them of a hunter. The same imagery led the Canadian Blackfoot tribe to call this constellation the 'Bull of the Hills' and the Navajo Indians to call it the 'Slender One'. On the other hand, the curved configuration of the Scorpio constellation looked like the tail of a Greek scorpion, whereas the seafaring Polynesians saw it as a Fish Hook and it appeared as an Azure Dragon to the Chinese.

By analysing the movements of the sun, moon and stars, astrologers and astronomers awed their countrymen with their predictions. Invoking connections between their gods and goddesses and the celestial objects gave these intellectuals power in royal courts. The earliest civilization that looked at the sky intelligently was perhaps in Mesopotamia, in the Fertile Crescent. The inhabitants of Sumer (~3500 BC) had certain mystical concepts[2] of the heavens and deified the sun, moon and Venus. By the time of Sargon of Akkad (2376–2294 BC), they had recognized a certain order in the movements of the stars and recorded crude qualitative ideas in cuneiform texts[3].

Meanwhile, Egypt had independently developed its own astronomy during the Old Kingdom (2900–2700 BC) and the Middle Kingdom (2100–1700 BC). Limited evidence on the subject is available from the diagonal calendars carved on coffin lids of the Middle Kingdom[4], perhaps to guide the noble dead to their heavenly destinations. The orientation of tombs and pyramids to the north and the temple decorations of the Ptolemaic period (from 300 BC) also point to the advanced state of Egyptian astronomy[5]. In 2773 BC, the Egyptians had introduced a very practical civil calendar[6] that began when the star Sothis (Sirius) rose at the same time as the sun (heliacal rising) and on the rising waters of the Nile. Every ten days, they chose a different star that rose with the sun and those ten days made a 'decan'. The sidereal year[7] was therefore divided into 36 decans, each of ten days, with five epagonal days.

Once every Sothic cycle (1460 years), the astronomical and civil years coincided. Coinciding with agricultural needs, the Egyptian solar year had three seasons of four months, each of 30 days, with five holidays at the end of the year. The day was split into 24 h, the length of the hour varying with the season. The zodiac was divided into 12 signs, each lasting three decans. Hellenic astronomers adopted the calendar because it was so elegant and, 3000 years later, Copernicus used it to formulate his lunar and planetary tables[8].

Astronomy originated much earlier than is generally believed in India and China. The Shujing (Book of Documents) records observations on stars and planets from 3000 BC to 2000 BC. As early as 2254 BC, the Shi Ji (Book of Records) states that the Emperor Yao got astronomers to calculate solstices and predict seasonal changes to assist farmers[9]. The Shangshu (Historical Classic) dated four cardinal asterisms to the 21st century BC. Stars and their movements were recorded during the Shang dynasty (1500 BC), and oracle bone writings from 1400–1200 BC indicate that the Yin used a year of 12 lunar months, each of 29 or 30 days, with an extra month every three years. Records and star catalogues of astonishing detail[10] were kept from 600 BC onwards.

Tracing the history of Indian astronomy poses serious problems. Vedic texts were not written, as they have to be recited according to certain modulations. In addition, writing and copying treatises[11] might lead to an addition or defect in the text. Knowledge was handed down by oral teaching, the student learning verses by heart. Only a chosen few had access to the explanation of the contained matter[12,13]:

> This mystery of the Gods is not to be
> imparted indiscriminately.
> It is to be made known to the well-tried pupil
> who remains a year under instruction.

Commentaries on these texts were published but, as these reflected contemporary styles, the original contents were altered over time and cannot be dated or quoted to any accuracy.

The oldest Indian reference to astronomy is in the hymn[14] to the god Indra in the Rig Veda:

> He like a rounded wheel hath in swift motion
> Set his ninety racing steeds with four
> Developed vast in form, with those who sing
>   forth praise
> A youth, no more a child, he cometh to our call.

A 14th century interpretation identifies Indra with Time, which had 94 periods—one year, two solstices, five seasons, 12 months, 24 half-months, 30 days, eight watches and 12 zodiacal signs. Fearing the alternate disaster that the Vedic knowledge might be forgotten in decadent days to come, Vasuka of Kashmir defied tradition and inscribed his version of the texts on birch bark[15].

Indian astronomers were known to be more advanced than the Chinese in their theoretical concepts. The Hindu system divided the sky into 27 nakshatras (asterisms or star groupings), later modified to 28 nakshatras, each with a yogatra (principal star). This model[16] dates to the Rig Veda (~1400 BC) and is codified in the Atharva Veda and the Yajur Veda (~800 BC). However, only records written after AD 470 are now available, when Varahamihira, son of Adityadasa and descended from the Zoroasters who fled Persia around the 1st century BC, compiled the Panchsiddhantika (Five Minor Texts).

The absence of earlier records has led Western historians to exaggerate the influence of Mesopotamia and Greece on Indian astronomy. Varahamihira did acknowledge the interaction with the science of Alexandria and used terms of Greek origin in his work. As well as referring to several treatises, his work comments on five great astronomical works of the time[17]. These were the Paitamaha Siddantha (Text of the Forefathers), Surya Siddantha (Text of the Sun), Vasishta Siddantha (written by the sage Vasishta), Romaka Siddantha (derived from Byzantine Rome) and the Paulisa Siddantha (ascribed to Paulus Alexandrinus). Of these, the Paitamaha Siddantha represents the Hindu astronomy not yet affected by imported technology and is regarded as representative of the ancient astronomical knowledge of India.

Commerce between the continents had been increasing over the centuries along with the development of science. Trading ships sailed between the Levant, Greece and North Africa for millennia. After expeditions by Nearchus (326 BC) and later by Hippalus, merchants exploited the monsoon winds and trade with the East Indies spurted[18]. Caravans had trailed across Asia from the Mediterranean to China along the Great Silk Route. Persian cities became trade junctions providing branch routes into the Indian subcontinent and over the Himalayas into Tibet and China. Exciting descriptions by sailors and traders led scholars, missionaries and entrepreneurs to travel in pursuit of knowledge, adventure and profit. These developments encouraged exchange of scientific ideas and inventions between civilizations of the world.

Diffusion of Knowledge

Clear documentary evidence relating to the diffusion of technology in the ancient world is scarce and perhaps this exchange of ideas was taken for granted.

Deductions about the origin of ideas in astronomy have to be based on the similarities of systems and the changes effected at various times. The absence of direct information on scientific exchanges has led to some controversy, such as the view that Indian astronomy was dominated by repeated transmissions from Mesopotamia and Greece[19]. A contrary view[20] has it that many ideas in Indian science believed to have been borrowed from Hellenic science had more ancient bases than the Greek works. Astronomical references from the Rig Veda, a text that probably originated in the 3rd millennium BC have been interpreted to prove that Indian astronomy travelled to Babylon and Greece[21].

The earliest migration of scientists occurred when the scholars at Sumer moved[2] to Babylon after the fall of their kingdom to continue their astronomical studies. Although their observations on star positions were rudimentary, the mathematicians of Babylon could predict eclipses. It seems probable that ~1500 BC, close interaction with Egyptian astronomers changed the Mesopotamian calendar. The very practical lunar calendar of the Babylonians began when the new crescent moon was sighted just after sunset[22]. To this day, Islamic priests follow the same system to announce the holy month of Ramadan. When the dissonance between the lunar day that began at sunset and the solar day asserted itself, the Babylonians switched to the Egyptian model. They divided the sky into three areas of 12 sectors, each named after a constellation or planet, very like the Egyptian design of 12 signs each of three decans.

Increasing Egyptian influence over the next three centuries meant that observations of stars became important and the movement of Venus was recorded in cuneiform writing for several years during the reign of Ammisaduqa[23]. By about 500 BC, mathematical astronomy had vastly improved in the Mesopotamian valley. The Babylonians realised that 235 lunar months spanned 19 solar years or 228 solar months, because each solar year had 12 months. As a result, seven intercalary months were introduced into their lunar calendar, based on the needs of the harvest. The zodiac was then divided into twelve sections, each being under the influence of the sun and the planets. Qualitative rules were framed to predict planetary and lunar phenomena and for the variation of night and day through the year.

Greece, too, felt the impact of advances in Egyptian and Babylonian astronomy. The early Grecian calendar was rather chaotic, with intercalations determined by lunar and even local political considerations.

The Greeks adopted Babylonian observations on asterisms and the movements of planets and used them to predict eclipses. Perhaps out of necessity, they imported the Egyptian calendar with hours of uneven length and remodelled it with equinoctial hours of constant length. The sexagesimal division of time and of angles was also of Babylonian origin. Inspired by Egyptian practices combined with Babylonian arithmetic, the Greeks divided the circle into six times 60 degrees and each degree into 60 minutes. Similarly, 60 minutes of time formed an hour. Thus, they laid a scientific base to the existing Babylonian astronomy. They split the ecliptic into 36 equal decans and astrologers turned these into powerful elements for computations in the mystical sciences. The concept returned to Arabia in a modified form through the works of Abu Mashar[22]. These texts were translated into Greek, Hebrew and Latin, finally appearing in Western Europe in imaginative aspects. These can be seen in the Room of the Months of the Palazzo Schifanoia in Ferrara, which is adorned with frescoes painted in AD 1460 by Francesco Cossa, Ercole dei Roberti and others; these frescoes are allegories of the months and scenes from the life of Borso d'Este.

Although little is known of early Persian astronomy, there were substantial astrological and astronomical texts in Persia. In 100 BC, the Persians had divided the heavens into four equatorial and one central palace, quite similar to the Coptic moon stations. Scientific exchanges between Asia and the Mediterranean affected Persian science. Greek and Indian treatises as well as the Roman Megesti were available in AD 250. The Zij-ash-Shah, revised under the Emperor Khosro I Anosharwan[23], incorporated Hindu texts with minor changes such as a shift of the zero meridian to Babylon. Persian scholars such as Ta-Mu-She of Jaghanyan travelled to China and translated a number of texts from Sogdian and Persia into Chinese.

Long before Western ideas came to the Asian continent, there were continual exchanges between Chinese and Indian scholars, who travelled to Tibet along the southern branch of the Silk Road. Along with other Hindu gods and beliefs, the Chinese adopted the ancient myth of two planets, Rahu and Ketu, who periodically 'ate' the sun and the moon to cause eclipses[24]. By 433 BC, Chinese astronomers[1] recorded a system of 28 hsiu (lunar mansions or moon stations) marked by a prominent star or constellation. The moon travelled past and lodged in each of these mansions. This system probably originated from the

Hindu system of 28 nakshatras (asterisms, marked by a prominent star or constellation) mentioned in the Rig Veda of 1400 BC. The Atharva Veda and the Yajur Veda (800 BC) give a complete list of these nakshatras. Whereas the Hindus named their nakshatras after their gods, the Chinese honoured a heavenly human society of emperors, queens, princes, royal courts and even bureaucrats and buildings.

As the two civilizations grew apart, only a quarter of the Indian yogatra (determinative stars) could be identified with the Chinese chuxing (prominent stars). The similarity between the information contained in the 670 BC cuneiform tablets from the time of king Assurbanipal of Nineveh and the Chinese moon stations suggests that a Mesopotamian model of the 4th century BC prevalent in Asia might have evolved into the Indian and Chinese systems[24]. Another pointer comes from an old Chinese theory[24] that there was a planet diametrically opposite Jupiter and moving counter to it. This concept is very similar to that of Philolaus of Tarentum who, in turn, might have derived it from old Babylonian sources. However, a Chinese casket lid from the 5th century BC depicted the 28 lunar mansions, indicating that Chinese and Indian astronomy dated back to times earlier than was believed, but how these two systems interacted then is uncertain. It is also uncertain[23] whether the Hindu nakshatras influenced the Chinese lunar mansions or the other way around, although the ancient Suryaprajnapathi text[26] states that system originated in India.

Indian astronomers appear to have interacted with other cultures from the earliest times. The prime motivation again was to reconcile the solar, lunar and sidereal years within a geocentric model of the universe. As the practice of introducing intercalary days and months was not satisfactory, the Indians looked elsewhere for guidance. The Surya Siddhanta[27] claims that it was revealed at the end of the Golden Age (2163102 BC) by the sun god Surya to Maya, an asura (demon), a possible reference to Assurbanipal of Nineveh. Over time, the treatise was modified into a more plausible planetary theory with inputs from Greece and Babylon. For instance, the text of AD 400 contains sections that are close to the Greek theory of epicyclic motion. It is believed[28] that Greek astronomy took three centuries to be incorporated into Indian texts, because it did not come directly but diffused through Persia. As a result, the theories of Hipparchus were absorbed without the refinements of Ptolemaic theory.

The Vasishta Siddhanta, according to the AD 499 version of the Panchasiddhantika of Varahamihira, is thought to have been taken from a 2nd or 3rd century text because its planetary theory is closely related to Babylonian material of the Seleucid period. Possibly, a Greek scholar brought the theory to India. The Vasishta Siddhanta marks a transition from the older, pure text of the Surya Siddhanta[17] to one adapted from Greek science. The differences from Alexandrian works might be due to the use of manuals on astrology rather than the original text. Hindu ritual practices might also have forced changes from the Syntaxis of Ptolemy and the astronomical works of Hipparchus.

The Romaka Siddhanta[19], written around AD 300, and the Paulisa Siddantha were clearly adaptations of imported science. Srisena recast the Romaka Siddantha, drawing from an older version with inputs from contemporary works[17]. This defines a yuga (age) that covers an integral number of revolutions of all the planets. Each yuga was divided into 150 periods, each of 19 years, with 235 months in a year. Meton of Athens had proposed such a calendar in 430 BC; this was improved by Claudius Ptolemy around AD 140. The yuga of the Romaka Siddhanta is taken from the Metonic calendar. The Romaka year has the same length as the tropical year of Hipparchus, as accepted by Ptolemy. Greek astronomy continued to spread into India over the years. Yativrasabha translated Greek texts into Sanskrit during the reign of the Kshetrapa king Rudraraman[19]. These were originally Egyptian texts composed around 250 BC.

Links between China and India strengthened over the years. Indian astronomers set up shop in Changan, the capital of Tang China in the 7th century AD. Chhutan Hsi-Yuan, an Indian monk compiled an astronomical text, the Khai-Yuan Chan Ching, in AD 718. This contained a translation of the Indian Navagraha calendar. Another Tantric Buddhist monk, I-Hsing, established a school of Indian astronomers resident in China and they translated Brahman astronomy into the Po-Lo-men Thien Wen Ching. Yixing, or Zhang Sui, the most outstanding astronomer of the era, helped monks to translate Indian sutras into Chinese. Three clans of Indian astronomers, the Siddartha, Kumara and Kasyapa, lived in Changan. Gautama Siddartha translated the Indian Navagraha (Nine Houses) calendar and compiled the Da-Tang Kaiyuan Zhanjing (Prognostication Manual of the Kaiyuan period of the Tang Dynasty) in AD 718. The Chinese never took to this calendar but the Koreans used it for some time. The Chinese system of mapping stars and lunar mansions spread to Korea, Japan and Vietnam but did not go beyond these borders.

The beginning of the 20th century saw the rise of Islamic science. Syria, once a centre of pre-Islamic astronomy was overwhelmed by the new culture. The Abbasid Caliphs were great sovereigns, aware of the importance of learning[29]. Astronomy was essential to locate and face Mecca for their prayers, to determine holy days and to design mosques. Although the Holy Quran frowned on foreign ideas, the Caliphs encouraged scholars to include Hindu and Greek astronomy in their learning.

By this time, Indian astronomy, the main source of information for the Arabs, was a mixture of modified Greek and Babylonian theories. The Moslems merely adapted this knowledge to their needs. They retained Ujjain as the zero meridian but they renamed it Arin. The present eon, which the Hindus called Kaliyuga, began at midnight of 17 February 3102 BC and was called the 'Era of the Flood'. Yaquib ibn Tariq (AD 772) of Baghdad collaborated with Hindu astronomers and was a key figure in the transmission of astronomy to Islam. 40 years later, Humayun Ibn Ishaq al-Ibadi established the Bayt al-Hikmah (House of Wisdom), where the Khandyakhadyaka, the astronomical compendium of Brahmagupta, was translated as the Zij-al-Arkand. The great astronomer al-Khwarizmi used this text, along with the Geographica and the Megale Syntaxis (the Almagest or al-Majisti), for his treatises.

The Mumtahan Zij calendar based on imported data was prepared at the Bayt al-Hikmah, as were Arabic versions of the Almagest. The great astronomers of the Baghdad centre included Abu Ma'shar, who propagated Greek astrology using a Persian translation of Sphaera Barbarica of Teukros the Babylonian in AD 542, and al-Battani, whose work became most important to European Renaissance astronomy. When the Caliphate fell apart, al-Battani migrated to ar-Raqqa in the Euphrates valley, Ibn Yunis went to Cairo and al-Beruni travelled to Afghanistan and India with Mohammed Ghazni. Al-Sufi listed the Arabic names for the stellar constellations and this formed the core of the Urano-metria of AD 1603, on which modern terminology is based.

The Byzantine Emperor Justinian, who built the cathedral of Sancta Sophia in Constantinople, closed the Neo-Platonic academy that had been founded by Plato 1000 years earlier[30], causing the academics to emigrate to Persia. However, three centuries later, in AD 773, the Caliph of Baghdad, al-Ma'mun and son of Haroun al-Raschid, acquired Ptolemy's Almagest in a peace treaty with the Byzantine emperor. He then gathered astronomers to his new centre. These scholars combined Greek astronomy with traditions from Persia and India. Their work spread to Sicily, Southern Italy and Spain. By the 12th century AD, Cordoba and Toledo became great centres of astronomy. After the Christian Church split in AD 1054, the Eastern Christians (centred on Byzantium) maintained cordial relations with the Caliphates and Islamic astronomy became fashionable under the emperor Comnenus. The talent of Byzantine scholars was such that their work was more advanced than the Alfonsine Tables of AD 1270.

A prisoner of the Mongols, the Moslem Nasiruddin al-Tusi became a trusted adviser of Prince Hulagu Khan and built another observatory in Maragha. He published the Persian Ilkhani Tables in AD 1271. The fame of this astronomer spread from Byzantium to China. Ibn-ash-Shatir worked on the theories of al-Tusi, and these appear in De Revolutionibus by Copernicus. Meanwhile Islamic science spread northwards to Persia. Under the supervision of Jalal-ad-Din Malik Shah, an observatory was built in Ispahan. The lunar year was abandoned in favour of the Jalali tropical year. Ulugh Beg, grandson of Timur, established the Samarkand observatory in AD 1420, planned by the astronomer al-Kashi. Their star catalogues and tables were part of the Historia Coelestis Britannica (AD 1725) of Flamsteed. This work was used by Gauss in AD 1799. Arab books on astronomy such as the Sanjari-Zij, Zij-al-Alai and various texts written by Chioniades were taken from Tabriz to Constantinople, where they were translated into Greek before reaching European libraries.

Arab Spain also contributed significantly to the spread of astronomy into Europe[4]. The Moslem scholar Majriti revised the Planisphaerium of Ptolemy into Arabic[31] and this was later rendered in Latin. Zargali (AD 1089) in medieval Spain took star data from old Graeco-Roman papyri for his Perpetual Almanac. He was also involved in the making of the Toledan Tables that Gerard of Cremona distributed in Europe. Ibn Ezra was another author whose treatises contributed to the spread of Islamic astronomy into Europe. The model of the heavens proposed by al-Bittrunji formed the core of a commentary by Maimonaides, and this was translated by Michael Scot from Toledo in AD 1217. Based on Islamic works of the time, Immanuel Bonfils composed the astronomical treatise Six Wings in Hebrew. Michael Chrysokokkes translated this into Greek along with a commentary, the Hexapterygon.

The reputation of Islamic science attracted scholars from Europe. Abelard of Bath (AD 1116–1142) travelled to Paris and then to Salerno and Sicily to learn Arabic before going to Spain. Boethius had translated parts of the Organon of Aristotle in AD 520. This new interest in Islamic science resulted in a flood of translations for two centuries. Toledo, in particular, proved to be a treasure house. The great scholar, Gerard of Cremona (AD 1175) translated the Almagest of Ptolemy and several other taxits of Thabit ibn Qurra and of Al-Farghani from Arabic to Latin. The earliest model of planetary motion by Euxodus of Cnidos (4th century BC) adapted by Aristotle was revived by Al-Bitrunji in 12th century Spain. This was translated by Michael Scot in AD 1217 as part of the De Motibus Celorum. Nasir ad-Din at-Tusi (13th century AD), a prominent astronomer at the Maragha observatory in Persia, modified Ptolemy's models in the Tadhkira fi ilm al-Haya (Memoir on Astronomy). A short treatise, translated by Gregory Chioniades, describes the lunar theory of at-Tusi. Scholars who travelled to Persia translated several Arabic and Persian astronomical works into Greek versions that Copernicus later used.

Arabic and Latin versions of Ptolemy's *Almagest* slowly filtered into Western Europe and stimulated interest in astronomy. Initially, Europeans merely wrote commentaries on the Almagest and calculated tables of planetary motions. George Trebizond, a Cretan emigrant in the Curia, translated the Almagest from the Greek for Pope Nicholas V in AD 1451 along with a long commentary, but this was never dedicated to the Pope after bitter criticism of the work. His son, Andreas Trebizond did, however, dedicate a more elaborate version to Pope Sixtus IV. Later, Johannes Regiomontanus produced an acceptable version of the Almagest that became a standard work for European astronomers.

An early Greek astronomer, Aristarchus, propounded a heliocentric theory but was shouted down by Pythagoras and Aristotle. Studying translations of Arab texts, Nicholas of Cusa (AD 1401–1464) and Leonardo da Vinci questioned the basic assumptions that the earth was the centre of the universe and was stationary. Nicolas Copernicus published the modern picture of the earth as a planet of the sun in the De Revolutionibus in AD 1530. No one paid any attention until AD 1539, when George Rheticus reissued the book. As it ran counter to the theories of the political powerful churchmen of the time, the De Revolutionibus was placed on the Papal Index in AD 1616 and stayed there for 200 years. The Inquisition penalized Galileo and Giordano Bruno for their acceptance of the Copernican theory. For suggesting that the solar system was only one of many in the universe, Bruno was tried before the Inquisition, condemned and burned at the stake in AD 1600. Galileo was forced to renounce Copernican theories in AD 1633 and was sentenced to life imprisonment.

Astronomy in China around AD 1250 was stagnant, discouraged by the Imperial Court. Arab astronomers arrived in China but they were overtaken by the Jesuit missionaries, in particular Adam Schall von Bell (AD 1591–1666), who convinced the Chinese that Galilean astronomy was superior to their own systems[24]. After he cured the Dowager Empress of a strange ailment, von Bell was appointed an official with vast influence at court. He then translated Western books and reformed the Chinese calendar. However the influence of the Jesuits waned after they were expelled in AD 1717. The Chinese thought that Western science could not coexist with traditional geomancy (feng shui). The expulsion of the missionaries triggered a return to their ancient models of the sky, much to the detriment of Chinese astronomy for the next two centuries.

The exchange of information between the various centres of astronomy led to a similarity, if not uniformity, of calendars and the view of the heavens. The Mayan civilization, totally isolated by the oceans, evolved a different and complex sacred calendar tailored to their agricultural economy[32]. They had 13 months in the year, each of 20 days. The solar year of 365 days was unavoidable but they divided it up into 18 months of 20 days, interposing five intercalary days. The Dresden Codex[33], one of only three records that survive on the astronomy of the Mayans, records that the planet Venus was most important. The Mayans identified Venus with Kukulcan, the Mayan equivalent of Quetzacoatl, and measured the rise and set of the planet to an accuracy comparable with modern astronomy.

Diffusion of Hardware

Ancient astronomers had various recipes to predict the movements of stars and strange events such as eclipses A measure of time was derived from these recipes, supported by observations of the night sky. Instruments were then built to ascertain the positions of celestial objects, primarily to time religious observances. Along with the spread of theoretical models of the universe and explanations of observed celestial phenomena, the design of instruments for the measurement of the positions of the sun, the moon, stars and planets were exchanged between people of different lands.

The technology of astronomical instruments and clocks spread between centres of civilization along with the interchange of theories and data on the movement of the sun, moon and stars. The earliest observations of the sky were of course naked eye sightings of the rise and set of heavenly bodies, primarily to determine agricultural practices: 'When the Pleiads, daughters of Atlas start to rise, begin your harvest, plough when they go down'[34]. The first instrument was the gnomon, which measured the passage of time by measuring the shadow cast by the sun from a pole. The gnomon provided more reliable measurements on the seasons. This device, a basic instrument of primitive societies and an essential feature of ancient Hindu astronomy, is still in use in the Malay Archipelago as a calendar[35].

The circle of stones at Namoratunga in Kenya[36], the sun circle at Cahokia in the Mississippi Valley[37] and the circle of stones at Stonehenge were all ancient instruments that observed star risings. They are so similar that one might suspect diffusion of technology in those ancient times. One of the oldest handheld instruments was the Egyptian 'merkhet', which was made from a palm leaf to measure the positions of stars. Gnomons or plumb lines measured the solar shadow during the day. From the 13th century BC, graduated sundials and water clocks were in vogue in Egypt[38]. The pharaoh Tuthmosis II is reported to have carried a portable sundial[38]. As well as the gnomon and the clepsydra, the Babylonians had the 'polos', a hemispherical bowl in which the shadow of a suspended bead revealed the time of day and the season.

As prominent stellar objects circled the earth, astronomers devised rings to represent their progression. The Greeks improved the polos by replacing the bowl by a skeleton sphere with a central belt for the zodiac and hoops perpendicular to it[39]. This was the 'armilla', the forerunner of the armillary sphere, a series of concentric rings that served as a stellar clock. When the analytical ability to project three-dimensional space on to a plane was achieved, the astrolabe came into use as a more practical and portable instrument. As civilization progressed, the obviously curved sky was modelled as a ball with the positions of prominent stars marked on it. Celestial globes or armillary spheres were made independently in China, Greece, Rome and the Levant. A Chinese silk scroll, Wuxingzhan (Astrology of the Five Planets) of the period 246–177 BC suggests that an armillary sphere was used to obtain data on the planets[10]. Centuries later, the tomb of Liu Sheng (AD 25–220) was found to contain a bronze clepsydra and a sundial. There is mention of such instruments in Vedic Hindu literature but the scarcity of recorded detail suggests, perhaps mistakenly, that only naked eye observations were made in the subcontinent before 500 BC.

By about AD 1200, the Alexandrian school of Claudius Ptolemy had made sophisticated armillary spheres probaly based on the designs that Hipparchus and Archimedes had used three centuries earlier[40]. Both the European and the Chinese centres added astronomical data and, in later models, incorporated some automatic movement. The early development of this instrument appears to have been independent as there was a basic difference between the Eastern and Western armillary spheres. Whereas the Chinese model rotated around the polar axis, the Europeans related the axis to the ecliptic. This was due to their divergent views of the universe. The Chinese and the Hindus believed the universe rotated around the Pole, or Mount Meru, whereas the Europeans laid more importance on the movement of the sun.

The Ptolemaic armillary spheres were so good that Islamic astronomers adopted their designs with hardly any change and took them to Spain in the 13th century AD. In the East, Chinese developments on the armillary sphere spread into the Orient. The Indian Surya Siddhanta[41] describes a wooden armillary sphere held along its equator, very close to the Chinese model in both construction and movement. Technology also diffused from India to China. The Indian astronomer monk Gautama Siddartha had set up an armillary sphere on the orders of Emperor Xuanzong[10]. Three years later, the Tantric monk Yixing, also known as Zhang Sui, made a water powered armillary sphere.

Other Indian instruments before AD 800 are very similar to those described by Ptolemy, Vitruvius and others, suggesting that, after the death of Claudius Ptolemy in AD 150, Alexandrian astronomers might have migrated to India[40]. The Arthashastra of Kautilya contains details of instruments for use in astronomy and in war. These include a circular chronometer and other mechanical clocks[42]. A later work, the Samaranga Sutradhara of Bhoja, describes an armillary sphere depicting the movements of the stars and planets, built by or with the assistance of Yavanas (Greeks). These astronomical instruments from Greece were imported into India, travelled West again into the Moslem countries and improved the technology existing there.

Around 180 BC, Hipparchus of Nicaea had caused a revolution in astronomy by his work on stereography, by which the three-dimensional heaven was projected on to two discs. This device, the astrolabe, was easily driven by water power and a clepsydra was ideal for this purpose. The Chinese missed out on this advance as they lacked the analytical background to make astrolabes. The astrolabe became known to Islamic science through translations of Greek texts around AD 900. The Moslems found it useful to determine prayer times as well as to find the direction of Mecca. As Islam spread and the Moslems conquered Spain, astrolabes were introduced through North Africa into medieval Europe in the 10th century AD.

Mensura Astrolabi, an 11th century manuscript by Hermanus Contractus, Abbott of Reichman Abbey, describes the use and construction of an astrolabe based on a 10th century manuscript of Lobel of Barcelona. By the 11th century AD, astrolabes and armillary spheres were established in Christian monasteries. The earliest specimens had Latin and Arabic words engraved on them. In 200 years, thanks to Christian monasteries in North Spain the astrolabe gained popularity in European centres of learning. Around AD 1267, Persian astrolabes were imported into China but they were not used for more than 300 years[40].

Water-powered clocks became popular in the Moslem world after AD 850, when Greek designs were adapted. Al Beruni (AD 1000) added gear mechanisms to show the calendar and the times of rise and set of the sun and moon. Developments in Indian astronomical instruments were closely allied to both Chinese and Moslem work, indicating a free flow of technology across the seas and the mountains. The astrolabe came to India during the reign of firoz Shah (AD 1351–1388) and this Yantra-raja (King of instruments) is described by Mahendra Suri in AD 1370. By AD 1300, clepsydra-powered clocks, far more complex than the original Greek models, were fabricated in India and exported to the Levant and to Europe. In this era, Ridwan and Al-Jazari describe armillary spheres, gnomons and water clocks very similar to those detailed in the Surya Siddhanta, along with clocks like those developed by Archimedes.

Some of these have innards that resemble Chinese gear wheels.

Clock technology travelled from Arabia into Spain in two waves[40]: first in AD 1087, when the Toledo Tables were first received; and then in AD 1274, with the Alfosine Tables and astronomical theories, techniques and instruments. However, Islamic water clocks were installed in Fez, Morocco and an Arabic book of AD 1150 describes the machinery of a water clock devised by Archimedes[43]. These instruments were Greek innovations that were imported into Arabia and then taken on to India, where several improvements were made incorporating Chinese technology. For instance, the Surya Siddhanta mentions perpetual motion machines.

Just two centuries after the Toledo Tables arrived, armillary spheres and suggestions for perpetual motion machines were mentioned[44] by Villard de Honnecourt in AD 1257 and by Leonardo da Vinci in AD 1507, among others. The connection of these devices with magnetism suggests a Chinese influence they are believed to have discovered and used lodestones. By AD 1300, a mass of Arabian texts on clock devices was available in Europe, but none of the designs incorporated the escapement, a vital component of mechanical clocks. However, within 50 years, this device had become commonplace. It is conjectured that the waterwheel escapement, the pioneering development in clock technology, came in garbled form from China through India and the Middle East to Europe[40].

Guo Shoujing (AD 1231–1316) was perhaps the last Chinese astronomer who used precise instruments, and made the most advanced and accurate traditional calendar. Chinese astronomy then went into a decline until Arab astronomers arrived in China and brought in their instruments. About this time, the Emperors of the Yuan dynasty (AD 1280–1367) welcomed Europeans who came to trade. In AD 1583, Chen Jui, Viceroy of Kuang Tang and Wang Pan, Governor of Chao-Ching in Kuang Tang, invited Jesuit missionaries who had established themselves in Macao to their courts. They sought the technology of the modern chiming clocks that the Jesuits possessed[40]. These devices were thought to be miraculous manifestations of the European mind, a view that the Jesuits enthusiastically endorsed. The Chinese had been measuring the sun's shadow by gnomons since 1500 BC and complex models had been developed for their use by the 12th century AD. All such astronomical instruments were made by the

Imperial Government and had perforce to be large and impressive. The advantage of size was realized later by Maharaja Jai Singh of Jaipur when he discovered that the large size of the apparatus tolerated errors in workmanship[45].

The Jesuit missionaries introduced the European horizontal sundial. They presented the Emperor with a magnificent spring clock and, in AD 1601, were entrusted with its maintenance. In turn, the Jesuits recruited European clock-makers and, in course of time, the Imperial Palace in Beijing acquired several thousand clocks, watches and celestial globes. However, when the expansion of Christianity threatened the monarchy, Emperor Sheng Tsu issued an Imperial Decree banning the spread of Christianity and expelled all missionaries except those at court[46]. The traditional geomancy of the Chinese could not be reconciled with Western science, which was rejected. The Chinese then claimed that the clocks were exclusively native inventions. After their expulsion, the Jesuits emphasized the decadent state of technology in Ming China and this influenced Western thought for a long time[40]. The Chinese had reverted to their ancient models, in which the stars were pasted on a mildly curved plane that revolved round the pole. However they could not resist the technological brilliance of the many clocks installed by the Jesuits.

The resurgence of science in 16th century Europe saw the spread of modern astronomy throughout the world. The radical ideas of Giordano Bruno, Copernicus and Galileo that the earth and other planets were part of a heliocentric system became acceptable and changed the nature of astronomy. At long last, the dissonance between the solar and lunar calendars that had plagued astronomers and calendar makers over four millenia was resolved. The invention of the mechanical escapement in Europe also revolutionized the design of clocks. These advances in astronomy and time keeping spread over the world encouraged by the aggressive trading policies of Portugal, Holland, England and France. Their colonial administrators, often scholars, built observatories from their private means and trained local people in astronomy and the observation of stellar events.

Throughout history, scholars have, irrespective of their origins, always exchanged information and ideas. The pace of these interactions, however increased rapidly over the centuries. The advent of electronic communication has only accelerated this trend to what might be an ultimate human limit.

References

1. Heggie, D.C. (1981) *Megalithic Science*, Thames & Hudson, London, UK
2. Bottero, J. (1973) *Ancient Empires* (Brandon, S.G.F., ed.), p.24, Readers Digest Association, London, UK
3. Negbauer, O. (1955) *Astronomical Cuneiform Texts*, Institute for Advanced Study, Princeton, USA; Lund Humphreys, London, UK
4. De Young, G. (1998) *Encyclopaedia of the History of Science, Technology, and Medicine in Non-Western Cultures* (Selin, H., ed.), p. 111, Kluwer
5. Spence, K. (2000) Ancient Egyptian chronology and the astronomical orientation of the pyramids. *Nature* 408, 320–324
6. Winlock, H.E. (1940) The origin of the ancient Egyptian calendar. *Proc. Am. Philos. Soc.* 83, 447
7. Negebauer, O. (1965) Transmission of planetary theories. *Scripta Math.* 22, 265
8. Whitrow, G.J. (1988) *Time in History*, p. 24, Oxford University Press
9. Debernadi, J. *Encyclopaedia of the History of Science, Technology, and Medicine in Non-Western Cultures* (Selin, H., ed.), p. 974, Kluwer
10. Yoke, H.P. *Encyclopaedia of the History of Science, Technology, and Medicine in Non-Western Cultures* (Selin, H., ed.), p. 108, Kluwer
11. Sachau, E.C. (1910) *Al-Beruni's Indica* (Vol. 1), p. 126, Kegan Paul, London, UK
12. Sarma, N. (1991) Measures of time in ancient India. *Endeavour* 15, 185–188
13. Burgess, E. (1935) *Translation of the Suryasiddantha, 1860* (Gangooly, P., ed.), p. 186, University of Calcutta
14. Chattopadhyaya, D., ed. (1982) *Studies in the History of Science in India*, Editorial Enterprises, Delhi
15. Sachau, E.C. (1910) *Al-Beruni's Indica* (Vol. l), p. 127, Kegan Paul, London, UK
16. Bose, D.M. *et al.* (1971) *A Concise History of Science in India*, Indian National Science Academy, Delhi
17. Thibaut, S. and Dwivedi, S. Pancha Siddanthika, Benares, 1889, Motilal Banarsidass, 1930.
18. Mookerji, R.K. (1912) *Indian Shipping*, Longmans, London
19. Pingree, D. (1978) *Dictionary of Scientific Biography: History of Mathematical Astronomy in India*, Scribners, New York

20. Filliozat, J. (1964) *Classical Doctrine of Indian Medicine*, Munshiram Manoharlal, Delhi

21. Kak, S. (1994) *The Astronomical Code of the Rig Veda*, Aditya Prakashan, Delhi

22. Negebauer, O. (1957) *The Exact Sciences in Antiquity*, Brown University Press

23. Negebauer, O. (1975) *A History of Ancient Mathematical Astronomy*, Springer-Verlag

24. Needham, J. (1954) *Science and Civilization in China* (Vol. 3), Cambridge University Press

25. Ho Peng Yoke (1998) Astronomy in China. In *Encyclopaedia of the History of Science, Technology, and Medicine in Non-Western Cultures*, p. 109, Kluwer

26. Wilson, H.H. (1864) *Vishnu Purana* (Hall, F., ed.), Trubna, London

27. Burgess, E. (1935) *Translation of the Suryasiddantha, 1860* (Gangooly, P., ed.), p. viii, University of Calcutta

28. Burgess, J. (1893) Notes on Hindu Astronomy. *J. Royal Asiatic Soc. Great Britain* 717

29. Nehru, J. (1934) *Glimpses of World History*, Lindsay Drummond, London

30. Whitrow, G.J. (1988) *Time in History*, p. 77, Oxford University Press

31. King, D.A., *Encyclopaedia of the History of Science, Technology, and Medicine in Non-Western Cultures* (Selin, H., ed.), p. 126, Kluwer

32. Whitrow, G.F. (1988) *Time in History*, p. 24, Oxford University Press

33. Thompson, J.E.S. (1972) A Commentary on the Dresden Codex, A Mayan Hieroglyphic Book, p. 62, American Philosophical Society, Philadelphia

34. Lamberton, R., trans. (1993) *Hesiod: Work and Days and Theogony*, Hackett

35. Ammarell, G. *Encyclopaedia of the History of Science, Technology, and Medicine in Non-Western Cultures* (Selin, H., ed.), p. 119, Kluwer, and references therein

36. Robbins, L.H. (1978) Namoratunga, The first archeoastronomical evidence in sub-saharan Africa. *Science* 220, 766

37. Fowler, M.L. (1975) A pre-Columbian urban centre on the Mississippi. *Sci. Am.* 233, 92

38. Breasted, J.H. (1936) *The Beginnings of Time Measurement and the Origins of our Calendar, Time and its Mysteries* (Series I), p. 80, New York University Press

39. Labat, R. (1957) *Ancient and Medieval Science* (Taton, R., ed.), Thames & Hudson, London

40. Needham, J. *et al.* (1960) *Heavenly Clockwork*, Cambridge University Press

41. Burgess, E. (1997) *The Surya Siddhanta* (Gangooly, P., trans.), p. 298, Motilal Banarsidass, Delhi

42. Raghavan, V. (1953) *Indian Hist. Quart.* 4, p. 256

43. Hill, D.R. (1976) *On the Construction of Water Clocks*, p. 9, Turner & Devereux

44. Direks: Perpetuum Mobile, Spon, London, 1861, Reprinted, B.M. Israel, Amsterdam, 1968

45. Proverbio. *Encyclopaedia of the History of Science, Technology, and Medicine in Non-Western Cultures* (Selin, H., ed.), p. 921, Kluwer

46. Yoke, H.P *et al.* (1987) *An Introduction to Science and Civilization in China*, University of Washington

Pyramids at Giza

Great Pyramid of Giza

Step Pyramid of Djose

Hieroglyphic alphabet

Technology in the Medieval West

Reading 5

Pamela O. Long

The western Roman Empire disintegrated in the fifth century after a series of migrations and conquests by so-called barbarians, including Goths, Vandals, Huns Franks, and Lombards. Especially in the Mediterranean regions, Roman legal, political, and social structures often persisted, as did some Roman technologies. Yet many characteristics of the empire were lost, and many others were overlaid by new ones. Although some towns remained, notably in the Lombard region of Italy, for the most part Roman urbanism disappeared.

Agriculture and the Work of the Countryside

The early medieval centuries in the west were profoundly rural. Two patterns of settlement and cultivation prevailed. Substantial villages, safer because more easily defensible, developed in regions where the soil was fertile enough to support larger concentrations of people. In regions of poor soil, such as Scotland, Wales, and the highlands of France, dispersed settlements were the rule. Peasants lived on isolated family farms or very small hamlets. They practiced what is called in-field, out-field agriculture. Each household had a small plot of land close to the dwelling (in-field) that was cultivated continuously and fertilized with the waste of humans and farm animals. It also cultivated a plot of land further away (out-field) until it became depleted of nutrients after a year or so. Then they let it lie fallow, using it for grazing (which yielded fertilizer in the form of manure), while they cultivated another plot.

Peasants in the larger villages usually practiced open-field farming. Each village was surrounded by a tract of land that was divided roughly in half. One was planted each year, while the second lay fallow. In northern Europe the field was divided into long narrow strips known as "balks." Each household held a number of strips scattered in each of the two fields.

In southern Europe cross-plowing made for rectangular or square fields. In both cases, peasants carried out biennial rotation, planting crops on some lands and letting others lie fallow. In the Mediterranean region peasants used a scratch plow or *sol ard*, discussed in chapter 1. Sometimes it was outfitted with a colter or knife attached to the beam of the plow in order to cut the soil in front of the plowshare, and it made shallow, crisscross furrows, thereby conserving moisture in the dry summers. In the north, the characteristically wet soil made a different sort of plowing advantageous. In the early medieval period, a heavy plow came into use that cut deep furrows and turned the soil, facilitating drainage. The heavy plow had a colter that cut the soil vertically, a plowshare that sliced it horizontally, and a mold-board that flung the slice aside. Heavy plows needed teams of four to eight oxen.

From the tenth century on, medieval Europe underwent a great increase in agricultural productivity. In his influential *Medieval Technology and Social Change*, Lynn White Jr. argued that this amounted to an "agricultural revolution" engendered by a series of technological innovations. White based his thesis on three related developments: the invention of the heavy, wheeled plow; a traction revolution entailing the invention of the rigid horse collar and the use of horseshoes, which allowed horses (faster and more efficient than oxen) to pull plows; and a shift from a two-field to a three-field system of crop rotation.

Many scholars disagree with White's thesis of an agricultural revolution. Historian of agriculture Karl Brunner noted that the renewed study of archaeological finds (particularly an important find of fifth-century tools and implements at Osterburken in Germany) and the reexamination of objects in museums support a view of gradual technological change in agricultural implements from Roman to early medieval times. Indeed, transitional kinds of plowshares that anticipated the heavy medieval plow have been found from late imperial Rome. With regard to scythes, long-handled cutting implements used for harvesting, Brunner suggested the same gradual development. When harvesting hay, a peasant swung the scythe, cutting a number of stalks with a single swing. But for much of the medieval era, peasants cut grain with a sickle, one sheaf at a time, to avoid losing any of the crop. Scholars disagree about when the scythe began to be used for harvesting grain. Michael Toch suggested a much later date than proponents of a revolutionary hypothesis have assumed, even as late as the sixteenth century. He argued that agricultural productivity in the early middle ages increased, not by virtue of technological change, but by "the more intensive application of human work" and by "diffusion, organisational adaptation and elaboration." Recently, Georges Raepsaet and Georges Comet have emphasized the complexity of technological change in agriculture and the importance of the social context to such change.

Whatever the rate of long-term change, the work of the peasants involved an exacting and largely unchanging routine. They plowed in the spring, sowed seeds by hand, and then smoothed the soil by dragging brushwood over it, or, later, with a wooden harrow. In the summer, fields had to be weeded and manured. Sheep and other livestock had to be tended, gardens cultivated, hay cut and stored. In the fall, grain was harvested, threshed, and stored. Winter was the season for repairing tools and making clothing. Each household would have several fruit trees, perhaps a small vegetable garden, and some animals, usually sheep and pigs, less often cattle. The village maintained a common pasture where animals could graze, and woods where pigs could root and where firewood, nuts, and berries could be gathered. A system prevailed wherein villagers plowed, sowed, and harvested crops together while a herdsman watched the animals in the pasture. Village women made simple clothing. Villagers often lived in mud huts with thatched roofs; they ate mostly bread and drank wine or ale. Meat was scarce, but was sometimes available after the slaughter of a pig, a tough animal that could fend for itself by rooting. Pork was preserved by salting, white pigskin was used to make shoes, harnesses, belts, and bags.

There is little doubt that agricultural productivity began to increase in the tenth century. Scholars have argued that changes were gradual, and that the causes of an increased food supply are a complex matter—for example, the decline in violence during the century after the worst period of raids by Vikings and other marauders, and also a small but significant warming of the climate. Rather than endorsing a technologically driven model of agricultural revolution, historians now take a comparative view. They ask what motivated one region to adopt technological changes while neighboring regions did not. They ask about the connections between rural and urban areas, agriculture and trade, connections that have been investigated especially in the Low Countries (present-day Netherlands and Belgium). Such questions signal a shift toward a conception of medieval agriculture as part of market economy rather than as a subsistence economy, particularly after the eleventh century.[7]

Whatever the other causes may be, the rising productivity of medieval agriculture can be attributed at least in part to improvements in agricultural technique. Very gradually, beginning in the eighth century, peasants changed from a two-field system of crop rotation to a three-field system. This new system emerged primarily in the north because it required spring planting, successful only in regions of wet soil. Available land would be divided into three parts. One-third lay fallow and one-third was planted in the fall with winter crops such as wheat and rye that would be harvested in the summer. The third part was planted in the spring with oats, barley, or nitrogen-fixing legumes. This could increase crop productivity by fully a third. But the changeover seems to have occurred far later than had once been assumed; one historian writes that "not until the mid-thirteenth century do we find a conscious and regular rotation." And rotation always varied according to region, soil, climate, demand, and custom.

There were innovations that resulted in the more efficient use of animal power, in particular the harnessing of horses as draft animals after the invention of a rigid horse collar. The traditional view was that the Romans had used a horse harness that put pressure on the windpipe, choking it when it tried to pull heavy loads. Although this view has been challenged as overly simplistic, it is nevertheless true that the rigid horse collar was a new medieval development. It rested on the animal's collarbone, allowing it to use its entire bodyweight to draw the load while at the same time breathing freely. While the rigid horse collar may have been invented in central Asia, it came into widespread European use after the tenth century. Another type of improved harnassing called the breast strap harness, which originated in ancient China, was also widely adopted in medieval Europe, especially in southern regions. Other improvements included the use of horseshoes, which helped to prevent lameness, and harnesses whereby horses could be hitched in tandem (one in front of the other) rather than side by side. Unlike oxen, horses require grain, such as oats, in addition to hay. Even though they are faster than oxen and have greater endurance, their employment as a draft animal was a gradual process, never universal, and varied from one region to another. In Flanders, for example, the horse was widely used from the twelfth century on, but this was "much more precocious use" than in regions such as Picardy in northern France, or England, where the-ox-to-horse transition did not occur until the sixteenth century.

In some areas, especially where dispersed settlements predominated, peasants cultivated their own land, called *allods*, and could consume all the products of their labor. In much of the rest of Europe, however, villages came to be ruled by lords in accord with a system called manorialism. While the origins of this system are controversial, it seems to have been generally in place by Carolingian times in the early ninth century. The land controlled by the lord, including one or more villages, was called a manor. Conditions and terms varied widely from one region to another, and even from one manor to another. Lords gained control of villages in a variety of ways, perhaps because of the peasants' need for protection from marauders, or simply by virtue of the disproportionate power of the lord. The lord and lady might live in the manor house with their children and supervise the peasants themselves, or they could employ a bailiff or overseer who supervised the work and collected the lord's revenues. Usually the lord took from a third to half of the produce, as well as livestock and fish caught in his streams. The lord's holdings, including the strips of the fields that belonged to him, were called the *desmesne*. His animals grazed on the common pasture. Male peasants supplied labor for building and digging ditches, while peasant women, supervised by the lady of the manor, often worked in the manor house spinning, weaving, and performing other chores.

Most of the tools, equipment, and food used by both lord and peasant were made on the manor or in the village. Flour had to be ground from grain, bread baked, beer, wine, cheese, and butter prepared by various methods. Animals had to be slaughtered and meat cured and made into sausages. If the laws of the manor allowed, peasants fished. From the forest, fruit and berries had to be gathered along with wood for fuel and other purposes such as making tools, barrels, and furniture. Peasants carried out their burdensome tasks cooperatively, but often divided according

to gender. While men did most heavy plowing and other heavy labor, women and older children may have goaded the oxen. Women bore and cared for children, raised poultry and livestock, milked cows, sheep, and goats, sheared sheep, tended cottage gardens, fetched water from wells, took grain to the mill, gathered firewood, and tended the fire. They spun yarn, made cloth from wool and from fibrous plants, sewed up clothing, did laundry, and prepared food and drink such as ale. In the fields they hoed, weeded, reaped, and tied hay into sheaves. At the end of the harvest, they gleaned, the backbreaking task of picking up stray grains from the field.

As the population increased, more and more land was cleared for cultivation, a process that medieval documents call *assarting*. First brush and light woods were cleared, then heavier forests cut and the troublesome tree roots removed. Marshland was drained as well, a difficult task but one which yielded rich, arable land for cultivation. Land reclamation was a gradual process that is largely undocumented, but is evident from place-names (*assart*, meaning clearing, is found in the names of many new villages), and from other scattered sources. The methods used to expand arable land depended in part upon geography. A special case is the Netherlands, where an extensive system of dikes was built between the twelfth and fourteenth centuries. At the same time, peat moors were drained and made suitable for cultivation by digging parallel ditches, burning the scrub, and then planting crops in the peat. Drainage was maintained only by scrupulous water management.

The emphasis that historians traditionally have placed on cultivation should not obscure the economic significance of uncultivated areas. Paolo Squatriti has noted the growing import of wetlands on the early medieval Italian peninsula. As populations declined and Roman systems of hydraulic management disappeared, marshland was increasingly utilized to obtain "fish, waterfowl, wood, twine, reeds, and pasture, as well as numerous mammals." Occasionally, fishing became a full-time occupation. In the tenth century, for example, the fishers of Pavia on the Po River in northern Italy maintained a fleet of at least sixty boats which they used to reach fisheries. This points to a highly developed enterprise of inland freshwater fishing.

Any generalized picture of medieval agriculture will be modified and vastly complicated by detailed investigation. Regional trends that have been observed include the transition from manorial system to lease-holding—a system in which the peasant simply leases a small plot from the lord—evident in the first half of the twelfth century in Flanders; the continued wage labor performed by peasants to supplement their income; and the increasingly close relationship of large and small farms "bound by the market." Careful studies of agriculture in Denmark and Sweden underscore the diversity between regions and among geographic areas within larger regions. Such studies enhance our sense of the complexity of medieval agriculture, and have ended the dubious practice of investigating one discrete region and then generalizing to all the rest of Europe.

An alternative agricultural system is exemplified by Cistercian farming begun in the twelfth century. The Cistercians were a reformed monastic order founded in Citeaux in southern France. Historians at one time credited the new order with pioneering land clearance and drainage in the twelfth and thirteenth centuries, and there is no doubt that they prospered on their new agricultural estates, called granges. More recent research has shown, however, that Cistercian successes were not the result of clearing land but rather of purchasing already cultivated land, thereby consolidating fragmented acreage and instituting new managerial practices. One of these practices entailed a new form of labor. The Cistercians took on able-bodied peasants as *conversi*, lay brothers who were celibate and entered the monastery as farm laborers. By becoming *conversi*, peasants gained their freedom from obligations to lords, avoided family responsibilities, and found economic security as well as a new religious vocation. The Cistercians recruited *conversi* primarily from the former tenants whose land they had acquired. Cistercian monks also became skilled at obtaining exemptions from obligations to lords, from customary tithes to the church, and from other fees and tolls. In the early decades of the new order, the monks themselves worked alongside the *conversi* and hired day laborers as well. They could move workers from one grange to another as needed. The result was a successful system of agriculture very different from both manorial cultivation and the farming of the free peasant.

In addition to the cultivation of fields, the Cistercians acquired much pastureland and they raised livestock, oxen, horses, cows, and especially sheep. They profited from trends that favored producers of animal products during the twelfth and thirteenth centuries, when the growth of towns created markets for leather, woolens, and parchment, as well as cheese, butter, and meat. In addition to acquiring pasture rights, the Cistercians instituted the practice of transhumance—moving their animals to the mountains in the summer and back to the plains in the winter. This practice was well established in southern France by the twelfth century, but the Cistercians practiced it on a much larger scale which enabled them to maintain herds of hundreds or even thousands of animals.

The Medieval Mill

The most important sources of power during the medieval centuries were human and animal muscle power. Yet, machines that harnessed the power of wind and water were also significant. The most important of these machines was the mill, first of all the grist mill for grinding grain into flour for making bread, and for grinding malted grains for making ale and beer. Various types of mills, including those powered by animals and humans, were used by the Romans and continued in use with regional variations during the medieval centuries. The ancients had also used water-powered mills. A magisterial essay by Marc Bloch, published in 1935, insured that the medieval watermill long would remain a focus for historians of technology. Bloch argued that the Romans failed to fully exploit water-powered mills because slaves could grind grain by hand. Historians such as Lynn White Jr. and Jean Gimpel posited an "industrial revolution" in medieval Europe based primarily on their acceptance of Bloch's view that the Romans neglected to exploit water-power, whereas medieval peoples embraced it, especially in the form of powerful mills driven by overshot and undershot waterwheels. More recently, this idea has been challenged, and it now appears that the Romans exploited waterpower to a greater extent than had been recognized, while medieval use developed very slowly in some areas.

The water-powered mill had a number of variations. Whether and how a particular type was used depended on geography as well as social organization and local custom. The simplest water-powered mill had a horizontal wheel that lay in a stream or river, or else was turned by water flowing from a millrace or funnel. At the center of the wheel was a vertical shaft that went through a fixed millstone and turned a second millstone. Grain poured into a funnel or hopper fell between the two stones. As the waterwheel turned, the shaft and upper millstone also turned. The grain, crushed between the millstones, emerged from the edges. Vertical water-wheels were more powerful than most horizontal wheels. Yet their gearing made them more complicated and expensive. In undershot mills, the water runs underneath the wheel, while in the overshot type the water pours over the top. Some vertical mills were mounted on barges floating in rivers, and mills were also mounted on the piers of bridges. In a very different type of mill, common on the Iberian peninsula, water was delivered under pressure, or "powered up." One such type was the tank or *arubah* mill, another the ramp mill. These powerful mills were sometimes used to turn multiple millstones.

Free peasants working their own land might use hand mills or querns and horse mills to grind grain. If they used a watermill, it would most likely be horizontal. On the manor, lords constructed heavier overshot or undershot vertical grist mills and forced peasants to have their grain ground for a fee amounting to a certain proportion of the flour. Because vertical mills were dependent on a reliable water flow, they were more prevalent in some areas than others. The rivers in parts of Italy, prone to floods in the winter and slack water in the summer, were not particularly suitable for vertical waterwheels. John Muendel has investigated archival documents, such as those pertaining to rentals, which mention parts of mills. He concludes that in the north Italian town of Pistoia most if not all watermills were horizontal. But in Florence, his investigations revealed a variety of types of mills, including horizontal, overshot, hybrid horizontal with gearing, floating, and undershot. Paolo Squatriti pointed out that manorial mills prevalent in Francia were not at all characteristic of the Italian peninsula, where urban control of mills and private control by individual peasants was more characteristic. He also demonstrated the diversity of types of mills in early medieval Italy, where querns and donkey mills existed alongside watermills.

Knowledge of medieval milling has been furthered by archeological investigations. For example, in his description of the archeology of English mills, David Crossley pointed to a new emphasis on the study of complete mill complexes, rather than just the remains of the mill itself. The study of mill-sites now entails investigation of watercourses, weirs, dams, and millponds, as well as the surrounding landscape, which may include surveys of entire valleys.

The post windmill was a medieval invention that appeared in northern Europe and England not long before 1185; Richard Holt called this "the most characteristic contribution of the age to the technology of exploiting natural power, and by far the most important". While Lynn White Jr. hailed the windmill as crucial to the technological dynamism of medieval Europe, Holt in contrast used the English case to argue that it was a new machine that was used primarily where conditions did not favor the watermill, and that it was adopted rather slowly.

Although grinding grain was by far the most important use, medieval mills were developed for other purposes as well. They were used for sawing wood and for grinding the bark used to tan leather. Water-powered forges were used to hammer the "bloom" in the process of working iron. Mills were used for grinding and sharpening the blades of tools and weapons. Most important, fulling mills were used in the stage of wool cloth-making that involved washing, soaking, and beating by means of hammers. Fulling mills featured the first European appearance of the cam, an eccentric lever that caught the hammer and lifted it, whereupon it would fall onto the cloth from its own weight.

The Medieval Castle

While village and field were traditional sites for the life and work of the peasant, the life of the nobility came to be centered on the castle. Nobles constructed these massive structures beginning in the ninth and tenth centuries, first as protection during the invasions of the Vikings, Magyars, and Slavic peoples. Later, castles evolved into multipurpose structures that served not only as protection but as residences for noble families, garrisons for feudal armies, administrative centers, and centers of craft production. They became more elaborate as time went on. Beginning as a simple protective enclosure made of earth and wood, and known as the motte-and-bailey castle, they became elaborate stone structures that symbolized the power of the nobles who used them both to defend the surrounding population and to control it.

The motte was an earthen mound topped by a wooden fortification and surrounded by a deep ditch. The bailey consisted of an enclosed yard built outside the ditch encircling the motte. During construction, earth from the ditch would be thrown into the center to create the mound. Then a wooden structure was built on top, sometimes just a tower to house soldiers, but, as time went on, more elaborate structures. Larger residential mottes housed the family and its retainers. The bailey was surrounded by a fence or palisade, usually constructed out of logs. Sometimes a ditch was dug around the bailey as well, which could provide refuge for the noble's horses and cattle, or for villagers living nearby. The ditch between the bailey and motte was crossed by one or more bridges which could be drawn up or destroyed in order to protect those inside from attackers.

The use of the motte-and-bailey castle as an instrument, not of defense, but of subjugation can be seen most clearly in England. In 1066 William and his army of Norman knights crossed the English Channel and defeated the Saxon army in the Battle of Hastings. After his victory William found it difficult to control the rebellious population, so he began entrusting his barons with the construction of castles at crucial defensive points and gave jurisdiction to each of these barons over the surrounding countryside. At the same time he also maintained control, because no baron could build a castle without his permission. By the time of William's death in 1087, there were more than 500 motte-and-bailey castles in England. The Normans were there to stay.

Gradually, beginning in the late tenth century, stone castles replaced motte-and-bailey structures. Although specific origins are controversial, within a hundred years they dotted the European countryside, many of them constructed on motte-and-bailey sites. They consisted originally of a stone tower, meant to garrison troops, that was surrounded by massive stone walls. As uses multiplied beyond mere fortification, castles became more and more elaborate, eventually becoming residences for noble families as well as administrative centers. The most important characteristic of castles were their walls—as much as six meters thick at the base—constructed by making two parallel walls out of ashlar blocks and filling the space with stone rubble. Castles usually incorporated immense towers as well as courtyards and rooms that housed family members, visitors, troops, and supplies.

In an influential study devoted to Lazio (the area around Rome in central Italy), Pierre Toubert investigated castle building as a phenomenon that was accompanied by changing patterns of social organization and settlement during the eleventh century. This process, which he called *incastellamento*, is the key structural element in a change of settlement patterns associated with the rise of feudalism. Toubert found that bishops and abbeys were important contributors to this movement. Dispersed habitats disappeared, while villages were formed near castles. The purpose of these castles was to dominate the countryside rather than protect it. Archaeologists working in the Iberian Peninsula, most importantly Pierre Guichard, have extended this investigation of *incastellamento*, focusing on its development in both Christian and Arab/Berber areas. Thomas Glick emphasized that the domination of a lord over the surrounding countryside could take a number of different forms. The lord could take rent in kind from peasants in a traditional manorial arrangement, for example, or he could extract taxes.

The Growth of Towns

Although life in the early Middle Ages was overwhelmingly rural, in some areas—especially northern Italy and southern France—remnants of Roman towns remained in greatly diminished form. Whether medieval towns were a development from Roman antecedents or were entirely new is a question that must be decided on a town-by-town basis. In general, however, towns in northern Italy and southern France were more likely to have had a Roman core than those that grew up in regions further from the center of the ancient empire. Remnants of Roman times included physical structures—walls, ruined buildings, and materials such as bricks and stones which medieval people often used to build new, more humble, structures. The study of towns, including their decline and subsequent medieval growth, has been substantially advanced by archaeological investigation of individual sites.

The specific origins of medieval towns usually involved a variety of factors. In part they grew up in response to increased trade and commerce. Occasionally they were created as marketing centers by powerful lords. They could develop around a monastery or the palace of a bishop, or near a lord's castle or a river crossing or trading post. In the eleventh century, kings and bishops began giving charters to towns, guaranteeing them a significant degree of autonomy and freedom. Whatever their origins, towns became a dynamic focus of medieval life after the eleventh century.

The effects of the revival of long-distance trade on urbanization deserves particular emphasis. Michael McCormick's monumental study of the "European economy" made it clear that the volume of travel and trade in the Carolingian Age between 700 and 900 was far greater than previously has been supposed. Travelers' destinations included numerous small ports as well as those on major trade routes. Venice, on the Adriatic Sea, developed an important trade with Constantinople; wheat, lumber, salt, and wine were shipped to the Byzantine capital in return for silk and spices. In the north, England began sending tin and raw wool to the European continent, bringing back Flemish wool cloth, silver from German lands, and luxury goods from northern Italy. In the early twelfth century, the counts of Champagne staged great trade fairs, providing security and safe roads for merchants from north and south. Fairs were also established in Lombardy, and the towns of Flanders traded their famous wool cloth for numerous other commodities. The Vikings gradually changed from marauders to traders who brought furs and slaves down the coast to French and German lands and down the Russian rivers to Constantinople, purchasing armor, gold, and silver. The Crusades, undertaken by western knights and merchants to capture the lands of the so-called infidel, boosted this long-distance trade; for example, when the First Crusade of 1097 resulted in the creation of the Latin Kingdom of Jerusalem, trade to the eastern Mediterranean expanded greatly. Four developments were intrinsically related: the growth of long-distance trade, the great expansion of the use of money, the proliferation of specialized crafts based in the new towns, and urbanization itself.

Cities were magnets for trade, but they also became centers of manufacture for everyday wares and luxuries, for saddles, pots and pans, glassware, jewelry, gloves, and consumables, most importantly bread. In the thirteenth century the most significant manufacture in western Europe was woolen cloth. And, as Steven Epstein noted, "perhaps the most distinctive, and relatively recent, feature of urban society was the large number of people who supported themselves through wage labour." Most cities were composed of neighborhoods, sometimes organized along craft lines. Usually, "dirty" trades that polluted the water or produced disagreeable smells—butchers, tanners, cloth fullers—were located at the edge of town. Many crafts were organized into guilds after the twelfth century. Boys and sometimes girls were apprenticed to masters in workshops. After some years, apprentices became journeymen or day laborers and were authorized to work for wages. Some journeymen became masters, setting up shops of their own. Guilds governed the rules of apprenticeship, set standards for the quality of the goods produced, and prevented outsiders from producing those goods. They also served social functions, burying deceased members, caring for indigent widows and children. Many scholars no longer accept the traditional generalization that guilds represented a conservative force, inhibiting invention and innovation. S. R. Epstein argues that the transmission of technologies throughout Europe was greatly facilitated by journeymen traveling from one town to another to get work (tramping), a well-established practice followed by many young artisans.

Crafts and Trades

In the early medieval village and manor, many material objects were made on site. Peasants made shoes, bags, and harnesses from pigskin; they fabricated wool or flax cloth and clothing; and they made wooden farm tools and cooking implements. The small amount of iron used for tools was mined elsewhere and worked by smiths, who at first constructed their forges and anvils in the forests to be near the necessary fuel. Eventually, however, the smith became a valued artisan in the village and manor, making horseshoes, plowshares, pots, pans, armor, and weaponry.

Another important site of early medieval craft production was the monastery. Many everyday items used by monks and nuns were made in monastic workshops, including wooden objects such as furniture and barrels, cloth and clothing, pottery, leather goods, and metal objects made by the smith. Some monasteries also supported skilled artisans who produced finely worked liturgical objects, from candlesticks to embroidered garments. This specialized work is vivid in the pseudonymous twelfth-century treatise, *De diversis artibus* (On Diverse Arts) by Theophilus ("Lover of God"). The author was probably Benedictine monk Roger of Helmarshausen, whose metalwork is known from three highly decorated objects—a jewel-studded book cover in Nuremberg and two portable altars in the cathedral treasury of Paderborn. In addition to providing numerous craft recipes on painting, glass, and metalwork, Theophilus discussed the ethics of monastic craftsmanship, stressing the importance of virtue, humility, piety, faith, and the open sharing with others of God-given craft skill and knowledge.

The growth of towns was characterized by the proliferation of specialized crafts. Tanners made leather out of skins which they provided to glovers, saddle-makers, and artisans who made parchment for books. Textile workers produced cloth, wool, linen, cotton, and silk. Metalworkers included goldsmiths, silversmiths, ironworkers, armorers, and smiths who worked with copper, tin, and pewter. Medieval crafts included glassmaking for vessels and windows, pottery and tile-making, painting, sculpting, and brick-making. Carpenters and joiners specialized in making a variety of objects from barrels to marriage chests to altarpieces.

The manufacture of wool cloth was the most important medieval industry. Most stages of production were carried out by specialists: shearing sheep and cleaning the wool, combing and carding, spinning, putting the yarn on the loom, weaving, dyeing, fulling, finishing. Wool cloth manufacture changed significantly over time. Early medieval cloth-making was a local, family or manor-based craft, carried out for the most part by women. Spinning was done by hand and cloth was woven on vertical warp-weighted looms. During the eleventh and twelfth centuries, wool became important in long-distance trade, and wool manufacture developed as a significant industry in many towns, especially in northern Italy and Flanders. Horizontal looms replaced vertical looms. The spinning wheel, introduced in the thirteenth century, was improved with foot treadles and a mechanism to control the tension. As textile manufacture changed from a local craft industry to one that produced commodities for long-distance trade, labor and power relationships changed. Women remained involved in spinning and other tasks within a "putting out" system in their own homes, and were paid (very poorly) by the piece. Most other parts of the process fell to the control of men, especially the drapers or wool merchants. The industry became organized into four major crafts: weaving, fulling, dyeing, and finishing.

Architecture and Building Construction

The architectural styles such as the Romanesque that emerged in the late tenth century and the Gothic that developed in the twelfth each entailed specific building techniques. All building construction, from humble to monumental, required the acquisition or fabrication of materials such as brick, stone, and wood, and their subsequent working by means of specialized tools and techniques. Aside from stone castles and great abbeys, the most prominent structures on the medieval landscape were churches and cathedrals. The new Romanesque style was seen mostly in cathedrals but also in other buildings such as abbeys. This style displayed a new interest in articulating and clarifying space. Elements such as side chapels were carefully integrated into the design as a whole. The creators of Romanesque buildings aimed to make coherent and unified spaces. This interest in articulating space led to the development of particular motifs, for example, a system of alternating columns and piers to divide and support a long wall (typical of central European Romanesque buildings) or a motif created by the Normans, the separation of one bay from another by tall shafts running from the floor to the ceiling. A further innovation, first achieved in Durham Cathedral in England, was to create a rib vault (an arched ceiling with sections separated by projecting bands or ribs) over the wide nave or center aisle of the church. At Durham the interest in articulation led to a technological innovation that would become intrinsic to a new style, the Gothic.

The Gothic cathedral may be traced to the mid-twelfth century and the new choir of St. Denis Abbey near Paris, built between 1140 and 1144. Three technical features accompanied the development of this new style—the pointed arch, the rib vault, and the flying buttress—each of which had developed separately in previous centuries. The flying buttress on the outside took the weight of the vaulting and the roof, enabling masons to build thin, high walls, filled with stained glass. As a result, Gothic buildings seem to soar and on bright days are flooded with light. Scholars continue to study these great structures, sometimes with new methodologies. Robert Mark has used the methods of structural engineering, creating plastic models and subjecting them to various stresses, in order to analyze how the cathedrals work as structures. He discovered, for example, that the pinnacles resting on the top of the buttresses of Rheims are not merely decorative but add to stability.

Of course, medieval masons did not have recourse to modern methods of structural analysis. Rather, they learned their craft by apprenticeship and thus acquired knowledge of previous construction. They built cathedrals bay-by-bay. Mark suggested that they tested a new idea on the first completed bay and, if it worked, carried it out in the succeeding bays. They experimented, he suggested, "by observing cracks developing in the tension-sensitive lime mortar used to cement the cut stones, and then by modifying the form of the structure in order to expunge the cracking." Designing the building as a whole, masons used "constructive geometry" in which they manipulated geometric forms, using drawing tools such as the straightedge and dividers, to determine the overall dimensions of ground plans, facades, and all the various elements of the building. When construction began, a "great measure" would be established to create the ground plan, and would then be employed throughout the construction. As Lon Shelby explained, the methods of the masons were traditional, empirical, and experimental. The great measure was part of the building. Once the dimension of one element had been fixed, all the other elements were "expressed in terms of the first." The unit of measure became a module for the entire building.

The master mason was in charge of both the design and construction, and usually stayed on-site during the construction process. He worked closely with a patron, and with masons, journeymen, and apprentices. He supervised construction involving stone and brick, but not that involving wood, glass, or lead, which were in the charge of carpenters, glaziers, and plumbers. The mason was actually concerned with three sites—the quarry where the stone was obtained; the lodge, a wooden building which served as his workshop; and the building site itself. Stone was hewn from the quarry and brought to the lodge where it was precisely cut according to where it was to be placed. To shape stones exactly, masons used numerous templates for piers, columns, vaults, doorways, and windows.

Masons were paid wages. Usually, they were not organized as urban guilds, because they traveled to worksites that could be in the countryside and or in distant towns. Some master masons became building contractors and supplemented their wages by profits acquired by supplying lumber, stone, or other materials. The lodge, usually constructed near the building site, functioned as the workshop of the masons and also as their organizational center.

The Value of Work

Scholars have argued that labor, including manual labor and skilled craftwork, gained more status in the medieval centuries than in the ancient world. And there is indeed much evidence for the ancient contempt for handwork, which was associated with slavery. Yet, recent scholarship has pointed to the positive ancient traditions alongside the negative. In the early medieval centuries, the growth of monasticism encouraged the appreciation of work. Monastic regulations treated labor as an essential part of the life of piety. In a different context, the growth of commerce and of towns led to an increase of skilled artisans and an increased appreciation for their skill and their wares. The emergence of merchant and craft guilds provided means of regulating particular trades, including rules for apprenticeship and quality control, but they also offered peer companionship and support for members, including funeral expenses for indigent members and support of indigent widows and their children.

With the growth of learning, the proliferation of schools, and the establishment of universities in the twelfth and thirteenth centuries, the "mechanical arts," as skilled craftwork was called, were sometimes included alongside the "liberal arts" in accounts of the disciplines. (The seven liberal arts comprised the trivium: grammar, rhetoric, and dialectic or logic, and the quadrivium, geometry, arithmetic, music, and astronomy.) Particularly notable was the *Didascalicon* of Hugh of St. Victor (c. 1096–1141) who gave seven mechanical arts—fabric-making, armament, commerce, agriculture, hunting, medicine, and theatrics—an important place. Nevertheless, during the medieval centuries, the skilled trades remained separate from the learned disciplines. The latter were communicated entirely in the language of learning, Latin. The medieval universities, first created in the early thirteenth century, admitted only men and all lectures and discussion, all books, were in Latin. In contrast, skilled artisans learned their trade within apprenticeship arrangements, and they did not learn Latin. The closing of the gap between the learned and the skilled, and interchange between the two groups occurred only in later centuries, beginning in the fifteenth and sixteenth.

Although work was often divided by gender, women participated in numerous productive activities and even controlled some of them. David Herlihy has investigated the evidence for a trend in which female work in public arenas was gradually curtailed in the late medieval and early modern centuries. In the early medieval centuries, however, women produced virtually all cloth and clothing. In the manor house and sometimes in monasteries, they worked in a workroom called the *gynaeceum*. Women are praised as fabric workers from the Carolingian age to the thirteenth century. Female participation in cloth-making cut across class lines. Noblewomen occupied supervisory roles and young girls, whether rich or poor, were often taught to spin and sew. From the eleventh century on, women gradually appear less frequently in the documentary evidence for cloth-making, while men appeal more frequently. By the thirteenth century men and women were working side-by-side in an industry once dominated by women. Their lack of legal autonomy and their exclusion from developing forms of credit prevented women from effectively engaging in long distance trade, while the growing importance of male ascetic monastic culture effectively excluded them, as David Noble has argued, from numerous other domains.

Work in a particular city, Paris, is illuminated by seven tax rolls from the late thirteenth and early fourteenth centuries (1292–1313) and by a copy of the principal guild statutes from the same period. The tax was called the *taille* and the tax rolls called the *Books of the Taille of the City of Paris*. It shows that women worked in 172 occupations in 1292 while men worked in 325. In 1313 women appear in 130 occupations and men in 276. Women appear "as drapers, money changers, jewelers, and mercers." Female mint workers, copyists, tavern-keepers, masons, artists, shoemakers, girdle makers, smiths, shield makers, archers, and candle makers are also found. That women dominated the silk cloth industry is evident in the records of female silk guilds. Only a very few occupations involving long-distance travel and heavy hauling were exclusively male. The woolen industry, however, was different. In 1292 there were 73 male weavers and only one female (who may actually have been working on silk or linen). In the same year there were 15 male dyers and again only one female. Here is an indication of a change that occurred all over Europe. When wool became important in long-distance commerce, it fell to the control of men. In all occupations, female workers earned about two-thirds as much as males.

There are sources that enable historians to gain some knowledge of women's work in other parts of Europe as well. For example, because ale-making was regulated in England, records survive from rural tribunals or manor courts. Making ale entailed soaking the barley for several days, draining it, germinating it to create malt, drying and grinding the malt, and then adding it to hot water to inculcate fermentation. The wort was drained from this mixture. Herbs and yeast could be added. Equipment for the process included large pots, vats, ladles, and straining cloths. Judith Bennett has shown that during the thirteenth and fourteenth centuries, the production of ale, the primary drink of the peasantry, was carried out entirely by women. When production later became more commercial, men controlled it.

In medieval Montpellier, in southern France, 30 of the 208 apprenticeship contracts extant from before 1350 concern females. Girls were apprenticed as bakers, spinners of gold thread, silver gilders, embroiderers, basket weavers, painters, and tailors; as in many medieval towns, women were extremely active in local retail markets. In Exeter, in southwest England, court records show that women were active in retailing, brewing, candle making, leather working, and cloth manufacture. Family occupation was important in determining the trade of both males and females. In both Montpellier and Exeter, craftswomen sometimes practiced trades similar to those of their fathers and husbands.

* * *

The brief discussion of women's work underscores a methodological issue pertinent to the study of all of medieval history. What can be known is based to a significant degree on written sources, and such sources always derive from one particular locality. The reason a source exists (whether it consists of court records or guild regulations or something similar) influences the nature of the material reported, on the one hand, but, on the other, fails to inform us about the many activities which were not included. The extent to which conclusions from any particular source can be generalized is always an issue. For the medieval centuries such problems are exacerbated by the relative paucity of sources. The development of medieval archeology and the incorporation of its findings by historians have brought about important new sources of information.

Scholars recently have explored particular regions in great detail, while at the same time treating their conclusions comparatively. In the matter of three technologies in particular—technologies pertaining to the military, to transportation, and to communication—the major geographic and political regions treated in this booklet were linked in important ways. Hence the following three chapters will underscore such interrelations topically rather than region by region.

The Needham Puzzle: Why the Industrial Revolution Did Not Originate in China*

Justin Yifu Lin

Introduction

One of the most intriguing issues for students of Chinese history and comparative economic history is, Why did the Industrial Revolution not occur in China in the fourteenth century? At that time, almost every element that economists and historians usually considered to be a major contributing factor to the Industrial Revolution in late eighteenth-century England also existed in China.

Chinese civilization, like the civilizations of Mesopotamia, Egypt, and India, originated from agriculture. The first unified empire, Qin, was formed in 221 B.C. By 300 B.C., Chinese society had developed into a form that had many characteristics of a market economy, with most land privately owned, a high degree of social division of labor, fairly free movement of labor, and well-functioning factor and product markets.[1]

This comparatively developed "market economy" probably created important attitudes toward profit and contributed to the swift diffusion of the best technology. In the Han dynasty (206 B.C.–A.D. 220), the iron-tipped plow, moldboard, and seed drill were widely used in the northern part of China, where the main crops were millet and wheat. The most significant improvements in Chinese agriculture came with the population shift from the north to the rice-growing areas south of the Yangtze River that started at the beginning of the ninth century, and especially after the introduction of a new variety known as "Champa rice" from Indochina at the beginning of the eleventh century.[2] This variety, characterized by better drought resistance and faster ripening, enabled farmers to extend the agricultural frontier from the lowlands, deltas, basins, and river valleys to the better-watered hill areas, and allowed production of two and even three crops a year.[3] The change from dryland crops to wetland rice led to a spurt of innovations in farm implements, including an improved plow that required less draft power, a share plow that could turn over sod to form a furrow, and the deep-tooth harrow.[4] Many of the elements of Arthur Young's scientific (conservation) agriculture, which led to the agricultural revolution in England in the eighteenth century, had become standard practice in China before the thirteenth century.[5] By the thirteenth century China probably had the most sophisticated agriculture and Chinese fields probably produced the highest yields in the world.

China's premodern achievements in science and technology were even more remarkable. Gunpowder, the magnetic compass, and paper and printing, which Francis Bacon considered as the three most important inventions facilitating the West's transformation from the Dark Ages to the modern world, were invented in China. Evidence documented in the monumental works of Joseph Needham and his collaborators shows that, except in the past 2 or 3 centuries, China had a considerable lead over the Western world in most of the major areas of science and technology.[6]

---

*An earlier version of this article was presented at seminars at the Australian National University, the Hong Kong University, the Chinese University of Hong Kong, and Peking University. I am indebted to the participants in those seminars for their insightful comments. I am especially grateful to Arman Alchian, Mark Elvin, Dean Jamison, E. L. Jones, James Lee, Joel Mokyr, Jean-Laurent Rosenthal, Scott Rozelle, Kenneth Sokoloff, and three anonymous referees for helpful suggestions. Robert Ashmore gave a helpful exposition review.

[1] Kang Chao, *Man and Land in Chinese History: An Economic Analysis* (Stanford, Calif.: Stanford University Press, 1986), pp. 2–3.

[2] Ping-ti Ho, "Early Ripening Rice in Chinese History," *Economic History Review* 2, ser. 9 (December 1956): 200–218.

[3] It is worth mentioning that, like many modern agricultural innovations, the promotion of Champa rice was sponsored by the government. The Song emperor Chen-Zong bought a large quantity of Champa rice from the south and made it available to farmers in the Yangtze delta (Chao, p. 200).

[4] Ibid., p. 224.

[5] Anthony Tang, "China's Agricultural Legacy," *Economic Development and Cultural Change* 28 (October 1979): 1–22.

[6] Before Joseph Needham's works on China's achievements in science and technology, few people in the West, or in China, for that matter, understood that many of the basic inventions and discoveries

---

It is no surprise that, based on this "advanced" technology, Chinese industry was highly developed. The total output of iron was estimated to have reached 150,000 tons in the late eleventh century. On a per capita basis, this was five to six times the European output.[7] Equally impressive was the advancement in the textile industry. In the thirteenth century, a water-powered reeling machine was adapted for the spinning of hemp thread, which was as advanced as anything in Europe until about 1700.[8]

High agricultural productivity and advanced industry facilitated the early development of commerce and urbanization. Peasants were linked to rural market fairs, which in turn were integrated in a national commerce network by canals, rivers, and roads. In addition to staples like rice, many local products, such as particular types of paper and cloth, became known and available nationwide.[9] Many cities flourished in the thirteenth century, astonishing even that sophisticated Venetian, Marco Polo. According to him, "Su-chou is so large that it measures about forty miles in circumference. It has so many inhabitants that one could not reckon their number"; and Hang-chou "without doubt the finest and most splendid city in the world, . . . anyone seeing such a multitude would believe it a stark impossibility that food could be found to fill so many mouths."[10] In short, China by the fourteenth century was probably the most cosmopolitan, technologically advanced and economically powerful civilization in the world. Compared to China, "the West . . . was essentially agrarian and . . . was poorer and underdeveloped."[11]

In retrospect, China had a brilliant start and remained creative for several thousand years of premodern history. Many historians agree that by the fourteenth century China had achieved a burst of technological and economic progress, and that it had reached the threshold level for a full-fledged scientific and industrial revolution.[12] However, despite its early advances

in science, technology, and institutions, China did not take the next step. Therefore, when progress in the West accelerated after the seventeenth century, China began to lag farther and farther behind. Needham put this paradox in the form of two challenging questions: first, why had China been so far in advance of other civilizations; and second, why isn't China now ahead of the rest of the world?[13] The goal of this article is to bring several relevant factors together that may provide a partial explanation to this puzzle.

Several hypotheses have been proposed by prominent scholars. These explanations can be classified into two categories: those based on failures of demand for technology and those based on failures of supply of technology. Section II reviews the existing demand-failure hypotheses. It is followed in Section III by a hypothesis of my own, which is essentially a supply-failure hypothesis. Section IV explores the factors inhibiting the development of modern science and technology in China and reviews other existing supply-failure hypotheses. A summary and some concluding remarks are in Section V.

The High-Level Equilibrium Trap

The most widely accepted hypothesis for China's later stagnation has been the "high-level equilibrium trap," first proposed by Mark Elvin and further expounded by Anthony Tang, Kang Chao, and other writers.[14] After reviewing China's many astonishing technological and institutional achievements before the fourteenth century, Elvin first refutes with convincing examples and evidence several conventional hypotheses, such as inadequate capital, restricted markets, political hazards, and lack of entrepreneurship in China, as explanations for the stagnation of China's technical creativity.[15] He then argues that the prime cause was unfavorable man-to-land ratio. Elvin's hypothesis, with Tang's and Chao's modifications, can be presented in a nutshell, as follows.[16]

upon which the modern world rests came from China. Needham's research started in 1937 as a result of his befriending some Chinese students in Cambridge who had profound knowledge of the history of Chinese science. His systematic research started in 1948, and the first volume of *Science and Civilization in China* was published in 1954 by Cambridge University Press. More than 10 volumes have been published and several more volumes are under preparation.

[7]E. L. Jones, *The European Miracle*, 2d ed. (Cambridge: Cambridge University Press, 1987), p. 202.

[8]Mark Elvin, *The Pattern of the Chinese Past* (Stanford, Calif.: Stanford University Press, 1973), p. 195; and Jones, p. 202.

[9]Elvin, p. 166.

[10]Quoted in ibid., p. 177.

[11]Carlo M. Cipolla, *Before the Industrial Revolution: European Society and Economy, 1000–1700*, 2d ed. (New York: Norton, 1980), p. 171.

[12]Chao; Elvin; Tang (n. 5 above); Joseph Needham, *Science in Traditional China: A Comparative Perspective* (Cambridge, Mass.: Harvard

University Press, 1981); and Wolfram Eberhard, "Data on the Structure of the Chinese City in the Pre-Industrial Period," *Economic Development and Cultural Change* 5 (October 1956): 253–68.

[13]Joseph Needham, "Introduction," in *China: Land of Discovery and Invention*, by Robert K. G. Temple (Wellingborough: Patrick Stephens, 1986), p. 6.

[14]Elvin, chap. 17; Chao; and Tang.

[15]Elvin's discussion is enlightening. However, I will not repeat his arguments here, for the sake of brevity. Readers interested in these arguments may refer to Elvin, pp. 286–98.

[16]Although their general views on China's unfavorable land-to-labor ratio and its implication for industrialization are similar, Elvin, Tang, and Chao have different emphases. While Elvin and Tang stress diminishing agricultural surplus and China's inability to support a sustained

China's early acquisition of "modern" institutions, such as family farming, fee-simple ownership, and the market system, provided effective incentives for technological innovation and diffusion. Therefore, the advancement of science and technology was initially much more rapid in China than in Europe. However, the Chinese family's obsession with male heirs to extend the family lineage encouraged early marriage and high fertility despite deteriorating economic conditions, resulting in a rapid expansion of population. The possibility for continued expansion of the amount of cultivated land was limited. At the end China stood at a position "where the level of living was subsistence and where the population was so large in relation to resources and the technological potentials were so fully exploited that any further advances in output would have required increases in population and consumption that would have out-stripped the resulting rise in food supply."[17] The rising man-to-land ratio implied that labor became increasingly cheap and resources and capital increasingly expensive. Therefore, the demand for labor-saving technology also declined. Moreover, the rising man-to-land ratio also implied a diminishing surplus per capita. As a result, China did not have a surplus to be tapped for sustained industrialization. Even though China had already approached the threshold of industrial revolution in the fourteenth century, "by that time population had grown to the point where there was no longer any need for labor-saving devices."[18] On the contrary, Europe enjoyed a favorable man-to-land ratio and a legacy of unexploited, traditional economic and technological possibilities, because of its hereditary feudal system. Although its scientific and technological development lagged behind China's in premodern ages, by the time sufficient knowledge was accumulated to the threshold of an industrial revolution, "a strong need to save labor was still acutely felt,"[19] and a large agricultural surplus was available to serve "as the principal means of financing industrialization."[20]

Although the above hypothesis is interesting, there are several reasons for abandoning this model as a valid explanation of China's failure to launch a full-fledged industrial revolution in the fourteenth century. I will first examine the implications of the man-to-land ratio for technological innovation and then will discuss the issue of "depletion of agricultural surplus."

The central assumption implicit in the above hypothesis is that of a bounded potential of agriculture in premodern ages. However, given the land, labor, and social institutions, the potential of agriculture, whether in modern or premodern ages, is a function of technology. If the development of technology is not inhibited, an "equilibrium trap" due to the adverse man-to-land ratio is not present. Therefore, the crucial issue is whether the lack of inventive creativity is a result of the rising man-to-land ratio.

It is true that up to the twelfth century, there was a steady flow of labor-saving innovations in plows and other farm implements, and that after that few labor-saving implements were invented, as shown by Chao.[21] However, changes in the orientation of invention were not due to the worsening of man-to-land ratio, as Chao claimed. China's population increased until about 1200, declined until approximately 1400, and recovered to the 1200 level at approximately 1500. It reached a new peak at about 1600, collapsed again

industrialization, Chao emphasizes the swelling labor surplus and its impact on demand for labor-saving technology. Additionally, Elvin does not explain the mechanism that led to the explosive population growth.

[17]Tang, p. 7.

[18]Chao, p. 227.

[19]Ibid.

[20]Tang, p. 19. In addition to the unfavorable man-land ratio, Elvin also viewed static markets as a factor inhibiting technical creativity in China. Although China had large, integrated national markets, e.g., for cotton cloth, Elvin argued, "It was not the size of the eighteenth-century British market (tiny compared to that of China) but the speed of its growth that put pressure on the means of production there to improve" (p. 318). However, this argument is misleading. Before the revolution in England's cotton industry, Europe had already imported cotton cloth from

Asia. However, "by hand methods Europeans could not produce cotton cloth in competition with the East. But the market was endless if cotton could be spun, woven, and printed with less labor, i.e., by machine" (R. R. Palmer and Joel Colton, *A History of the Modern World,* 6th ed. [New York: Knopf, 1984], p. 429). Therefore, as in China, the market size for the British cotton industry was initially static. The growing potential came from the cost-competitiveness of mechanized production. If there had been an industrial revolution in China's cotton industry that made costs of production lower than those of the traditional technology, the market potential in China for the "modern mechanized" firms would have been endless also. Elvin's explanation is also not consistent with the fact that during the first century of the Ming dynasty (1368–1644) and the Qing dynasty (1645–1911), the Chinese economy was booming. A similar view on the role of market expansion was advanced by Jones, although he attributed the stagnation of markets in China as well as in Turkey and India to the invasion of Manchu, Ottoman, and Mughal—all minorities from the steppes. He argued that "the economies became command hierarchies imposed on customary agricultures. These weakened investment in human and physical capital, slowing and diverting for the duration of the empires much further growth of the market" (pp. xxiv–xxv [n. 7. above]). If Jones's explanation for the slowdown in the Manchu empire (1645–1911) is valid, China should have had brisk development under the Ming empire (1368–1644), which was ruled by Han Chinese. However, the slowdown in technological development was already under way during the Ming dynasty.

[21]Chao points out that among the 68 major farm implements that appeared after 221 B.C. and before modern times in China, 35 were invented between 961 and 1279, and only four were invented in 1369–1644. Among these four, two were labor using rather than labor saving in nature (p. 195).

Table 1

Per Capita Acreage of Cultivated Land, A.D. 2–1887

Cultivated Land		Population		
Year	Amount (Million Mu)	Year	Number (Million)	Per Capita Acreage (Mu)
2	571	2	59	9.67
105	535	105	53	10.09
146	507	146	47	10.78
976	255	961	32	7.96
1072	666	1109	121	5.50
1393	522	1391	60	8.70
1581	793	1592	200	3.96
1662	570	1657	72	7.92
1784	886	1776	268	3.30
1812	943	1800	295	3.19
1887	1154	1848	426	2.70

Source—Kang Chao, *Man and Land in Chinese History: An Economic Analysis* (Stanford, Calif.: Stanford University Press, 1986), p. 89.

by about 1650, and thereafter has grown continuously. Due to the decline in population, the estimated per capita acreage at the end of the fourteenth century was actually about 50% higher than that at the end of the eleventh century and was even about 10% higher than that at the end of the tenth century (see table 1). The per capita acreage in the mid-seventeenth century was also higher than that at the end of the eleventh century. If the man-to-land ratio were the valid explanation for the burst of labor-saving innovations up to the twelfth century, then that rate should have been even higher in the fourteenth and fifteenth centuries and again in the mid-seventeenth century.

Moreover, even if we take the man-to-land ratio in the early twentieth century as the point of discussion, the claim that there was "no need for labor-saving devices" is tenuous. Because of widespread double-cropping, labor shortages have always existed during the peak season when farmers have to simultaneously reap the first crop and prepare the land and sow or transplant the second crop. According to John Buck's survey in the 1920s, there was on average only one and a half months free of field labor for the whole of China. Most of this period was accounted for by winter unemployment in the dryland farming areas of northern China. In the irrigated parts of southern China there were hardly any periods during the year in which farm households were not fully occupied in agricultural activities.[22] Therefore, the relatively low rate of labor-saving inventions after the twelfth century cannot be explained by the fact that the population had grown to

the point where there was no longer any need for labor-saving devices, as the hypothesis claimed.[23]

The other reason, implied in the above hypothesis and emphasized by Elvin and Tang, for why the demand for technology might have been dampened is an "inadequate" agricultural surplus arising from the adverse man-to-land ratio. However, this explanation has several problems. First, from the preceding discussion of demographic dynamics and per capita acreage, we can conclude that, given the technological level and social institutions, the surplus per capita in the fourteenth and fifteenth centuries should have been higher than that in the twelfth century, especially after the period of peace ushered in by the founding of the Ming dynasty in 1368. What we find, however, is a deceleration of labor-saving innovations.

Second, even if we take the twentieth century as a reference point for discussion, the claim that the high man-to-land ratio had depleted the agricultural surplus as a source for capital formation cannot be supported empirically. According to Carl Riskin's estimates, 31.2% of China's net domestic product was available for "nonessential" consumption in 1933.[24] Such

---

[22]John L. Buck, *Land Utilization in China* (Nanjing: Nanjing University Press, 1937).

[23]The above discussion does not imply that the change in man-to-land ratio has no consequences for technological innovation in an economy. Since relative scarcities of resource endowments in an economy affect relative prices of those resources, for the purpose of cost minimization, the optimal intensities of capital, labor, and land usage embodied in a technology for performing a certain task will be different from economy to economy and from region to region within an economy. For example, when Japan imported textile machinery from Britain in the late nineteenth century, most of the innovations involving the imported textile machinery made the machinery more labor-intensive; as a result, there was a substantial decline in the capital-labor ratio between 1886–90 and 1891–95 in the firms using the British machinery (Keijiro Otsuka, Gustav Ranis, and Gary Saxonhouse, *Comparative Technology Choice in Development: The Indian and Japanese Textile Industries* [New York: St. Martin's Press, 1988], p. 21). The reason for that type of modification was that the capital intensity of British machinery was tailored to Britain's capital and labor endowments, which differed from Japan's. A similar phenomenon was also found when foreign technology was transferred to China. Robert F. Dernberger's study of China's modern sector in the early twentieth century shows that Chinese industry tended to be small and had a high labor-capital ratio. However, the unfavorable man-to-land ratio did not inhibit the demand for new technology, as "Chinese-owned factories in the modern manufacturing sector outnumbered foreign-owned factories by more than ten to one" ("The Role of the Foreigner in China's Economic Development," in *China's Modern Economy in Historical Perspective,* ed. Dwight H. Perkins [Stanford, Calif.: Stanford University Press, 1975], p. 41). In the above examples, the new technology, even after the adjustment in labor-using innovations, was still more capital-intensive than the indigenous technology. As pointed out by W. E. G. Salter, "the entrepreneur is interested in reducing cost in total, not particular costs such as labor costs or capital costs. . . . Any advance that reduces total costs is welcome, and whether this is achieved by saving labor or capital is irrelevant" (*Productivity and Technical Change* [Cambridge: Cambridge University Press, 1960], p. 43).

[24]Carl Riskin, "Surplus and Stagnation in Modern China," in Perkins, ed.

estimates certainly depend on how "essential consumption" and "essential government expenditures" are defined. Riskin's findings indicate that the income flow in 1933 could provide for a rate of investment above 11% of national income, cited by W. W. Rostow and other economists as a threshold level for sustained economic development.[25] Moreover, the average rate of national income used for capital accumulation during the first 5-year plan period (1953–57) under the socialist government was 24.2%.[26] At that time, agricultural technology was still essentially traditional.[27]

From the above discussions, I find that the fact that the Industrial Revolution failed to occur in China in the fourteenth century cannot be attributed to a lack of demand for new technology, as asserted by the "high-level equilibrium trap" hypothesis.

Population, Science, and Invention

Given a set of inputs, a technological innovation must bring with it an increase in output. So long as humans' material desires are not satiated, the demand for new, better, and more cost-effective technologies is always present, though changes in the relative scarcity of labor and land in an economy may alter the patterns of invention. If technological change fails to take place, the problem does not stem from a lack of demand but from a failure on the supply side.[28] To address the Needham puzzle, we thus need to turn our attention to the supply side of technology.

Britain's Industrial Revolution in the eighteenth century is often identified with the mechanization of the textile industry, the exploitation of iron and coal, and Watt's invention of an atmospheric steam engine. However, what really distinguished the Industrial Revolution from other epochs of innovation bursts in human history, such as the one in the eighth- to twelfth-century China, was its sustained high and accelerating rates of technological innovation. The problem of China's failure to initiate an industrial revolution in the fourteenth century, therefore, is not simply a question of why China did not take a further step to improve its water-powered hemp-spinning machine. Rather, the question is why the speed of technological innovation did not accelerate after the fourteenth

century, despite China's high rate of technological innovation in the pre-fourteenth-century period.

The key to this question may lie in the different ways in which new technology is discovered or invented. The hypothesis I propose as a likely explanation to the Needham puzzle is as follows: in premodern times, technological invention basically stems from experience, whereas in modern times, it mainly results from experiment cum science. China had an early lead in technology because in the experience-based technological invention process the size of population is an important determinant of the rate of invention. China fell behind the West in modern times because China did not make the shift from the experience-based process of invention to the experiment cum science–based innovation, while Europe did so through the scientific revolution in the seventeenth century.

To support the above hypothesis, I will first present a simple stochastic model of technological invention à la Robert S. Evenson and Yoav Kislev and then will use it to analyze the historical development in China and Europe.[29]

A Model of Technological Invention

A technology can be defined as a body of knowledge about how to combine a set of inputs for producing a certain product. The net output, measured in a value term, produced by a given technology is defined as the productivity of the technology. A better technology means one with higher productivity. The supply of technology comes from inventive activity, which can be described as "trial and error" or "hit or miss" performed by the potential inventors, including farmers, artisans, tinkers, and researchers in the fields or in the laboratories. Each trial produces a technology with a certain productivity level, which is represented as a point under an invention distribution curve.[30] A trial can thus be perceived as a random draw from the invention distribution. Figure 2A portrays the basic features of the invention distribution curve.[31] If a draw results in a technology with a higher productivity than the existing technology, a better technology is invented. The probability of inventing a better technology by a random draw can be measured by the shaded area in figure 2A.

[25]W. W. Rostow, *The Stages of Economic Growth* (Cambridge: Cambridge University Press, 1960).

[26]State Statistical Bureau, *China Statistical Yearbook, 1988* (Beijing: Zhongguo tongji chubianshe, 1988), p. 60.

[27]Dwight H. Perkins and Shahid Yusuf, *Rural Development in China* (Baltimore: Johns Hopkins University Press, 1984), chap. 4.

[28]Theodore W. Schultz, *Transforming Traditional Agriculture* (New Haven, Conn.: Yale University Press, 1964).

[29]Robert E. Evenson and Yoav Kislev, "A Stochastic Model of Applied Research," *Journal of Political Economy* 84 (April 1976): 265–81.

[30]This assumption implicitly assumes that numerous technologies, each with a different combination of inputs, can have the same productivity level.

[31]In the figure, we assume that the distribution function is standard normal. However, the arguments in this article are independent of the type of the distribution function.

The adoption of a better technology to production is called technological innovation, which requires a diffusion process and time. For simplicity in describing the model, I will assume that once a better technology is invented, it is adopted by the whole economy.[32]

The mean and variance of the invention distribution function for an inventor is a function, among other things, of the inventor's stock of scientific knowledge and ingenuity, the material available for invention, and the surrounding physical environment. An increase in an inventor's stock of scientific knowledge increases the mean of his invention distribution function, shown in figure 2B as a rightward shift of the distribution curve.[33] Different inventors may have different invention distribution functions because of differences in their stock of scientific knowledge. Therefore, with a given technological level in an economy, the increase of an inventor's scientific knowledge improves the probability of his inventing a better technology. It is also possible for an inventor with a low stock of scientific knowledge to make big inventions, although the probability of such events is low.

It is worth mentioning that scientific knowledge itself is a result of the trial and error of scientific research, which can be described in a similar way to the invention of technology. However, science and technology have several different characteristics. Technological knowledge is used directly for the production of outputs, while scientific knowledge is used to derive testable hypotheses about the characteristics of the physical world, which may or may not facilitate the production of technology. New technology can be discovered by a veteran farmer or an artisan as a result of casual work, while scientific progress, especially in modern times, is more likely to be made by scientists following a rigorous scientific method. This scientific method is characterized by a "mathematization" of hypotheses about nature combined with relentless experimentation.[34] Because scientific knowledge must be acquired by

technological inventors before it can affect the outcome of invention, there is a time lag between progress in science and progress in technology.

An inventor's ingenuity affects his invention distribution function. That is, the better an inventor's ingenuity, the greater the likelihood of his inventing a better technology. However, the distribution of innate ingenuity is assumed here to be the same across nations and times. Change in the available materials can also change the mean, and probably also the variance, of the invention distribution function. One salient example is the progress from the Stone Age to the Bronze Age, and then to the Iron Age. Taking the example of plows, the productivity of an iron plow is in general higher than that of a bronze plow, which in turn is higher than that of a stone plow.

The model used here assumes that the source of invention is trial and error. It is important for our discussion to distinguish two types of trial and error: one is experience based and the other is experiment based. Experience-based trial and error refers to spontaneous activity that a peasant, artisan, or tinker performs in the course of production. Experiment-based trial and error refers to deliberate, intense activity of an inventor for the explicit purpose of inventing new technology. New technology obtained from experience is virtually free, while that obtained through experiment is costly. However, in a single production period an artisan or farmer can have only one trial, while an inventor can perform many trials by experiment. Since experience-based invention involves no cost and is a spontaneous result of production, cost-return calculations are not involved in experience-based invention. On the other hand, economic considerations are a key factor in determining the undertaking of experiment-based invention.[35]

It is possible to extend the above model in several directions. However, it is sufficient to suggest a new perspective on the Needham puzzle by the several implications that can be drawn from this simple model: (1) The likelihood of inventing a better

---

[32]This assumption is harmless because of the long time span in the article. A better technology will eventually prevail in an economy.

[33]In the figure, we assume that the increase in the inventor's stock of scientific knowledge results in a rightward shift in the mean of distribution of the invention distribution function without changing the variance. However, the increase in the stock of scientific knowledge can also result in an increase in the variance of the distribution without changing the mean. The increase in both the mean and the variance will increase the likelihood of inventing a better technology. For a mathematic proof of the statement, see Evenson and Kislev. Of course, the mean and variance of the distribution can also be changed at the same time.

[34]Joseph Needham, *The Grand Titration: Science and Society in East and West* (London: Allen & Unwin, 1969), p. 15. For further discussions on scientific discoveries, see Sec. IV of this article.

[35]The Schmookler-Griliches hypothesis of market demand–induced invention and the Hicks-Hayami-Ruttan hypothesis of relative factor scarcity–induced invention are relevant only for analyzing experiment-based invention. However, for the adoption of technology, economic considerations should have been relevant since antiquity (Zvi Griliches, "Hybrid Corn: An Exploration in the Economics of Technological Change," *Econometrica* 25, no. 4 [October 1957]: 501–22; Jacob Schmookler, *Invention and Economic Growth* [Cambridge, Mass.: Harvard University Press, 1966]; John R. Hicks, *The Theory of Wages* [London: Macmillan, 1932]; Yujiro Hayami and Vernon W. Ruttan, *Agricultural Development: An International Perspective* [Baltimore: Johns Hopkins University Press, 1971]).

technology is a positive function of the number of trials. (2) The probability of inventing a better technology is a negative function of the highest productivity of previous draws from the invention distribution—the level of existing technology. (3) Increases in the stock of scientific knowledge and improvements in the quality of available materials raise an inventor's likelihood of finding a better technology.

Technological Change in Premodern and Modern Times

Technological innovation, by definition, is an improvement in productivity. What distinguishes technological innovation in the modern age from that in premodern ages is the higher rate of innovation in the modern age, which is a consequence of the shift in the method of technological invention. Although in some cases systematic experimental methods were used in premodern times, as, for example, in the discovery of magnetic declination in China,[36] it is an accepted view that technological invention was predominately derived from experience. Inventions were made by artisans or farmers as minor modifications of existing technology as a result of experience obtained from the production process.[37] The experimental method became the predominant way of finding new knowledge only after the scientific revolution in the seventeenth century. The use of science to guide experiments came even later.

The first hypothesis from the above model predicts that, when experience is the major source of technological invention, the size of population in an economy is an important factor in determining the rate of invention and the level of technology in that economy. A larger population implies more farmers, more artisans, more tinkers, and so on and, therefore, more trial and error. Moreover, given the assumption that the level and distribution of the innate ingenuity tends to be the same statistically for large and small populations, a larger population thus implies that there would be a larger pool of gifted people in that economy. From the model described in Section *A*, we can conclude that in premodern times a large population contributed positively, in a probabilistic sense, to the level of technology and the rate of technological invention, ceteris paribus.[38] This may explain why

the great civilizations of antiquity were located in Mesopotamia, Egypt, and India, where fertile river valleys were favorable for agriculture and could support a large population.

Chinese civilization originated on the Loess Plateau in northwestern China, later than the great river civilizations. During the Former and Later Han dynasties (206 B.C.–A.D. 220) China's population was concentrated on the North China Plain and to the west of the gorges of the Yellow River. The principal grain was millet, but wheat, barley, and rice were also grown. During the fourth and fifth centuries A.D., Chinese settlers began to migrate in large numbers into the Yangtze River valley. Initially the method of farming was very crude, mainly slash-and-burn. As more people moved south, farming became settled, and wetland rice cultivation began to dominate. The pattern of Chinese agriculture that was practiced up to modern times essentially was established by the Song dynasty (960–1279).

Figure 1 shows that China had about twice the population of Europe until about 1300.[39] The aforementioned invention model predicts that, with experience as the principal source of technological invention, China had a higher probability than Europe of discovering new technology. In the eighth to twelfth century, the burst of inventions in China probably was due partly to the increase in population and partly to the shift of population from the north to the south.[40] Accompanying this shift was the transition from dryland crops to wetland rice, which with

[36] Needham, *Science in Traditional China* (n. 6 above), p. 22.

[37] Cipolla (n. 11 above), A. E. Musson, ed. *Science, Technology, and Economic Growth in the Eighteenth Century* (London: Methuen, 1972), p. 58.

[38] The contribution of population to invention is emphasized by William Petty, Simon Kuznets, F. Hayek, and Julian Simon. See the discussion by Julian L. Simon, *Theory of Population and Economic Growth*

(New York: Blackwell, 1986), chap. 1. It should be emphasized that the above arguments are true only in a probabilistic sense. The model does not imply that a small country cannot make a major contribution to technological invention. It only implies that, when experience is the major source of technological invention, the probability of such event is smaller for a small country than for a large country.

[39] In antiquity, a determinant factor of the population size in an economy was the carrying capacity of the agriculture, which depends on physical environment and technology. The physical environment in China was not particularly favorable for agriculture, compared to the fertile river valleys (Ping-ti Ho, *Huangtu yu zhongguo nongye de qiyuan* [Loess and the origin of Chinese agriculture] [Hong Kong: Chinese University Press, 1969]). However, probably due to luck, China was early in inventing the technology of row cultivation and intensive hoeing (sixth century B.C.), the iron plow (sixth century B.C.), the efficient horse harness (fourth century B.C.), the multitube seed drill (second century B.C.), and so forth (Robert K. G. Temple, *China: Land of Discovery and Invention* [Wellingborough: Patrick Stephens, 1986]). Therefore, agricultural productivity in China was relatively high compared to that in Europe at that time. This might have contributed to China's larger population.

[40] In the first century A.D., more than 75% of the Chinese population was in the north; by the end of the thirteenth century, the opposite was true. Beginning with the fourteenth century, the population in the north recovered gradually to reach about 45% in the twentieth century (Elvin [n. 8 above], chap. 14).

suitable technology brought a much higher yield than dryland crops.[41] However, with the original dryland farming technology, the yield of rice was still much lower than its potential. This shift in crops amounted to a rightward shift of the invention distribution function arising from the change in material available to inventors. Therefore, there was a burst in technology related to rice farming, including new tools, new crop rotations, and hundreds of types of new seed. The many other technological innovations in this period, such as water transportation, which Elvin ably documented in his celebrated book, also could be explained by the same line of reasoning.

Conventional wisdom has often argued that China's achievements in ancient times were due to its early acquisition of "modern" socioeconomic institutions, including the unified nation-state, family farms, free labor migration, and so forth, which should have facilitated a more rapid diffusion of technology once invented.[42] However, to the extent that technological inventions in premodern ages were fundamentally experience based and independent of economic calculations, the impact of socioeconomic institutions on technological invention was at most indirect, based on the possibility that fast diffusion of better technology might have allowed the economy to sustain a larger population than it would have otherwise.

After a decline of population during the twelfth to fourteenth centuries, China's population started to grow exponentially, except during the short period from 1600 to 1650. From the first hypothesis in the invention model, a larger population implies that there is more trial and error and, therefore, more invention. However, the second hypothesis predicts that, given an invention distribution curve, the marginal returns to the probability of invention from a larger population will eventually diminish. The post-fourteenth-century experience in China seems to support the implication of the second hypothesis. After the burst of technological invention from the eighth to twelfth centuries, the technological level moves to the right end of the experience-based invention distribution curve. Invention was still possible and actually continued to appear, but the probability of big breakthroughs became smaller and smaller. Most inventions took the form of minor modifications. Technological change could not recover

to a higher rate until there was a rightward shift in the invention distribution curve, made possible by applying Galilean-Newtonian physics, Mendelian genetics, and contemporary biological, chemical, plant, animal, and soil sciences.[43]

During the period of experience-based technological invention, Europe was at a comparative disadvantage due to its smaller population—a smaller population means a smaller number of trials. However, this disadvantage was countered by the shift to experiment-based technological invention and the closer integration of science and technology arising from the scientific revolution in the seventeenth century. Of course, as mentioned above, the experimental method had been used to invent technology even in ancient times. However, the popularization of the experiment as a vehicle for inventing new technology was a phenomenon that emerged only after the scientific revolution.[44] The experimental method removes the constraints of population size on technological invention. The number of trials that an inventor can perform in a laboratory within a year may be as many as thousands of farmers or artisans could perform in their lifetimes. However, if only experimental methods had been applied, the result would have been a single burst of technological inventions, as in the eighth to twelfth centuries in China. Soon thereafter, Europe would have faced the gradual exhaustion of invention potential and a slowdown in the rate of innovation, as China did after the fourteenth century. Therefore, more important than the popularization of the experimental method is the continuous shift to the right of the invention distribution function by the increasing integration of science with technology.

As in any society, science and technology in Europe initially were separate and distinct: science was viewed as philosophy, while technology was the practice of artisans. Scientists had no interest in, or inclination toward, technological affairs, and technological developments were mostly the results of the toil of unlettered artisans. It was only by the time of Galileo that "sciences concerned with utilitarian technology had found spokesmen capable of winning attention and commanding respect."[45] At the beginning, the contribution of science to technology was sporadic; in fact,

---

[41]The average yield of millet, wheat, and rice in the fourteenth century is estimated to be 104, 108, and 310 catties, respectively (Chao [n. 1 above], p. 215).

[42]Ibid.; Elvin; Needham, *Science in Traditional China* (n. 12 above); and Tang (n. 5 above).

[43]As predicted by the third hypothesis of the model, as the Chinese acquired modern methods and better materials in the form of Western-designed machinery with contact with the West in the late nineteenth century, creativity reappeared (Elvin, p. 315).

[44]Peter Mathias, "Who Unbound Prometheus? Science and Technical Change, 1600–1800," in Musson, ed. (n. 37 above).

[45]Cipolla (n. 11 above), p. 244.

whether or not science was a major contributing factor to the Industrial Revolution in the eighteenth century is still subject to debate.[46] However, at least by the mid-nineteenth century, science had already begun to play an important role in technological invention.[47] The sustained acceleration in the rate of technological innovation that is a major characteristic of modern economic growth is made possible only by the continuous rightward shift of the invention curve brought about by the continuous progress in science.[48] Needham found evidence that China began losing ground to Europe in the technological race only after the scientific revolution had occurred in Europe.[49]

Why a Scientific Revolution Did Not Occur in China

Any discussion of the Needham puzzle is incomplete without an explanation of why modern science did not arise in China. As stated above, science, in essence, is a body of systematic knowledge about nature that is expanded through a mechanism similar to that of technological invention, that is, the process of trial and error. China's large population gave it a comparative advantage in developing science in pre-modern times. However, since the advent of the scientific revolution, scientific discoveries have been made primarily by a new and more effective method that is the combination of two elements: (*a*) mathematization of hypotheses about nature, and (*b*) using controlled experiments or replicable tests to examine the validity of hypotheses. The Chinese were not historically unreceptive to the experimental method. In fact, in ancient times, they had conducted more systematic experimentations than did the Greeks or the medieval Europeans.[50] The question, then, is why the many gifted of China's large population, with the advantages of superior early achievement, did not make the transition to the new methodology in the fourteenth, fifteenth, seventeenth, or eighteenth centuries. The key to this problem lies in various factors that inhibited the growth of modern science in China.

Considerable research, including some by Needham himself, has been done in an attempt to identify the inhibiting factors in China's politico-economic institutions. A survey of all existing hypotheses is beyond the purview of this article. I will only comment on two of them. Needham's explanation is that China had a "bureaucratic system," which arose from the need of maintaining its vast array of irrigation systems, while Europe had an "aristocratic feudalism," which was relatively more favorable to the emergence of a mercantile class. When the aristocracy decayed, it gave birth to capitalism and modern science. The bureaucratic system in China at first was favorable to the growth of science. However, it inhibited the emergence of mercantilistic values and thus "was not capable of fusing together the techniques of the higher artisanate with the methods of mathematical and logical reasoning which the scholars had worked out, so that the passage from the Vincian to the Galilean stage in the development of modern natural science was not achieved, [and was] perhaps not possible."[51]

In a similar vein, though with different emphasis, Wen-yuan Qian and others argue that it was China's imperial and ideological unification that prohibited the growth of modern science.[52] In their view, intolerance was common to all premodern societies. In Europe, however, there were competitions between church and state, between church and church, and between state and state, which made the resistance to new basic ideas less effective. Therefore, Europe's cluster of more or less independent states created favorable conditions for scientific development. China, on the other hand, was ruled by one dominant ideological system backed by absolute political power, and no genuine public dispute was allowed. As a result, despite the fact that "the Chinese people have been innovative in mechanical skills and technologies, traditional China's politico-ideological inhibitions kept Chinese people from making direct contributions to the theoretical infrastructure and methodological foundations of modern science."[53]

[46]Musson, ed.

[47]Rondo Cameron, *A Concise Economic History of the World* (Oxford: Oxford University Press, 1989), p. 165.

[48]Hicks (n. 35 above), p. 145; Simon Kuznets, *Modern Economic Growth: Rate, Structure, and Spread* (New Haven, Conn.: Yale University Press, 1966), pp. 10–11; Nathan Rosenberg and L. E. Birdzell, Jr., *How the West Grew Rich* (New York: Basic Books, 1986), p. 23.

[49]Needham, *Science in Traditional China*, p. 122. That the lack of modern science limits the possibility of technological invention is acknowledged but not developed by Elvin (p. 297); Tang; and Chao (p. 227).

[50]Needham, *The Grand Titration* (n. 34 above), p. 211.

[51]Ibid.

[52]Wen-yuan Qian, *The Great Inertia: Scientific Stagnation in Traditional China* (London: Croom Helm, 1985); Kenneth Boulding, "The Great Laws of Change," in *Evolution, Welfare, and Time in Economics*, ed. Anthony M. Tang, Fred M. Westfield, and James S. Worley (Lexington, Mass.: Lexington Books, 1976); Albert Feuerwerker, "Chinese Economic History in Comparative Perspective," in *Heritage of China: Contemporary Perspectives on Chinese Civilization*, ed. Paul S. Ropp (Berkeley and Los Angeles: University of California Press, 1990).

[53]Qian, p. 91.

The above explanations improve our understanding of the issues in some ways. However, discrimination against merchants and artisans in ancient China was probably not as serious as Needham makes out. As legally defined, traditional China was at once a Confucian and a "physiocratic" state; merchants were the lowest social class within a four-class scheme. However, there was a discrepancy between legal texts and social realities. Historical data reveal that successful merchants, moneylenders, and industrialists of the Former Han period (206 B.C.–A.D. 8) were treated almost as social equals by vassal kings and marquises.[54] By the medieval period, big business and financial organizations had already appeared and flourished in China, most of them owned by members of gentry families. Therefore, young men who were not interested in books and learning but had an adventurous personality could find socially approved outlets in commerce.[55] Furthermore, during the Ming-Qing period, the discriminatory laws forbidding merchants to take civil service examinations were formally removed. After 1451, the channel for purchasing offices and even academic degrees was opened. Thus money could be directly translated into position and became one of the determinants of social status.[56]

It is also true that, as Qian argues, through the civil service examinations China was able to effectively impose a state ideology. However, Qian may have overstressed the shackling effects of ideological and political uniformity on intellectual creativity. One counterexample to Qian's assertion was the challenge of Wang Yangming (1472–1529) to traditional philosophy and social order. Wang's teaching stressed heterodox intuitive knowledge, the intrinsic equality of all men and the unity of knowledge and conduct, all in sharp contrast to the official neo-Confucian philosophy that emphasized academic conservatism and social status quo. His teaching initiated a powerful social movement and numerous followers and admirers established hundreds of private academies (shuyuan) to disseminate Wang's philosophy. Although the Ming court proscribed his teaching in 1537, 1579, and 1625, Wang's disciples were able to continue the movement, and they left a permanent imprint on the nation's educational system.[57] Admittedly, the political environment was not

conducive to unorthodox thinking. However, if a revolutionary philosophy such as Wang Yangming's was able to emerge and take root, the effects of ideological rigidity on intellectual creativity in premodern China must not have been as inhibitive as Qian believes. Revolutionary movements often have to emerge in settings unfavorable to their existence. Copernicus, Kepler, Galileo, and other pioneers of the scientific revolution in Europe had to contend with schoolmen who upheld the dogma of the authority and omniscience of the classics and even had to risk their lives in religious courts. In fact, it may be that the pioneers of the scientific revolution in premodern China had to battle harder than their European contemporaries for social recognition and acceptance, due to such factors as pointed out by Needham, Qian, and others; however, it is also fair to say that politico-ideological authority in premodern China was not absolute and that the Chinese system did not in itself preclude the possibility for geniuses to make revolutionary breakthroughs.[58]

I agree with Needham, Qian, and others that China's failure to make the transition from premodern science to modern science probably had something to do with China's sociopolitical system. However, the key to the question is not so much that this system prohibited intellectual creativity, as they argued, but rather that the incentive structure of the system diverted the intelligentsia away from scientific endeavors, especially from the mathematization of hypotheses about nature and controlled experimentation.

In premodern times, many scientific findings were made spontaneously by geniuses with innate acumen in observing nature. Individual ingenuity is, of course, important for the progress of modern science. However, the advance of modern science from its inception has relied on the mathematical systematization of hypotheses about the external universe and on tests by controlled experiment. To be able to accomplish this, a scientist must have updated knowledge about the universe, as well as training in abstract mathematics and controlled experimental methods. This knowledge and training gives scientists a stock of acquired human capital with which to observe nature to determine what can be added to science by empirical

---

[54]Ping-ti Ho, *The Ladder of Success in Imperial China: Aspects of Social Mobility, 1368–1911* (New York: Columbia University Press, 1962), p. 42.

[55]Eberhard (n. 12 above).

[56]Ho, *The Ladder of Success in Imperial China*, p. 51.

[57]Ibid., pp. 198–200.

---

[58]I want to emphasize that I do not mean to belittle the importance of cultural pluralism, political tolerance, and so on, which are emphasized by Qian and others, for the progress of modern science. However, what I am discussing here is the scientific revolution at the crucial point of its birth, which was Europe in the seventeenth century. From the historical evidence cited by William Monter, one can conclude that records of tolerance at that time were not much better in Europe than in China. It would be a fallacy here to compare the Western system after the rise of modern science with the premodern system in China.

observation and experiments. A larger population means more geniuses, and therefore, in premodern times, implied probabilistically more achievements in premodern science. However, even though there may be many geniuses in a society, without the necessary acquired human capital, the society will not be able to launch a scientific revolution. This special human capital, a necessary requirement for membership in the club of modern science, is expensive and time-consuming to acquire.

For several reasons, which are embedded in China's historical and political legacy, the gifted in ancient China had fewer incentives than their Western contemporaries to acquire the human capital required for "modern" scientific research. In the West, the states were governed by hereditary feudal aristocrats. In China, after the Qin unification in 221 B.C., the state was ruled by bureaucrats. Civil service examinations were instituted during the Sui dynasty (589–617), and after the Song dynasty (960–1275), all bureaucrats were selected through competitive civil service examinations. Government service was by far the most honorable and in every sense the most worthwhile occupation in premodern China. Therefore, traditional Chinese society considered entry into the ruling bureaucracy the final goal of upward social mobility.[59] Naturally, the gifted were attracted to these jobs, and they had ample incentives to invest their time and resources in accumulating the human capital required for passing the examinations.[60] The basic readings for the examinations, which students had to memorize by heart, were the Confucian classics, with a total of 431,286 characters.[61] That required 6 years of memorization, at the rate of 200 characters a day. After memorizing the classics, students were required to read commentaries several

times the length of the original texts and to carefully scan other philosophical, historical, and literary works, which were needed as a basis for writing poems and essays in the examinations.[62] Because of the strictly defined curriculum for the examinations, most people, including most of society's geniuses, would not have had the incentive to devote time and resources before passing the examinations to the type of human capital required for scientific research. Moreover, once they passed the examinations, they would be occupied with the demands of officialdom and with official ladder climbing, and thus would for the most part still have no time or incentive for acquiring those types of human capital.[63]

In premodern times China's population was larger than that of Europe. This would indicate that China had more geniuses than Europe in premodern times. However, because of the incentive system created by the specific form of civil service examination and officialdom, fewer of the gifted in China than those in Europe were interested in acquiring the human capital essential for the scientific revolution. Therefore, despite her early lead in scientific achievement, China failed to have an indigenous scientific revolution.[64]

---

[59]Ho, *The Ladder of Success in Imperial China*, p. 92.

[60]This is best reflected by a famous poem by the Song emperor Chen-Zong:

To enrich your family, no need to buy good land:
Books hold a thousand measures of grain.
For an easy life, no need to build a mansion:
In books are found houses of gold.
Going out, be not vexed at absence of followers:
In books, carriages and horse form a crowd.
Marrying, be not vexed by lack of a good go-between:
In books there are girls with faces of jade.
A boy who wants to become a somebody
Devotes himself to the classics, faces the window, and reads.

Quoted in Ichisada Miyazaki, *China's Examination Hell: The Civil Service Examinations of Imperial China* (New Haven, Conn.: Yale University Press, 1976), p. 17; published in Japanese in 1963 and translated into English by John Weatherhill.

[61]Ibid.

[62]The examinations not only enabled the emperors to select the most talented people for the civil service but also instilled a moral system, through Confucian teaching, which greatly reduced the costs of governing. Jones (n. 7 above) wondered why China was able to maintain unity despite wide regional diversity and the backwardness of premodern means of control and communication (p. 221). As suggested by Ray Huang, "The dynasty stood upon its moral character, which was its strength. Otherwise it would never be able to govern the people. The secret of administrating an enormous empire . . . was not to rely on law or power to regulate and punish but to induce the younger generation to venerate the old, the women to obey their menfolks, and the illiterate to follow the examples set by the learned" (*1587, a Year of No Significance: The Ming Dynasty in Decline* [New Haven, Conn.: Yale University Press, 1981], p. 22). These principles of loyalty and filial piety were the essence of Confucian teaching. After the thirteenth century, a military coup was almost unheard of in China, and China proper was never for long under more than two administrations. The civil service examination and the stress on Confucian classics probably was the ingenious institutional innovation that contributed to this phenomenon.

[63]The comment by Sung Ying-Hsing (Song Yingxing), author of the famous 1637 technology book *T'ien-Kung K'ai-Wu* (A volume on the creations of nature and man; Chinese technology in the seventeenth century) to his book is the best footnote to this point. He wrote: "An ambitious scholar will undoubtedly toss this book onto his desk and give it no further thought: it is a work that is in no way concerned with the art of advancement in officialdom" (trans. E-tu Zen Sun and Shiou-Chuan Sun [University Park: Pennsylvania State University Press, 1966]).

[64]During Song times, mathematics and astronomy were studied in the state university (*guo zi jian*) and were required for examinations. It is a pity that in 1313 such subjects were dropped and the required readings officially limited to a set of Confucian texts. As a result, by Ming times many of the mathematical writings of the Song were lost or had become incomprehensible.

Concluding Remarks

In this article I have attempted a hypothesis for the puzzle: Why was China's science and technology so far in advance of other civilizations historically, only to fall in modern times? Many causes may have contributed to this paradoxical phenomenon. In this article I postulate a simple hypothesis with a few relevant factors that are ignored by most students of Chinese history as well as of comparative economic history. In premodern times, most technological inventions stemmed from the experiences of artisans and farmers, and scientific findings were made spontaneously by a few geniuses with innate acumen in observing nature. In modern times, technological inventions mainly result from experiment cum science. Scientific discovery is made primarily by the technique of mathematized hypotheses and models about nature tested by controlled experiment or replicable tests, which can more reliably be performed by scientists with special training. Under the premodern model of technological invention and scientific discovery, the larger the population in a society, the greater the number of experienced artisans, farmers, and geniuses in the society. Therefore, other things being equal, more advance in technology and science would be more likely to occur in a larger society. China had comparative advantages in premodern times because of its large population but fell behind the West in modern times because technological invention in China continued to rely on happenstance and experience, while Europe changed to planned experiment cum science in the scientific revolution of the seventeenth century. The reason that China failed to have a scientific revolution I have attributed here to the contents of civil service examinations and the criteria of promotion, which distracted the attention of intellectuals away from investing the human capital necessary for modern scientific research. Therefore, the probability of making a transition from primitive science to modern science was reduced.

To the extent that the above hypothesis is valid, several policy implications for economic development are in order. In premodern times the large population size in an economy is potentially an asset for economic growth, as a large population is likely to contribute to a higher rate of technological innovation and scientific discovery in that economy. Experience-based invention is still an important source of technological change in modern times, especially with respect to minor modifications of existing technology. However, if this large population is ill equipped with the acquired human capital necessary for undertaking modern scientific research and experiment, the likelihood that the economy will contribute to modern technological invention and scientific discovery is small. For a developing country in modern times, many technologies can certainly be imported from developed countries at a much lower cost than the cost of inventing them independently. However, many empirical studies have found that the success or failure of technology transfers crucially depends on the domestic ability to follow up with adaptive innovations on the imported technology, which in turn depends on domestic scientific research capacity.[65] Therefore, in modern times a large population is no longer an endowment for economic development. More important than the size of the population is education with an emphasis on modern curriculum.

[65]Gustav Ranis, "Science and Technology Policy: Lessons from Japan and the East Asian NICs," in *Science and Technology: Lessons for Development Policy*, ed. Robert E. Evenson and Gustav Ranis (Boulder, Colo.: Westview, 1990); David C. Mowery and Nathan Rosenberg, *Technology and the Pursuit of Economic Growth* (Cambridge: Cambridge University Press, 1989).

Reading 7

Historiographical Layers in the Relationship between Science and Technology

Sungook Hong

Abstract

This paper aims to examine historiographical layers in the historical narrative on the relationship between science and technology, a topic which has been exhaustively discussed without consensus by both historians of science and technology. I will first examine two extreme positions concerning this issue, and analyze the underlying historiographical standpoints behind them. I will then show that drawing implications for the relationship between science and technology from a few case-studies is frequently misleading. After showing this, I will move to the "macrohistory" of the relationship between science and technology, which reveals a long process in which the barriers between them gradually became porous. Here, I will examine the historical formation of three different kinds of "boundary objects" between science and technology, which facilitated their interactions by making their borders more permeable: instruments as a material boundary object; new institutions, laboratories, and departments as a spatial boundary or boundary space; and new mediators as a human hybrid.

Science/Technology Relationship: A Historiographical Introduction

No other issue in the history of science and technology has been so exhaustively debated without consensus than the relationship between science and technology. In these debates, two extreme positions have been clearly articulated. The first position is that science and technology have long been intimately connected and are almost fused into one, to such an extent that any distinction between them is now meaningless. It even argues that the separate terms—science and technology—should be replaced with one term, "technoscience." The second position, in contrast to the first, is that science and technology were, and largely still are, distinct activities, each with its own norms, methods, and communities. Between these two extreme opinions, there are a myriad of different opinions, constituting a wide spectrum on this issue.

Who takes up these two ends of the spectrum? Marxist historians and philosophers of science have favoured the idea that science and technology have been intimately connected. Benjamin Farrington, for one, argued that the beginnings of science in ancient Greece were closely associated with material production as can be seen among the pre-Socratic philosophers such as Thales, Anaximander, and Anaximenes, who were engaged in technical or manual activities. The relationship between them became remote as a result of the inception of the idealist natural philosophy of Plato and Aristotle. It was the Scientific Revolution in the sixteenth and seventeenth centuries that remarried science with technology. The philosophical foundation of the remarriage, Farrington argued, was best reflected in Francis Bacon's "industrial" philosophy of science. Other Marxists had similar opinions. In his very influential article, Borris Hessen claimed that Newton's new mechanics in *Principia* was caused by the technological and economic needs of the seventeenth-century European society, such as navigation and ballistics. J.D. Bernal's *Science in History* and *Science and Technology in the Nineteenth Century* also stress the intimate connections between science and technology throughout history. According to him, the contributions of science to technological developments, and the reverse contributions of technology to the progress of science became numerous since the Industrial Revolution. This made science an important part of the productive force of a society.

The Marxist idea was influential in the 1970s, but the recent notion of "technoscience" has little to do with the Marxist tradition. One of its intellectual origins is in Bruno Latour's and Michel Callon's actor-network theory, according to which scientific facts are inseparably linked with technological artifacts. The link between them has constructivist, and hardly Marxist, grounds. The actor-network theory suggested an extended definition of agencies to include non-human, as well as human, actors. According to this theory, scientific facts are a sort of artifact, in the sense that they are stabilized by the alliance between humans and non-humans (instruments and other non-human actors in the laboratory). The alliance is also an essential ingredient for the expansion of the technoscientific network outside the laboratory. The actor-network is thus formed among different groups of humans and non-human agencies. This renders a strict distinction between science and technology meaningless. That scientists are thinkers and engineers are tinkers, or that science and technology are two different human activities is an outmoded notion which should now be discarded.

Who takes the other extreme? A majority of historians of technology in North America has moved, either consciously or unconsciously, to this side. The establishment of the "Society for the History of Technology" (SHOT) in 1958 was initiated by a discord between some historians of technology and the then editor of ISIS, the official journal of the History of Science Society (HSS). During its early years, the history of technology, as a discipline, faced the task of defining its boundary between two relatively well-established fields, the history of science and economic history. For this, the SHOT historians of technology should undermine two prevalent arguments: 1) technological development is determined by science and 2) technological development is determined by economic need. Of the first, there was the "applied-science thesis" which argued that technology is merely a product of the application of scientific laws and methods. In connection with this thesis, the "linear model" of technological development proposed that basic science would always bring about technological invention and innovation.

Against this applied-science thesis and linear model, historians of technology presented two novel ideas. First, they maintained that scientists and engineers have different communities. Further, the communities of civil, mechanical, electrical, and chemical engineers have developed different, and inverted, values and norms than those of the scientific community. In the scientific community, for example, the abstract understanding of natural phenomena was given the highest credit, while abstract understanding was least valued in the engineers' community. Edwin Layton expressed that science and technology are the "mirror-image twins." They look like twins, but their appearances (their norms and values) are inverted, reflecting their "seek-to-know" versus "seek-to-do" imperatives. Secondly, historians of technology argued that the essence of technology lies in technological knowledge, rather than in the hardware. It has its own method, data, and content, but its essence consists in its pursuit of efficiency and its emphasis on design ability. On these two grounds, historians of technology proposed that the interaction between science and technology should not be understood as an application of scientific knowledge to technological hardware, but as an interaction between two equal sets of knowledge.

Historians of science never reached a consensus on this issue. However, historians like Alexandre Koyré, Rupert Hall, Richard Westfall, and Charles Gillispie, have pointed to the absence of direct, synergetic influence between science and technology until quite recently. Koyre characterized the Scientific Revolution by intellectual innovations such as "the destruction of the cosmos" and "the geometrization of space," for which the role of experiment and instruments, let alone technology, were minimized. Hall asserted that the impact of science upon technology during the Scientific Revolution was minimal. Technology and art infiltrated into science, not because the boundary between them became porous, but "scholars" in universities attended to and adopted the method of the "craftsmen" outside the university. Westfall criticized Farrington's concept of "industrial science," because, he says, what happened to the relationship between science and technology had little to do with economic sectors that can be properly categorized as "industrial." Hall argued that the direct interaction between technology and scientific knowledge was almost absent even during the Industrial Revolution. Important technological developments of the day, such as the textile machines, the steam engine, railway, the use of charcoal for the production of iron, and the chemical production of soda, were not made possible by science. Taking examples from the French *Encyclopaedia* movement, Gillispie asserted that science provided technology with a rational classification and systematization of existing techniques and industrial procedures—a process which he called the "natural history of industry."

There are, however, interesting differences between historians of science like Hall and Gillispie, on the one hand, and the SHOT historians, on the other. Hall and Gillispie took for granted that *modern* technology is deeply indebted to science. Hall apparently thought that modern technology was a sort of applied-science when he listed the four contributions of science to technological knowing: 1) mathematical analysis; 2) controlled experiments; 3) natural laws such as thermodynamics and genetics; and 4) new natural phenomena. Thus, modern technological knowledge is "closely approximate to" modern science in basic ideas. This is what most, perhaps not all, of the SHOT historians could not have accepted. According to the SHOT historians, it was not the passive adoption by technology of high scientific theories, but an interactive process between them that led to the birth of a new engineering science. Engineers have formulated systematic theories even when no scientific guidance was available. On the basis of detailed studies into aeronautical engineering, Walter Vincenti has unmasked the differences in, for example, the uses of models and experiment in science and engineering. While scientists use scale models and do experiment to build or test hypotheses, engineers use models and experiment to obtain data for design, with which to bypass the difficulty caused by the absence of an exact quantitative theory.

The issue of the relationship between science and technology has become more complicated due to conflicting interpretations of historical case studies. The Marxist historians, the SHOT historians, and the historians of science used historical examples and case studies to make their points more convincing. These case studies have not apparently settled the long controversy over the relationship between science and technology. But why? Why such examples and case-studies cannot settle the controversy? Why do historians make different interpretations out of the same historical event? The next section will examine these issues. I will show that the microhistorical examinations of case studies will lead to different, and even conflicting, conclusions on science/technology relationship.

Historical Case-Studies and the Limit of Implications

Those who supported the idea of the distinctiveness between science and technology preferred to refer to Project Hindsight—a project conducted in the early 1960s with support from the Department of Defence of the United States. Project Hindsight probed the contribution of science and technology to the development of the modern military defence system. Investigators had first sorted out "events" that led to the system, and apparently discovered that only "0.3% of these events" were caused by science. More than 90% of these events turned out to be of technological origin, and 8% of applied-science origin. Yet, people who disagreed with the findings of the Project Hindsight preferred Project TRACES, a counter-project sponsored by the National Science Foundation. Project TRACES traced the origins of several innovations like contraceptives, the electron microscope and video tape recorders, and showed that they were rooted in basic (non-mission oriented) scientific research. As we can easily see, these two different conclusions had different policy implications.

Each group tends to emphasize different aspects of the same historical event. Thomas Edison's inventions, and in particular his electric lighting system, have been stressed as showing the autonomy of technology from science. However, discontents have stressed the crucial contribution of the physicist F. Upton, Edison's employee, to the development of Edison's electric lighting system. The engineering method of W.J.M. Rankine (1820–1872), one of the pioneers of modern engineering, has been interpreted either as evidence showing the application of science to engineering or as evidence showing the formation of a new field of engineering distinct from science. The invention of the Diesel engine by the German engineer Rudolf Diesel has been depicted either as showing the contribution of science (thermodynamic principles such as Carnot's cycle) to technology or the uselessness of scientific ideals (again Carnot's cycle) in the real world of technology. The invention of the laser was regarded as showing either a fruitful interaction between science (quantum physics) and microwave engineering or a huge gap between scientific ideas on stimulated emission (suggested in the 1910s) and its application to the laser (made in the late 1950s). Faraday's discovery of electromagnetic induction in 1831 has been interpreted as a crucial piece of evidence showing the contribution of science to technology, since it made possible the electric motor and dynamo. Yet, those who are inclined to oppose this implication would say that it was not Faraday, but many obscure technologists, who essentially contributed to the invention of the commercial dynamo which was only available at the end of 1860s, thirty years after Faraday's discovery. Likewise, the entire history of electricity in the nineteenth century has been described in two different ways. A. Keller asserted that "without Volta, Davy, Oersted, Ohm, Ampere and Faraday, there would have been no telegraph, telephone, electrolysis, electroplating, no batteries, generators, illumination or trams." Contrarily, Derek de Solla Price characterizes the same period as the period when a "revolutionary change in the technology of science caused repercussions all over science and technology and created entire sets of new methodologies."

Why do these historical interpretations diverge? There are several reasons for this. First, a disciplinary gap between the history of science and technology certainly contributed to the worsening of the situation. Historians of science seldom examine "inside technology," while historians of technology sometimes hold an outmoded conception of science. As de Solla Price once complained, "our parascientific professions happen to have been professionalized quite separately so that most historians of science get a certificate of ignorance in the history of technology and vice versa." Secondly, there is a semantic problem. As Otto Mayr pointed out, the term science or technology is a vague "umbrella term", which indicate "general areas of meaning without precisely defining their limits." Other terms such as (pure or practical) engineering, engineering science, scientific engineering, technological knowledge, and theory and practice, as well as various concepts used to denote their relationships such as interaction, application, flow of information, mediation, and transformation, cannot be precisely defined. Metaphors and models such as "dancing pair," "mirror-image twin," "symbiosis," and "dialectic interactions" distorts as much as reflects the relationship between science and technology. Third, historians' general propositions as to the relationship between science and technology, consciously or not, reflect both their analysis of specific past case-studies and their present concerns. I think that Layton's and other SHOT historians' ideals on the distinctiveness between scientific and technological knowledge reflects the ideas of some nineteenth-century and twentieth-century American engineers who stressed for the professionalization of engineering the independence of technology from science. Let me give another example. In the history of radio, Marconi developed his own version of the history of wireless telegraphy in which technology was largely independent of science, while the British Maxwellians such as Oliver Lodge undermined it by stating various contributions of science to early radio. The actors' histories were inscribed in their scientific articles and monographs, memoirs, autobiographies, obituaries, and patents, and in archives such as letters and unpublished diaries. The interpretations of later historians were heavily influenced by these sources, which they regarded as "authority."

So, among these competing conclusions, which one does reflect the true historical relationship? Can we draw an "objective" picture that truly represents the relationship between science and technology? What I have tried to show in this section is that drawing a general conclusion on the relationship between science and technology out of a single or a series of specific historical case-studies is often risky, for one can draw multiple, and even conflicting, implications from the same historical event. In the next section, I will make one attempt to bypass this difficulty. For this, I will propose a "macrohistory" of the relationship between science and technology, in which long-term trends and patterns are analysed. The macrohistory that I will present will emphasize the historical formation of three different kinds of boundary objects between science and technology, which made their boundary more porous, and which facilitated their interactions. They include: instruments as a material boundary object; new institutions and laboratories as a spatial boundary object or boundary space; and new group of people who mediated between science and technology as a boundary person or a human hybrid.

Boundary Objects and Macrohistorical Relationships between Science and Technology

Science (or natural philosophy) and technology (or arts) were rather distinct and separate fields in ancient times and in the Middle Ages. There were several reasons for this separation. First, natural philosophy and technology were carried out by different groups of people. Natural philosophy was done by scholarly intellectuals, while technology was carried out by uneducated artisans who usually belonged to the lower class. Because of this, we do not know the names of the inventors of the grinding stone, siphons, water pumps, the mechanical clock, and the wind-mill, all designed in ancient and medieval times. Secondly, natural philosophy and technological activity were considered to be quite different in nature. The goal of natural philosophers was to discuss natural phenomena for philosophical or theological purposes, while the goal of artisans invent useful things. Aristotelian natural philosophy, which prevailed in the medieval ages, took for granted a strict dichotomy between the natural and the artifactual, which justified natural philosophers' low appreciation of technological activity.

The *macrohistory* of the relationship between science and technology reveals a long process in which the barriers that had blocked the interaction between science and technology gradually became porous and partially removed. The trigger for this was prepared in the Renaissance when the social status of engineers rose considerably. The so-called Renaissance engineers made fortune and established social prestige by constructing buildings, ships, and canals. Further, they educated themselves to be more erudite. They contacted university scholars, learned classical languages like Greek and Latin, read Euclid and Archimedes, and wrote books on engineering topics. This brought university scholars to become interested in engineering problems. Simon Stevin and Galileo tackled statical and mechanical problems such as those in the lever, buoyancy, and the projectile. Pascal and Torriceli tried to explain hydraulics and phenomena in water pumps. Others examined geometrical problems in military fortification, navigation and cartography. This naturally created several contact points between scholars and engineers.

Material Boundary Object: Instruments

Another change during the Scientific Revolution was the introduction of an experimental method into science. Aristotle considered observation to be important to science, but was strictly against the employment of experimental methods, because he thought that experimental intervention would disturb the true course of nature. Once experiment was introduced into science by Francis Bacon, Galileo, and Robert Boyle, and accepted as the proper scientific method, technologies rushed into science in the form of instruments. Scientists used and constructed various instruments such as inclined planes, pumps, pulleys, levers, lenses, prisms, and clocks for their scientific research. Telescopes and microscopes greatly widened human senses, generating a series of epistemological controversies over the authenticity of the images they produced. Robert Boyle's air pump created a vacuous space which did not normally exist in nature. The natural world that natural philosophers were interested in was no longer "naked" nature, but a sort of artifactual world—or "nature in the wider sense"—created or mediated by instruments. By doing so, instruments shortened the epistemological gap between science and technology.

I would argue that instruments such as the microscope, telescope and air-pump, technologies such as the pendulum clock, and demonstration machines such as Atwood's machine (eighteenth-century)

provided the first interface for natural philosophers, instrument makers, and technologists. In other words, they became the first boundary object which lay overlapped across science and technology. Instruments mediated the interaction between science and technology in four ways. First, they constituted a major inflow of technology into science. Second, new artifacts created a new, strange effect that drew the attention of scientists to provide a scientific explanation. Scientists sometimes transformed such an effect into a standardized scientific technique or a marketable instrument. Lamps exhibited a strange effect named the Edison effect, and the Edison effect was transformed into the thermionic valve. Another example is a strange electrical property of some crystals. This was used as the crystal rectifier (the cat's whisker) in early wireless telegraphy. It then attracted the endeavour of several theoretical and experimental physicists in the first half of the twentieth century, which eventually led to the birth of the contact semiconductor in 1947. Third, a scientific theory designed for instruments could later be applied (with modifications) to machinery, as Maxwell's electromagnetic theories on electrical instruments were later used for heavy AC machineries. Finally, as Layton has asserted, measuring instruments (like the Prony dynamometer) designed to measure the efficiency of a technology (like the water turbine) can incorporate relevant scientific principles. This kind of instrument plays a crucial role for the diffusion and improvement of the technology.

Spatial Boundary Object or Boundary Space

Let's move to the second boundary object. The second boundary object was the social space in which scientists and engineers could meet and discuss their common interests. Since the late seventeenth century, coffeehouses, salons, and pubs in London were developed as a place where Newtonian natural philosophers like Desagulier and businessmen like James Brydges met and discuss on Newtonianism, navigation, commerce, insurance, and the newly invented Savory engine. In the last quarter of the eighteenth century, "Philosophical and Literary" Societies were established in various industrial cities such as Birmingham, Manchester, Leeds, Liverpool, Newcastle, Sheffield, and Derby. The members of these societies included notable scientists, engineers, businessmen, and industrialists of the cities. The Lunar Society of Birmingham, the most famous of these societies, had as its members James Watt, Josiah Wedgewood, Joseph Priestley, Erasmus Darwin, and local industrialists and engineers. They discussed diverse scientific and technical issues such as the phlogiston theory of combustion, thermodynamics, and the steam engine. They believed that scientific knowledge would bring about technological and industrial advances.

Debates have taken place over whether scientific knowledge discussed in these societies really contributed to technological development. Neil McKendrick argued that Josiah Wedgewood, a successful ceramic industrialist and a member of the Lunar Society, was in fact much indebted to the knowledge and techniques of chemistry for his success in making china. But his critics would argue that Wedgewood's chemical knowledge was based on Joseph Priestley's theory of the phlogiston that proved to be non-existent, and further that Wedgewood simply legitimized his success in terms of the current scientific theory. This issue becomes more complicated, since historians have implicitly projected their present standard of science upon the past. Historians tend to think that an incorrect theory such as Priestley's phlogiston theory cannot result in successful applications. Wedgewood's knowledge, if successful, must have come from somewhere other than faulty science— for instance, from his accumulated experiences in china-making in the field. This kind of logic leads to the conclusion that sciences can be genuinely applied to technology only when their theories are correct. When did scientific theories obtain a self-correcting mechanism? One is likely to say that physical sciences in the second half of the nineteenth century, armed with mathematics and experiment, were quite different from those in the second half of the eighteenth century. Take Kelvin's thermodynamics, Maxwell's electromagnetic theory, or Liebig's organic chemistry and compare them with the eighteenth-century caloric theory, Galvanism, and the phlogiston theory. No wonder the new thermodynamics, electromagnetism, and organic chemistry—all of which constitute what we now call "science"— found their ways into the industrial and technological applications of science.

I have discussed the creation of new social spaces such as the literary and philosophical societies in which scientists, engineers, and industrialists interacted and talked about their common interests. These interactions became an intellectual motive force for the establishment of several provincial colleges such as Owens College in Birmingham and Mason College in Manchester which specialized in training local elites including industrialists and engineers. The social gap between university-based scientists, and self-employed inventors and consultant engineers, became almost negligible in these newly created colleges. In 1840, the first chair for an engineering professorship was established in the University of Glasgow, which accelerated the already existing interactions between scientists and engineers. In 1832, the British Association for the Advancement of Science (BAAS) was established, and in the 1840s it set up a committee for the purpose of exploring the magnetic field of the world. In 1860, telegraphers and physicists who had contacted in both the BAAS and submarine telegraphy companies agreed on establishing the Committee for Electrical Standards in the BAAS. Under this committee, theorists like James Clerk Maxwell worked with the practical telegrapher Fleeming Jenkin to find a way to determine the new unit and standard system. Through this collaboration, Maxwell obtained a new insight into the ratio of electrostatic to electromagnetic charge of electricity, which later became a basis of his electromagnetic theory of light.

Another important boundary space, the research laboratory in industry, was established as the belief in the mutual benefit between science and technology became popular and the actual contact between academics and industrialists became increasingly commonplace. In the field of dyestuff chemical industry, the university-trained chemists had established successful business of their own, as can be seen in the successful career of William H. Perkin who, as a student of A. Wilhelm Hofmann, synthesized the first chemical dye, aniline purple, in 1856 and became a successful businessman. In the 1870s, chemical industry already hired university-trained chemists for the problems of cost and quality. In the 1880s, the Bayer Company hired a few Ph.D. chemists to conduct research only. This became the germ for the Bayer Company's well-known research division, which soon found new colours for synthetic dyes. This led to the expansion of the division into a research laboratory. This chemical laboratory became a model for the electrical industry. In 1900, the General Electric Company in the United States established the first research laboratory among electrical industries under the directorship of W.R. Whitney, and hired physicists and chemists for long-term research. These scientists essentially employed a hybrid method characterized by the "natural study of the artificial" and "technological theories." Langmuir in the lab was even awarded the Nobel Prize in chemistry for his research that he performed in the laboratory. The Bell Telephone laboratory, which was established for solving serious technical problems in long-distance telephony, became another very successful company laboratory. Its contributions to pure science include C.J. Davisson and L.H. Germer's discovery of electron diffraction in 1927, and A. Penzias and R. Wilson's discovery of cosmic background radiation in 1962, each of which essentially contributed to physics and astronomy, respectively.

Boundary People and Engineering

The proliferation of the boundary spaces went hand in hand with the formation of hybrid people who were interested in, and contributed to, both science and technology. James Joule, one of the discoverers of the principle of energy conservation, was an early protocol. His experience in engineering research on electric battery and the steam engine made him adopt the language of engineers such as "duty" or "work," and this linguistic resource became an important ground for his discovery of the mechanical equivalent of heat. From the 1840s on, hybrid figures appeared frequently. William Thomson (later Lord Kelvin) worked on both electromagnetic theory and submarine telegraphy, combining theory and practice into one. His mathematical theory on the transmission of telegraphic signals became a model for scientific electrical engineering. In the field of civil engineering, W.J.M. Rankine integrated both theory and practice. In telephone engineering of the 1880s, Oliver Heaviside, who discovered the mathematical condition for the distortionless transmission of signals, followed the route that Thomson had paved thirty years previously to combine new mathematical techniques with the real world of technology. In AC power engineering, John Hopkinson, John Ambrose Fleming, and William Ayrton built the new discipline. These pioneers mediated between scientific theory and practical knowledge in the field, constructing the core of engineering knowledge.

The historical evolution of various engineering disciplines in the nineteenth and twentieth centuries is beyond the scope of this paper. However, I will briefly discuss three features of engineering which have been paid less attention by historians of both science and technology. First, historians have pointed out that engineering (or scientific engineering or engineering science) has mediated between science and technology, but they have implicitly assumed that science represents a high theory and technology represents practical know-how, and that the mediation has therefore taken place as the transfer of information from high to low. This assumption

is sometimes not compatible with the historical fact that many scientists were neither interested nor contributed much to technology or that, at many times, their contributions were entirely neglected by practical engineers. In my articles on the history of electrical engineering, I proposed that mediation between science and technology in the early phase of electrical engineering was made possible by two kinds of "mediating people"—the scientist-engineers who, on the basis of their scientific training, understood the operation of machinery in scientific terms, and the practicing-engineers who systematized their accumulated knowledge in the field and in the workshop. I also stressed that the mediation took place on four different levels: that is, between a scientific and a "workshop theory"; between a scientific theory and a practice in the field and in the workshop; between a scientific practice in the lab and a workshop theory; and between a scientific practice in the lab and a practice in the field and the workshop.

Secondly, if we consider the creation and the development of, for example, electrical power engineering as a product of intense interaction between science and technology (each of which had its own theories and practices) through the mediation by hybrid engineers, we can divide its history into two qualitatively different periods. During 1880–1900, which I want to call the "formative period," the boundary of power engineering was highly flexible with no clearly defined principles and practices. The boundary was continuously redefined by the extensive interaction between the scientist-engineers and the practicing engineers. In this process, they actively drew their theories and practices from many resources including scientific theory, tacit knowledge acquired from craft skill, operating knowledge of machinery, and workshop as well as laboratory practices, and moulded them into electrical engineering. Science penetrated the real world of technology through this truly multidimensional interaction. However, during the period from 1900 on, which I want to call the "stabilized period," during which major problems were largely solved and the boundary of power engineering became rather rigid, engineers were more concerned with the increase in the efficiency of the system and the perfection of it. At this stage, they drew mainly on existing, well-established body of engineering theory and practice. The interaction between science and technology becomes a routine interaction between two well-defined sets of knowledge, that is, between science and engineering knowledge.

The third feature is more speculative. Historians have pondered over why the so-called science-based industry, which emerged in the second half of the nineteenth century, started with chemical and electrical science and engineering, which were then newly established, rather than with older, well-established engineering fields such as mechanical, civil, architectural, or hydraulic engineering, or with traditional sciences such as mechanics and astronomy. Think of the early phase of electrical engineering. In this case, the major resource that engineers could draw on was science. In the late 1870s and early 1880s, for instance, various scientific theories on direct and even alternating current phenomena were available. Various theories and hardware of measuring instruments, which had been devised for laboratory purposes, were at the hands of engineers. Science was the most readily available resource for engineering problems, since there was no established engineering theory and practice. It was not the epistemological superiority of science, but the efforts of hybrid engineers to mobilize science as a resource to solve particular problems, that started the science-based industry or engineering.

Concluding Remarks

The emergence and the spread of three boundary objects—instrumental, spatial, and human—that gradually blurred the distinction between science and technology was not smooth or straightforward. When we turn to *microhistories*, the apparently regular patterns and trends in the macrohistory dissolve into fuzzier, chaotic configurations. These microhistorical configurations seems to have been conditioned by various contingent and local factors which were seldom taken into account in the macrohistory. Tensions and conflicts between science and technology can always be found in microhistorical analyses. Even in such a hybrid space as the BAAS's Electrical Standards Committee, there was tension between engineers and scientists. Scientific engineering appears less clear when we examine it through a "microscope". William Thomson's theoretical work on signal transmission were in fact less essential to the success of the submarine telegraphy than innovations of a technological kind such as the improved insulation of the cables. Heaviside's innovative idea on distortionless signal transmission was almost completely ignored by engineers at the time. The status of research in the corporate laboratory has been unstable due to the tension between managers who wanted to direct scientists to particular problems in the industrial production and scientists who wanted to maintain their autonomy over research.

These microhistories, however, cannot overrule the macrohistorical patterns and trends, any more than the macrohistory should not make us blind to local and temporal irregularities and contingencies. What these microhistorical irregularities and contingencies reflect is the essential instability inherent in boundary objects. The flexibility of a boundary object induces various, and even conflicting, elements as its constituents, but the same flexibility contributes to its instability. However, this instability can be a resources for its dynamics. A flexible, unstable boundary object is like a motion of the bicycle: the instability remains constructive as long as it is pushed forward. As long as a boundary object begets other boundary objects, it can maintain a dynamic instability. Some economists described the history of the capitalism in terms of dynamic instability in the sense that the inherent instability in capitalist economy remains constructive as long as it remains dynamic. This dynamic instability is a homology between the science-technology hybrid and the capitalist context in which this hybrid evolved.

Reading 8

The Scientific Revolution and The Death of Nature

Carolyn Merchant*

Abstract

The Death of Nature: Women, Ecology, and the Scientific Revolution, published in 1980, presented a view of the Scientific Revolution that challenged the hegemony of mechanistic science as a marker of progress. It argued that seventeenth-century science could be implicated in the ecological crisis, the domination of nature, and the devaluation of women in the production of scientific knowledge. This essay offers a twenty-five-year retrospective of the book's contributions to ecofeminism, environmental history, and reassessments of the Scientific Revolution. It also responds to challenges to the argument that Francis Bacon's rhetoric legitimated the control of nature. Although Bacon did not use terms such as "the torture of nature," his followers, with some justification, interpreted his rhetoric in that light.

In 1980, the year *The Death of Nature* appeared, Congress passed the Superfund Act, ecofeminists held their first nationwide conference, and environmentalists celebrated the tenth anniversary of Earth Day. *The Death of Nature*, subtitled "Women, Ecology, and the Scientific Revolution," spoke to all three events. The chemicals that polluted the soil and water symbolized nature's death from the very success of mechanistic science. The 1980 conference "Women and Life on Earth: Ecofeminism in the '80s" heralded women's efforts to reverse that death. Earth Day celebrated a decade of recognition that humans and ecology were deeply intertwined. The essays in this *Isis* Focus section on the twenty-fifth anniversary of *The Death of Nature* reflect the themes of the book's subtitle, and I shall comment on each of them in that order. I shall also elaborate on my analysis of Francis Bacon's rhetoric on the domination and control of nature.[1]

I

Charis Thompson's provocative, well-argued paper deals with the connections between women and nature and the foundations of and responses to ecofeminism. When *The Death of Nature* appeared in 1980 the concept of ecofeminism was just emerging. The 1980 conference organized by Ynestra King and others seemed to me to offer an antidote to the death of nature and the basis for an activist movement to undo the problems that the Scientific Revolution had raised for contemporary culture in the form of the environmental crisis. Moreover, it connected the effects of nuclear fallout and chemical pollutants on women's (and men's) reproductive systems to the relations between production and reproduction I had discussed in the book.[2]

*Department of Environmental Science, Policy, and Management, University of California, Berkeley, California 94720.

Earlier versions of this essay were presented at the conference "The Scientific Revolution: Between Renaissance and Enlightenment," University of Florida, Gainesville, February 2005 (Pts. II and III), and as part of the session "Getting Back to *The Death of Nature*: Rereading Carolyn Merchant" at the annual meeting of the History of Science Society, Minneapolis, Minnesota, 4 November 2005, sponsored by the Women's Caucus. I thank the reviewers, commentators, and participants at both conferences and Robert Hatch, Roger Hahn, David Kubrin, Wilber Applebaum, and Bernard Lightman for their insights and suggestions.

[1] Carolyn Merchant, *The Death of Nature: Women, Ecology, and the Scientific Revolution* (1980; San Francisco: HarperCollins, 1990) (hereafter cited as **Merchant**, *Death of Nature*). For a list of reviews and commentaries on the book from 1980 to 1998 see Merchant, "*The Death of Nature*: A Retrospective," in "Symposium on Carolyn Merchant's *The Death of Nature*: Citation Classics and Foundational Works," *Organization and Environment*, 1998, *11*:180–206 (the retrospective is on pp. 198–206); this symposium featured commentaries by Linda C. Forbes, John M. Jermier, Robyn Eckersley, Karen J. Warren, Max Oelschlaeger, and Sverker Sörlin. See also Kevin C. Armitage, "A Dialectic of Domination: Carolyn Merchant's *The Death of Nature: Women, Ecology, and the Scientific Revolution*, 2000, online, reviewed for H-Ideas' Retrospective Reviews of "books published during the twentieth century which have been deemed to be among the most important contributions to the field of intellectual history." See also Noël Sturgeon, Donald Worster, and Vera Norwood, "Retrospective Reviews on the Twenty-fifth Anniversary of *The Death of Nature*," *Environmental History*, 2005, *10*:805–815.

[2] Sherry Ortner's foundational article, "Is Female to Male as Nature Is to Culture?" in *Woman, Culture, and Society*, ed. Michelle Rosaldo and Louise Lamphere (Stanford, Calif.: Stanford Univ. Press, 1974), pp. 67–87,

Thompson notes that ecofeminism linked the domination of women with the domination of nature and recognized the values and activities associated with women, including childbearing and nurturing. She correctly points out that during the 1980s and 1990s ecofeminism faced a critique by academic women that it was essentialist in its conflation of women with nature, implying not only that women's nature is to nurture but also that women's role is to clean up the environmental mess made by men. Women who, as ecofeminists, came to the defense of nature were actually cementing their own oppression in the very hierarchies that (as the anthropologist Sherry Ortner had argued) identified men with culture and women with nature.[3]

My own efforts to deal with the problems of essentialism and nature/culture dualism led me to develop a form of socialist ecofeminism rooted not in dualism but in the dialectics of production and reproduction that I had articulated in *The Death of Nature*. There I had argued that nature cast in the female gender, when stripped of activity and rendered passive, could be dominated by science, technology, and capitalist production. During the transition to early modern capitalism, women lost ground in the sphere of production (through curtailment of their roles in the trades), while in the sphere of reproduction William Harvey and other male physicians were instrumental in undermining women's traditional roles in midwifery and hence women's control over their own bodies.[4]

During the same period, Francis Bacon advocated extracting nature's secrets from "her" bosom through science and technology. The subjugation of nature as female, I argued, was thus integral to the scientific method as power over nature: "As woman's womb had symbolically yielded to the forceps, so nature's womb harbored secrets that through technology could be wrested from her grasp for use in the improvement of the human condition."[5]

The dialectical relationships between production and reproduction became for me the basis for a socialist ecofeminism grounded in material change. I also addressed the related problem of the depiction of nature as female, and its conflation with women, by advocating the removal of gendered terminology from the description of nature and the substitution of the gender-neutral term "partner." This led me to articulate an ethic of partnership with nature in which nature was no longer symbolized as mother, virgin, or witch but instead as an active partner with humanity.[6]

I don't believe, however, that Thompson's statement that "by the early to mid 1990s ecofeminism had largely been relegated to a marginal position in feminist theory in the academy" is quite accurate. During the 1990s and 2000s, ecofeminists dealt with the problem of essentialism by articulating new theories that acknowledged the variable, gendered, raced subject and the socially constructed character of nature. All were deeply cognizant of the critiques of essentialism and identity politics and moved beyond them to argue for ethically responsible, situated, relational subjects engaged in ecofeminist political actions.[7]

The role of ecology in the Scientific Revolution was the second of the three themes in *The Death of*

influenced my thinking about women's relationships to nature and culture. I was also influenced by Rosemary Radford Ruether, "Women's Liberation, Ecology, and Social Revolution," *WIN*, 4 Oct. 1973, 9:4–7; and Ruether, *New Woman/New Earth: Sexist Ideologies and Human Liberation* (New York: Seabury, 1975). Susan Griffin consulted me on some of her ideas while writing *Woman and Nature: The Roaring Inside Her* (New York: Harper Collins, 1978). Although Françoise d'Eaubonne had used the term "ecofeminism" in 1974 in "The Time for Ecofeminism," few scholars in the United States had heard the word at that time: Françoise d'Eaubonne, *Le féminisme ou la mart* (Paris: Horay, 1974), pp. 215–252. Ynestra King taught a course on "Ecofeminism" at the Institute for Social Ecology in Plainfield, Vermont, about 1976.
[3]Ortner, "Is Female to Male as Nature Is to Culture?" For a history of theories associated with ecofeminism see Carolyn Merchant, *Radical Ecology: The Search for a Livable World* (1992; New York: Routledge, 2005), Ch. 8. Thompson's own recent work shows why issues of reproduction so important to the origins of early modern science continue to be vitally significant today. See Charis Thompson, *Making Parents: The Ontological Choreography of Reproductive Technologies* (Cambridge, Mass.: MIT Press, 2005).
[4]Harvey argued that the semen of the male, as the most perfect animal, was the efficient cause of conception, while the egg was mere matter. In fact, he held that the male semen was so powerful that impregnation of the egg could occur without contact with the sperm. "How," he wrote, "should such a fluid [the female's] get the better of another concocted under the influence of a heat so fostering, of vessels so elaborate, and endowed with such vital energy? —how should such a fluid as the male semen be made to play the part of mere matter?" William Harvey,

Works (London: Sydenham Society, 1847), pp. 298, 299, quoted in Merchant, *Death of Nature*, p. 159. For a recent assessment of scholarship on midwifery see Monica H. Green, "Bodies, Gender, Health, Disease: Recent Work on Medieval Women's Medicine," *Studies in Medieval and Renaissance History*, 3rd Ser., 2005, 2:1–46 (I thank Katharine Park for this reference).
[5]Merchant, *Death of Nature*, p. 169. See also p. 172: "For Bacon as for Harvey, sexual politics helped to structure the nature of the empirical method" as power over nature.
[6]On socialist feminism see Merchant, *Radical Ecology* (cit. n. 3), Ch. 8; on partnership with nature see Carolyn Merchant, *Reinventing Eden: The Fate of Nature in Western Culture* (New York: Routledge, 2003), Ch. 11.
[7]These theoretical works included Val Plumwood's *Feminism and the Mastery of Nature* (New York: Routledge, 1993), which proposed a form of social ecofeminism that dealt with problems of domination and difference by positing the relational self, and Noël Sturgeon's *Ecofeminist Natures: Race, Gender, Feminist Theory, and Political Action* (New York: Routledge, 1997), which dealt explicitly with the argument of the rejection of ecofeminism by the academy while validating women's on-the-ground activism. Likewise, Mary Mellor's *Feminism and Ecology* (New York: New York Univ. Press, 1997) and Ariel Kay

Nature's subtitle, "Women, Ecology, and the Scientific Revolution." In his well-argued theoretical paper on the intersections between environmental history and the history of science, Gregg Mitman raises the critical question of the linkages between the two fields, represented professionally by the American Society for Environmental History, founded in 1976, and the History of Science Society. Indeed, Mitman's own work has been at the forefront of these linkages.[8]

In *The Death of Nature*, a bridge between the history of science and environmental history was developed most explicitly in Chapter 2, "Farm, Fen, and Forest," on the ecological and economic changes taking place in Western Europe during the period of the rise of mercantile capitalism and the nation-state.[9] The chapter argues that ecological and technological changes in the late sixteenth and early seventeenth centuries helped to create material conditions that made new ideas plausible. As both Thompson and Mitman point out, I do not argue that material or ecological changes *cause* or *determine* ideological changes. Rather, they make some ideas prevalent at a given time seem more plausible than others. Some ideas die out or become less compelling (in this case those associated with natural magic and the organic worldview), while others are developed and accepted, in particular (in this case) those that led to mechanical explanations for phenomena and the mechanistic worldview. *The Death of Nature* moved back and forth between material and social conditions and ideas about nature and science. Thus ecological and material changes are seen as fundamental to understanding the rise of mechanism and to the argument for the links between environmental history and the history of science.[10]

Mitman states that "*The Death of Nature* presents us with a materialist history of environmental change that pointed toward, but never quite embraced, an ecological history of material, cultural, and social relations through which nature became not universal, but many." While it is true that in *The Death of Nature* I focused on nature symbolized as female, I do not believe that nature is necessarily a universal force. Rather, nature is characterized by ecological laws and processes described by the laws of thermodynamics and by energy exchanges among biotic and abiotic components of an ecosystem. Any of these components can become an actor or actors in an environmental history of a particular place. In my 1989 book *Ecological Revolutions* I developed a theory of ecology, production, reproduction, and consciousness in which, as Mitman puts it, "material, cultural, and social relations" are all interacting parts of ecological history. While I would still argue that the drivers of change are material (bacteria, insects, plants, and animals—including humans) and economic (explorations, colonization, markets, and capital), new ideas can support and legitimate new directions and actions taken by groups of people, societies, and nations.[11]

The Scientific Revolution is the third theme in the book's subtitle and the one addressed most cogently by Katharine Park's essay. *The Death of Nature* in general had an arresting impact in many fields and was used widely in courses; why, Park asks, was it not embraced more warmly by historians of early modern science? From evidence over the years, it would seem to me that the book did indeed have a substantial audience among historians of science and was read in numerous classes.[12] But if it was not awarded accolades by more of the field's heavyweights (although I would take Everett Mendelsohn, Walter Pagel, and Frances Yates as fully sufficient and most satisfying), I think its reception had less to do with hyperprofessionalism than with the book's challenge to the pedestal on which historians had tended to place the Scientific Revolution. The book questioned the grand narrative of the Scientific Revolution as progress and undermined the valorization of the most revered fathers of modern science—such as Harvey, Bacon, Descartes, and Newton. It argued that seventeenth-century mechanistic science itself contributed to the most pressing

Salleh's *Ecofeminism as Politics* (London: Zed, 1997) proposed socialist feminist approaches to ecofeminism as political positions. Chris Cuomo's *Feminism and Ecological Communities* (New York: Routledge, 1998) dealt with issues of race and ecofeminism, while Karen Warren's *Ecofeminist Philosophy* (Lanham, Md.: Rowman & Littlefield, 2000) proposed a multicultural, relational ethic of care.

[8]See esp. Gregg Mitman, *The State of Nature: Ecology, Community, and American Social Thought, 1900–1950* (Chicago: Univ. Chicago Press, 1992).

[9]I elaborated on these connections at the History of Technology meeting (a 4S meeting) in Toronto in 1980 and in a 1982 article: Carolyn Merchant, "Hydraulic Technologies and the Agricultural Transformation of the English Fens," *Environmental Review*, 1982, 7:165–177.

[10]Merchant, *Death of Nature*, p. 68: "As European cities grew and forested areas became more remote, as fens were drained and geometric patterns of channels imposed on the landscape, as large powerful waterwheels, furnaces, forges, cranes, and treadmills began increasingly to dominate the work environment, more and more people began to experience nature as altered and manipulated by machine technology. A slow but unidirectional alienation from the immediate daily organic relationship that had formed the basis of human experience from

earliest times was occurring. Accompanying these changes were alterations in both the theories and experiential bases of social organization which had formed an integral part of the organic cosmos."

[11]Carolyn Merchant, *Ecological Revolutions: Nature, Gender, and Science in New England* (Chapel Hill: Univ. North Carolina Press, 1989).

[12]Merchant, "*Death of Nature*: A Retrospective" (cit. n. 1), pp. 198–206.

ecological and social problems of our day and dared to suggest that women were as much the victims as the beneficiaries of the progress of science. The book contributed to a growing body of scholarship that led to the historian of science's interest in the social construction of nature and authority and the importance of the role of women in science and to the questioning of grand narratives and the ways that science was implicated in ideologies of progress.

Park is correct that I did not challenge the idea of the Scientific Revolution itself. I focused on the major transformations in science and society that occurred during the sixteenth and seventeenth centuries (1500–1700), from Copernicus to Newton, from Renaissance natural magic to the mechanical worldview, and from the breakup of feudalism to the rise of mercantile capitalism and the nation-state. I could well have emphasized the explorations of the New World (depicted as female) as a source of natural resources for the emerging European economies, connections I later developed in *Ecological Revolutions* and *Reinventing Eden*. Our understanding of the ways that "early modern science" engaged with the everyday world has been enriched by Park's own work on metaphors and emblems of female nature and the body, as well as studies of scientific patronage and practice and the witnessing of experiments.[13]

Yet the notion of a "Scientific Revolution" in the sixteenth and seventeenth centuries is part of a larger mainstream narrative of Western culture that has propelled science, technology, and capitalism's efforts to "master" nature—a narrative into which most Westerners have unconsciously been socialized and within which we ourselves have become actors in a storyline of upward progress. Demoting the "Scientific Revolution" to the mere nomer of "early modern science" obscures the power of the dominant narratives of colonialism and imperialism that have helped to shape Western culture since the seventeenth century at the expense of nature, women, minorities, and indigenous peoples. This move hides the political power of scientific narratives

in remaking the earth and its natural resources as objects for human use.[14]

But not only did *The Death of Nature* invoke mechanistic science in the destruction of nature; it further suggested that the scientific method as power over nature, exemplified in the rhetoric of Francis Bacon, implied the constraint and even the torture of nature.[15] The most heated critiques of the book have come from those whom Park has called the FOBs (Friends of Bacon). These critics have argued that the feminist project to reframe Bacon's thought has seriously misread his intentions and his accomplishments. I shall spend the rest of this essay looking at their arguments.

II

Francis Bacon's influence and reputation as a founder of modern science have been the subject of debate in recent years. Here I revisit Bacon's impact as portrayed in *The Death of Nature*, responding to the defenders of Bacon who question feminist readings of his rhetoric, absolve him of advocating the torture of nature, and maintain that he was not a slave driver but a humble servant of nature.[16] I argue that Bacon's goal was to use constraint and force to extract truths from nature. His choice of words was part of a larger project to create a new method that would allow humanity to control and dominate the natural world.

In *The Death of Nature*, I stated that "much of the imagery [Bacon] used in delineating his new scientific objectives and methods derives from the courtroom, and, because it treats nature as a female to be tortured through mechanical inventions, strongly suggests the interrogations of the witch trials and the mechanical devices used to torture witches," and I quoted a passage from Bacon's *De Dignitate et Augmentis Scientiarum* (*Of the Dignity and Advancement of Learning*) (see Table 1, col. 2). I also suggested that "the strong

[13]Merchant, *Death of Nature*, pp. 131–132, 288; Merchant, *Ecological Revolutions* (cit. n. 11), pp. 55–56; Merchant, *Reinventing Eden* (cit. n. 6), pp. 117–123; Katharine Park, "Nature in Person: Medieval and Renaissance Allegories and Emblems," in *The Moral Authority of Nature*, ed. Lorraine Daston and Fernando Vidal (Chicago: Univ. Chicago Press, 1994), pp. 50–73; Daston and Park, *Wonders and the Order of Nature, 1150–1750* (Cambridge, Mass.: MIT Press, 1998); Mario Biagioli, *Galileo, Courtier: The Practice of Science in the Culture of Absolutism* (Chicago: Univ. Chicago Press, 1993); and Steven Shapin and Simon Schaffer, *Leviathan and the Air-Pump: Hobbes, Boyle, and the Experimental Life* (Princeton, N.J.: Princeton Univ. Press, 1985).

[14]Merchant, *Reinventing Eden*, pp. 1–8.

[15]Merchant, *Death of Nature*, pp. 168, 172.

[16]Alan Soble, "In Defense of Bacon," *Philosophy of the Social Sciences*, 1995, *25*:192–215, rpt. with additions and corrections in *A House Built on Sand: Exposing Postmodernist Myths about Science*, ed. Noretta Koertge (New York: Oxford Univ. Press, 1998), pp. 195–215, esp. pp. 203–206 (subsequent references to the essay will be to this later version); William R. Newman, "Alchemy, Domination, and Gender," *ibid.*, pp. 216–239; Nieves H. De Madariaga Mathews, *Francis Bacon: The History of a Character Assassination* (New Haven, Conn.: Yale Univ. Press, 1996), Chs. 24, 33; Mathews, "Francis Bacon, Slave-Driver or Servant of Nature? Is Bacon to Blame for the Evils of Our Polluted Age?" http://itis.volta.alessandria.it/episteme/madar1.html; Peter Pesic, "Nature on the Rack: Leibniz's Attitude towards Judicial Torture and the 'Torture' of Nature," *Studia Leibnitiana*, 1997, *39*:189–197; Pesic, "Wrestling with Proteus: Francis Bacon and the 'Torture' of Nature," *Isis*, 1999, *90*:81–94; Iddo Landau, "Feminist Criticisms of Metaphors in Bacon's Philosophy of Science," *Philosophy*, 1998, *73*:47–61; and Perez Zagorin, *Francis Bacon* (Princeton, N.J.: Princeton Univ. Press, 1998), pp. 121–122.

sexual implications of the last sentence can be interpreted in the light of the investigation of the supposed sexual crimes and practices of witches." I summed up Bacon's approach to the domination of nature with the sentence: "The interrogation of witches as symbol of the interrogation of nature, the courtroom as model for its inquisition, and torture through mechanical devices as a tool of the subjugation of disorder were fundamental to the scientific method as power."[17]

Bacon did not use the phrases "torture nature" or "putting nature on the rack" (nor did I claim in *The Death of Nature* that he did so). He believed that everything in nature should be studied, including those valid things that witches might indeed know about nature. But nature was nevertheless to be studied through interrogation. The goal, as Peter Pesic argues, was to extract the truth. The critics read the methods of interrogation Bacon advocated as a benign means of obtaining knowledge, whereas I read them as legitimation for the domination of nature.

The passage in Table 1 was just a small part of the larger argument I made that Bacon's treatment of nature as female legitimated the control of nature through science and technology. It is nevertheless instructive to reexamine the context out of which that passage arose and the views of James VI of Scotland (who in 1603 became James I of England [1566–1625]) and Francis Bacon (1561–1626) on torture.

Table 1 compares the relevant passage from the original 1605 English edition of *The Advancement of Learning* with the same passage from the 1875 (and identical 1870) English translation of the expanded version of the essay, *De Dignitate et Augmentis Scientiarum* (1623); the original Latin edition of 1623 (republished in 1858); and the French translation of 1624. With regard to the 1858 Latin edition of *De Dignitate et Augmentis Scientiarum*, James Spedding, Robert Leslie Ellis, and Douglas Devon Heath state in their note to the phrase "quod et Majestas tua exemplo proprio confirmavit" ("as your Majesty has shown in your own example"): "The allusion is to King James' *Daemonologie*, a work in three books, consisting of dialogues between Philomathes and Epistemon; the latter of whom represents the king's opinions on witchcraft."[18]

In an effort to exonerate Bacon and James I of any negative implications for science, nature, and women that a reader might draw from their writings, Alan Soble states, "Bacon is not alluding to cruel methods of inquisition, but is pointing out that James I was willing to get his hands dirty by studying witchcraft. What James I 'show[ed] in his own example,' says Bacon, is that everything in nature is an appropriate object for scientific study—one of Bacon's principles—not that science should torture nature as if it were a witch." In the *Daemonologie* James did indeed distinguish between "Astronomie and Astrologie" and noted the differences between "naturall reason" and "unlawful charmes, without natural causes." But he did not, as Soble claims, "study witchcraft" to see what within it might have been "an appropriate object for scientific study." On the contrary, the book reveals James's involvement in both the torture of witches and the sexual aspects of the witch trials.[19]

Although torture was officially banned in English common law from the time of the Magna Carta, it was nevertheless used during the reigns of the Tudors (Henry VII, Henry VIII, Mary, and Elizabeth I) and Stuarts (James I and James II). Under those monarchs, the Court of the Star Chamber ordered hangings, whippings, mutilations, and the pillory. James I believed that witches had powers over people and nature, knew secrets, and could be forced to confess those secrets if interrogated under torture or shown the instruments of torture. In the *Daemonologie* he denounced witchcraft and advocated the death of witches by fire. The devil, he wrote, "makes them to renunce their God and *Baptisme* directlie, and giues them his marke vpon some secreit place of their bodie." Witches could be detected by probing for that insensible part on the body in order to find the devil's mark, ducking them in water to see if they would float (if they floated they were guilty, since the water would receive "in her bosom" those who had been baptized but not those whose impiety had caused them to renounce baptism, hence

[17]Merchant, *Death of Nature*, pp. 168–169, 172.

[18]Francis Bacon, *De Dignitate et Augmentis Scientiarum* (1623), in *Works*, ed. James Spedding, Robert Leslie Ellis, and Douglas Devon Heath, 14 vols. (London: Longmans Green, 1857–1874, 1875–1881) (hereafter cited as Works, in parentheses, with volume and page numbers), Vol. 1, pp. 496, 498. The note is inserted by the editors at the end of the quoted passage; they refer to King James the First, *Daemonologie* (1597) (New York: Dutton, 1924).

[19]Soble, "In Defense of Bacon" (cit. n. 16), p. 203; and James I, *Daemonologie*, pp. 11, 33 (quotation), 77 – 81. Soble argues that inserting the words I omitted in the passage on sorceries, witchcrafts, charms, etc., from *De Dignitate et Augmentis Scientiarum* changes the meaning of the passage quoted in Table 1, col. 2—i.e., the [bracketed] words "For it is not yet known in what cases, and how far, effects attributed to superstition participate of natural causes, and therefore" and "(if they be diligently unravelled)." Inserting these words does strengthen the idea that witches, sorcerers, alchemists, and natural magicians might have valid knowledge of nature, but it docs not change Bacon's goals, as stated in the passage, of finding this knowledge by "hound[ing] nature in her wanderings" and of "further disclosing the secrets of nature."

Table 1

Francis Bacon's *Advancement of Learning*

The Advancement of Learning, (1605)	De Dignitate et Augmentis Scientiarum, English, 1875	De Dignitate et Augmentis Scientiarum, Latin, 1623	Le Progrez et advancement aux sciences, French, 1624
History of Nature is of three sorts; of nature in course, of nature erring or varying, and of nature altered or wrought; that is, history of Creatures, history of Marvels, and history of Arts.	The division which I will make of Natural History is founded upon the state and condition of nature herself. For I find nature in three different states, and subject to three different conditions of existence. She is either free, and follows her ordinary course of development; as in the heavens, in the animal and vegetable creation, and in the general array of the universe; or she is driven out of her ordinary course by the perverseness, insolence, and froward-ness [*sic*] of matter, and violence of impediments; as in the case of monsters; or lastly, she is put in constraint, moulded, and made as it were new by art and the hand of man; as in things artificial. . . . Of these the first treats of the Freedom of Nature, the second of her Errors, the third of her Bonds.	Partitionem *Histariae Naturalis* moliemur ex statu et conditione ipsius Naturae, quae in triplici statu posita invenitur, et tanquam regimen trinum subit. Aut enim libera est natura et cursu consueto se explicans, ut in coelis, animalibus, plantis, et universo naturae apparatu: aut a pravitatibus et insolentiis materiae contumacis et ab impedimentorum violentia de statu suo detruditur, ut in monstris; aut denique ab arte et opera humana constringitur et fingitur, et tanquam novatur, ut in artificialibus.	L'Histoire Naturelle, est de trois sortes: De la Nature en son cours, de la Nature errante et variante, et de la Nature alteree et travaillee, c'est à dire l'Histoire de creatures, l'Histoir de merveilles, et l'Histoire des Arts. La premiëe d'icelle est son double manifeste & en bonne perfection: les deux dernieres sont traitées si faible-ment, que je suis constraint de les noter comme défectueuses.
. . . from the wonders of nature is the nearest intelligence and passage towards the wonders of art: for it is no more but by following and as it were hounding Nature in her wanderings, to be able to lead her afterwards to the same place again. Neither am I of opinion, in this History of Marvels, that superstitious narrations of sorceries, witchcrafts, dreams, divinations, and the like, where there is an assurance and clear evidence of the fact, be alto-gether excluded. For it is not yet known in what cases, and how far, effects at-tributed to superstition do participate of natural causes; and therefore howso-ever the practice of such things is to be condemned, yet from the speculation and consideration of them light may be taken, not only for the discerning of the offences, but for the further disclosing of nature. Neither ought a man to make scruple of entering into these things for inquisition of truth, as your Majesty hath shewed in your own example; who with the two clear eyes of religion and natural philosophy have looked deeply and wisely into these shadows, and yet proved yourself to be of the nature of the sun, which passeth through	. . . from the wonders of nature is the most clear and open passage to the wonders of art. For you have but to fol-low and as it were hound nature in her wanderings, and you will be able, when you like, to lead and drive her afterward to the same place again. Neither am I of opinion in this history of marvels, that superstitious narratives of sorceries, witchcrafts, charms, dreams, divina-tions, and the like, where there is an assurance and clear evidence of the fact, should be altogether excluded. [For it is not yet known in what cases, and how far, effects attributed to super-stition participate of natural causes, and	. . . Harum prima *Libertatem Naturae* tractat; secunda *Errores;* tertia *Vincula.* . . . quod a miraculis naturae ad miracula artis expeditus sit transitus et pervius. Neque enim huic rei plus inest negotii, praeterquam ut naturae vestigia persequaris sagaciter, cum ipsa sponte aberret; ut hoc pacto postea, cum tibi libuerit, eam eodem loci deducere et compellere possis. Neque vero prae-ceperim ut ex historia ista mirabilium superstitiosae narrationes de malefi-ciis, fascinationibus, incantationibus, somniis, divinalionibus, et similibus, prorsus excludantur, ubi de facto et re gesta liquido constet. Nondum enim innotuit quibus in rebus, et quousque, effectus superstitioni attributi ex causis naturalibus participent. Ideoque licet hujusmodi artium usum et praxim merito damnandum censeamus, tamen a speculatione et considerations ipsa-rum (si strenue excutiantur) notitiam	. . . L'autre à cause que des merveilles de la nature, l'intelligence & le passage vers les merveilles de l'art en sont plus proches: car ce n'est autre chose qu'en suivant & comme chassant la nature en ses fourvoiements, etre par après ca-pable de la conduire en la même place. Je ne suis pas aussi d'opinion en cette *Histoire de merveilles,* que les supersti-tieuses narrations de songes, de divina-tions & d'autres choses semblables, où il y a une assurance & Claire evidence du fait, soient du tout exclues. Car on ne sçait pas encore en quell cas & de combine les effets attribués à la super-stition participent des causes naturel-les: Et partant encore que la pratique de telles choses soit à condemner, toutefois de la spéculation & considéra-tion d'icelles, l'on peut prendre de la offences, mais pour d'avantage de sec-ourir la nature: Ny l'on ne dôit pas faire scrupule d'entrer en ces choses pour la recherché de la verité, come votre Majesté a montrée par son exemple, qui avec les yeux clairs de la

pollutions and itself remains as pure as before. But this I hold fit, that these narrations which have mixture with superstition be sorted by themselves, and not to be mingled with the narrations which are merely and sincerely natural.[1]

therefore] howsoever the use and practice of such arts is to be condemned, yet from the speculation and consideration of them [(if they be diligently unravelled)] a useful light may be gained, not only for a true judgment of the offenses of persons charged with such practices, but likewise for the further disclosing of the secrets of nature. Neither ought a man to make scruple of entering and penetrating into these holes and corners, when the inquisition of truth is his [sole] object—as your Majesty has shown in your own example; who, with the two clear and acute eyes of religion and natural philosophy, have looked deeply and wisely into those shadows, and yet proved yourself to be truly of the nature of the sun, which passes through pollutions and is not defiled. I would recommend however that those narrations which are tinctured with superstition be sorted by themselves, and not mingled with those which are purely and sincerely natural.[2]

hand inutilem consequemur, non solum ad delicta in hoc genere reorum rite dijudicanda, sed etiam ad naturae secreta ulterius rimanda. Neque certe haesitandum de ingressu et penetratione intra hujusmodi antra et recessus, si quis sibi unicam veritatis inquisitionem proponat; quod et Majestas tua exemplo proprio confirmavit. Tu enim duobus illis clarissimis et acutissimis religionis ac naturalis philosophiae oculis, tales umbras prudenter ac perspicaciter perlustrasti; ut te Soli simillimum probaveris, qui polluta loca ingreditur, nec tamen inquinatur, Caeterum illud monuerim, narrationes istas cum rebus superstitiosis conjunctas seorsum componi, neque cum puris et sinceris naturalibus commisceri oportere.[3]

Religion & de la Philosophie naturelle, a regardé profondement & sagement dans ces ombres, & toutefois s'est montrée être du naturel du soleil, qui passe par toutes les ordures, & demeure aussi pur que devant. Or je tiens qu'il est convenable que les Narrations qui ont un mélange avec la superstition, soient assorties d'elles mêmes, & non pas pour être mêlées avec les narrations, qui sont purement & sincèrement naturelles, pour les narrations qui concernent les miracles des Religions, ou non pas naturelles, & partant impertinentes pour l'Histoire de la nature.[4]

[1] Francis Bacon, *The Advancement of Learning* (1605), in *Works*, ed. James Spedding, Robert Leslie Ellis, and Douglas Devon Heath, 14 vols. (London: Longmans Green, 1870), Vol. 3, pp. 330–331. Note Bacon's use of nature in the female gender in his own English first edition, written in 1605.

[2] Carolyn Merchant, *The Death of Nature: Women, Ecology, and the Scientific Revolution* (1980; San Francisco: HarperCollins, 1990), p. 168; and Francis Bacon, *De Dignitate et Augmentis Scientiarum* (1623) (*Works*, Vol. 4, pp. 294, 296). Words in brackets omitted in *The Death of Nature*; "whole" in *The Death of Nature*; "sole" in *The Death of Nature*, corrected to "sole."

[3] Bacon, *De Dignitate et Augmentis Scientiarum* (*Works*, Vol. 1, pp. 496, 498). Note inserted at penultimate sentence of quoted passage by Spedding, Ellis, and Health: "The allusion is to King James's *Daemonologie*, a work in three books, consisting of dialogues between Philomathes and Epistemon: the latter of whom represents the king's opinions on witchcraft."

[4] Francis Bacon, *Le Progrez et avancement aux sciences diuines & humaines* (Paris: Pierre Billaine, 1624), Bk. 2, Ch. 2, pp. 197, 199–201 (French the modernized, with the exception of the title). I thank Roger Hahn for assistance with the transcription.

God), and threatening or torturing them to see if they would repent (crocodile tears indicating a false repentance).[20]

James VI's personal involvement with the questioning of the accused and his obsession with witchcraft are revealed in the 1591 tract *Newes from Scotland, Declaring the Damnable Life and Death of Doctor Fian, a Notable Sorcerer Who Was Burned at Edenbrough in January Last*. The trial came about when Fian (John Cunningham), along with Agnis Sampson and Agnes Tompson of Edinburgh, was accused of causing a devastating storm during the return passage of James and his fiancé from Norway to Scotland. Agnis Sampson was brought before the king and other nobility, where she was interrogated and refused to confess. She was taken to prison and searched for the devil's mark on her private parts. According to the author of the *Newes,*

> It has lately been found that the Devil do generally mark them with a private mark, by reason the Witches have confessed themselves, that the Devil do lick them with his tongue in some private part of their body, before he receives them to be his servants, which mark commonly is given them under the hair in some part of their body, whereby it may not easily be found out or seen, although they be searched: and generally so long as the mark is not seen to those which search them, so long the parties that have the mark will never confess anything. Therefore by special commandment this Agnis

Sampson had all her hair shaven off, in each part of her body, and her head thrown with a rope according to the custom of that country, being a paine most grievous, which she continued almost an hour, during which time she would not confess anything until the Devil's mark was found upon her privates, then she immediately confessed whatsoever was demanded of her, and justifying those persons aforesaid to be notorious witches.[21]

To convince James that she spoke the truth, Agnis Sampson took him aside and revealed the very words that he and his wife had uttered on the first night of their marriage. James acknowledged that her words were accurate and believed the rest of what she told him. Following that, Agnes Tompson was questioned and confessed that a cat had been the cause of the storm. The women also "confessed that when the Devil received them for his servants, and that they had vowed themselves unto him, then he would carnally use them, albeit to their little pleasure, with respect of his cold nature."[22]

Then Doctor Fian, alias John Cunningham, was examined. According to Leonard A. Parry, in his *History of Torture in England*:

> under the most terrible torture, he confessed his guilt, though he immediately afterwards retracted his confession. The bones of his leg were broken into small pieces in the boot. This was not enough.

---

[20]L. A. Parry, *The History of Torture in England* (1934; Montclair, N.J.: Patterson Smith, 1975), pp. 1–3, 7; and James I, *Daemonologie*, pp. 11, 33 (quotation), 77–81. According to James: "There are two other good helpes that may be vsed for their trial; the one is the finding of their marke, and the trying the insensiblenes thereof. The other is their fleeting on the water: for as in a secret murther, if the deade carcase be at any time thereafter handled by the murtherer, it wil gush out of bloud, as if the blud wer crying to the heauuen for reuenge of the murtherer, God hauing appoynted that secret super-naturall signe, for tryall of that secrete vnnaturall crime, so it appeares that God hath appoynted (for a super-naturall signe of the monstruous impietie of the Witches) that the water shal refuse to receiue them in her bosom, that haue shaken off them the sacred Water of Baptisme, and wilfullie refused the benefite thereof; No not so much as their eyes are able to shed teares (thretten and torture them as ye please) while first they repent (God not permitting them to dissemble their obstinacie in so horrible a crime) albeit the women kinde especially, be able other-waies to shed teares at euery light occasion when they will, yea, although it were dissemblingly like the Crocodile" (p. 81).

[21]Anonymous, *Newes from Scotland, Declaring the Damnuble Life and Death of Doctor Fian, a Notable Sorcerer Who Was Burned at Edenbrough in January Last* (London: John Lane, 1591), bound with King James the First, *Daemonologie* (New York: Dutton, 1924), pp. 12–13. The original English reads: "it hath latelye been found that the Deuill dooth generallye marke them with a priuie marke, by reason the Witches haue confessed themselues, that the Diuell dooth lick them with his tung in some priuy part of their bodie, before hee dooth receiue them to be his seruants, which marke commonly is giuen them vnder the haire in some part of their bodye, whereby it may not easily be found out or seene, although they be searched: and generally so long as the marke is not seene to those which search them, so long the parties that hath the marke will neuer confess any thing. Therefore by special commaundement this Agnis Sampson had all her haire shauen of, in each parte of her bodie, and her head thrawen with a rope according to the custome of that Countrye, beeing a paine most greeuous, which she continued almost an hower, during which time she would not confesse any thing vntill the Diuels marke was found vpon her priuities, then she immediatlye confessed whatsoeuer was demaunded of her and justifying those persons aforesaid to be notorious witches."

[22]*Newes from Scotland*, p. 18. Original English: The women also "confessed that when the Diuell did receiue them for his seruants, and that they had vowed themselues vnto him, then he would Carnallye vse them, albeit to their little pleasure, in respect of his cold nature."

The King himself suggested a new device. "His nailes upon all his fingers were riven and pulled off with an instrument called in Scottish, a turkas, which in England we call a payre of pincers, and under everie nayle there was thrust in two needels over, even up to the head." Notwithstanding all this, "so deeply had the devil entered into his heart, that hee utterly denied all that which he had before avouched." He was burnt alive.[23]

Throughout the proceedings, James was both a witness to and a full participant in selecting the means of torture. Both sexual torture and physical torture were integral components of the interrogation process.

When James VI became James I, King of England, in 1603, he instituted stricter death penalties for offenses attributed to witchcraft than had his predecessor Elizabeth I. His 1604 Witchcraft Act (in effect until 1736) repealed the milder law of Elizabeth and instituted more severe treatment. Individuals convicted of practicing witchcraft, enchantment, and sorcery or of harming the cattle or goods of any other person would be imprisoned for a year without bail and pilloried in a public place once a quarter for six hours. Those convicted of causing death or injury to another person would suffer the pain of death as felons and lose the privilege of clerical blessing. Parry notes, "This act was passed at a time when Coke was Attorney-General, Bacon a member of parliament, and twelve Bishops sat on the Commission to which it was referred! James I was a confirmed and wholehearted believer in witchcraft."[24]

In summary, it is abundantly clear that in 1597, when James VI wrote the *Daemonologie*, he advocated torture to reveal the truth and condoned the examination of the private parts of the accused for evidence of witch marks. In 1604, when, as James I, he instituted his witchcraft law, he believed that witches should be imprisoned or put to death.

What were Bacon's views about the torture of nature and of witches? It would be naive to believe that Bacon was ignorant of the most severe means of torture or of the methods of examining women's bodies for evidence that they had consorted with the devil— or of James I's early obsession and involvement with

those methods. The European Inquisition, torture practices, and death were part of the context of his life and world and were certainly known by that widely read and influential man. In addition to the rack, the instruments and methods of torture included the breast strip, breast press, witches chair, ducking stool, judas cradle, expanding vaginal pears, wheel, ladder, strangle, hanging strap, and funnel and water torture.[25]

Bacon did not advocate the practice of torture or use of the rack on human beings. He nevertheless used imagery drawn from torture in his writings and believed that witchcraft and sorcery could reveal useful information. The use of torture rhetoric condones a transfer of methodological approaches used to extract information from the accused to extracting secrets from nature. The method of confining, controlling, and interrogating the human being becomes the method of the confined, controlled experiment used to interrogate nature. Torture should be used not on witches but on nature itself. The experimental method is superior to that developed by magicians to control nature. A question must be asked and an experiment designed to answer it. For the experimental method to succeed, the experiment must be a closed, isolated system in which variables are controlled and extraneous influences excluded. Witnessing is critical to the process. The trial—that is, the experiment—must be witnessed by others. Indeed, it was one of Bacon's singular contributions to realize that, to be understood, nature must be studied under constrained conditions that can be both witnessed and verified by others. Bacon used metaphor, rhetoric, and myth to develop his new method of interrogating nature. As Peter Pesic notes, "Since he was describing something not yet formed, he used a rich variety of rhetorical figures to express his vision."[26]

Bacon promoted the study and interrogation of sorcerers and practitioners of the occult arts for clues as to how nature worked and how the devil worked through nature. In endeavoring to gain power over nature, he drew heavily on the alchemical and magical traditions for clues that would lead to the human control of nature. He accepted the goal and idea of control, but he sought new methods of extracting knowledge. What was true should be sorted out from

---

[23]Parry, *History of Torture in England* (cit. n. 20), p. 180; and *Newes from Scotland*, pp. 27, 28, as quoted by Parry.

[24]"An Acte Against Conjuration Witchcrafte and Dealinge with Evill and Wicked Spirits," 1604—1 Jas. I, c. 12; and Parry, *History of Torture in England*, p. 180. The last witch trial in England took place in 1712.

[25]Parry, *History of Torture in England*, pp. 76–87, 162–177, 182; George Ryley Scott, *The History of Torture throughout the Ages*, 2nd ed. (London; Kegan Paul, 2003), pp. 168–255; and Merchant, *Death of Nature*, pp. 168–172.

[26]Pesic, "Wrestling with Proteus" (cit. n. 16), p. 81. Pesic's goal in this article, however, is to argue that Bacon did not advocate torture or use torture as a model for the experimental method.

what was erroneous. The problem with magic was that it was rooted in individual knowledge and judgment, rather than being subjected to a set of universal rules and agreements. As Paolo Rossi put it: "According to Bacon, magic endeavours to dominate and to improve nature; and for this it should be imitated. Where it needs revising is in its claim to use one man's inspiration instead of the organised efforts of the human race, and to make science serve individual ends rather than mankind."[27]

Even though Bacon opposed the practice of torture, his rhetoric and metaphors for the interrogation of nature under constraint come from the devices of torture that were part of his cultural milieu, including the rack. Being tortured on the rack was referred to as being "put to the question." The rack was introduced into England during the reign of Henry VI but was used only for cases of high treason, such as that ordered by James I after the Guy Fawkes Gunpowder Plot of 1605. It consisted of an oak frame three feet above the ground, on or under which the prisoner was placed on his back, with hands and feet bound to rollers and levers on each end. The levers were then moved to exert force on the joints and sockets until the prisoner responded to the interrogation.[28]

Concerning the rack, Bacon wrote that Elizabeth I consulted him about a plagiarized text by Sir John Haywarde that was dedicated to her mortal enemy, Lord Essex (Bacon's initial benefactor); he reports her saying, "with great indignation, that she would have him racked to produce his author; I replied nay, madam, he is a doctor, never rack his person, but rack his style: Let him have pen, ink, and paper, and help of books, and be enjoined to continue the story where it breaketh off, and I will undertake, by collating the styles, to judge wether he were the author or no." Bacon thus opposed using the rack literally yet advocated using it stylistically. In promoting his experimental method he used rhetoric that implied and even condoned torture—verbs such as "vex," "hound," "drive," "constrain," "straiten," "mold," "bind," "enslave," "spy on," and "transmute" were applied to nature. Such words were metaphors for the interrogation of nature (putting nature to the question), intended to reveal the truths of nature through experimentation.[29]

In the time between the 1605 and 1623 editions of Bacon's *Advancement of Learning*, witch trials served as models of interrogation to reveal hidden secrets that could be used to convict the accused and levy the death sentence. On the Home Circuit between 1605 and 1626—during James's reign and Bacon's association with the Court—nineteen witches (fifteen women and four men) were convicted, twelve of whom (ten women and two men) were hanged. In the Lancashire (Pendle Forest) witch trials of 1612, in which ten witches were publicly hanged, several confessed under interrogation to have allowed the devil (in the form of a familiar) to suck on their body parts. James I, who continued his earlier interest in interrogating witches, intervened in two subsequent cases. In 1618 he interviewed John Smith, a boy who had accused nine witches who were subsequently hanged, decided that he was an impostor, and stopped the hanging of the remaining women the boy had accused. In 1621 he likewise interviewed Katherine Malpas, who had accused two women of bewitching her, an accusation later revealed to be a fabrication. Although James came to believe that many witches were either deluded or prevaricators, he did not repeal his 1604 witchcraft law, and it remained in effect until 1736. By the time of James's involvement in the Malpas case in 1621, however, Bacon had fallen out of favor with the king, having been accused earlier that year of accepting bribes, sentenced to the Tower of London (where he served only two days), and banished both from holding office and from Parliament. When he wrote the expanded version of *The Advancement of Learning* in Latin in 1623 he was again hoping to curry favor with the king.[30]

In the passage from the 1605 *The Advancement of Learning* quoted in Table 1, Bacon had written, "Neither ought a man to make scruple of entering into these things for inquisition of truth, as your Majesty hath shewed in your own example; who with the two clear eyes of religion and natural philosophy have looked deeply and wisely into these shadows, and yet proved yourself to be of the nature of the sun,

[27]Paolo Rossi, *Francis Bacon: From Magic to Science* (Chicago: Univ. Chicago Press, 1968), pp. 32–33, on p. 32.

[28]Parry, *History of Torture in England* (cit. n. 20), pp. 180–181 (on the use of the rack see pp. 41, 54, 76); Scott, *History of Torture throughout the Ages* (cit. n. 25), pp. 168–180; and Bacon, *De Dignitate et Augmentis Scientiarum* (*Works*, Vol. 4, p. 298).

[29]Bacon, quoted in Parry, *History of Torture in England*, p. 40.

[30]R. H. Robbins, *Encyclopedia of Witchcraft and Demonology* (New York: Crown, 1959), pp. 277–279; Christina Hole, *Witchcraft in England* (New York: Scribner's, 1947), p. 140; C. L'Estrange Ewen, *Witchcraft in the Star Chamber* (Privately printed, 1938), pp. 13–14, 26–29, 33–34, 56; Ewen, *Witch Hunting and Witch Trials: The Indictments for Witchcraft from the Records of 1373 Assizes Held for the Home Circuit A.D. 1559–1736* (London: Kegan Paul, Trench, Trubner, 1929), chart on p. 106 (the Home Circuit covered Essex, Hertfordshire, Kent, Surry, and Sussex counties); and Rachel A. C. Hasted, *The Pendle Witch-Trial, 1612* (Lancashire: Lancashire County Books, 1993), p. 2.

which passeth through pollutions and itself remains as pure as before." In 1623 he modified the first part of that sentence to read: "Neither ought a man to make scruple of entering and penetrating into these holes and corners, when the inquisition of truth is his sole object, —as your Majesty has shown in your own example." He added "and acute" to the phrase "two clear [and acute] eyes" and "truly" in front of the phrase "[truly] of the nature of the sun." He also changed "for the further disclosing of nature" to "the further disclosing of the secrets of nature." These changes may refer to James's interrogations in the 1618 and 1621 cases, as well as his 1597 *Daemonologie*, or, alternatively, they may be meant to emphasize more strongly his continuing interest in investigating witchcraft. They may also represent Bacon's renewed efforts to regain James's favor. In any case, I would still maintain that as metaphors they reflect the sexual aspects of the witch trials, including the practice, originally condoned by James VI, of interrogating and identifying witches by sticking needles in their private parts to identify their "insensible" witch marks.[31]

If Bacon did not explicitly state that nature should be put on the rack, however, where did that phrase come from? The rack and its association with Bacon and the torture of nature seem to have been present in cultural exchanges at least by the late seventeenth century. Peter Pesic details the history of the association of Bacon's ideas with the torture of nature and of putting nature on the rack.[32] He points out that its connection to Bacon may first have been put in writing by Gottfried Wilhelm Leibniz. In 1696 Leibniz wrote about "the art of inquiry into nature itself and of putting it on the rack—the art of experiment which Lord Bacon began so ably." Four years later, Jean Baptiste du Hamel, secretary of the Paris Academy of Sciences, wrote, "We discover the mysteries of nature much more easily when she is tortured [*torqueatur*] by fire or some other aids of art than when she proceeds along her own road."[33] The Latin verb "*torqueo*"

means "to turn, twist, wind, or wrench" and "of torturing on the rack, etc.: to rack, wrench," as well as "to rack, torture, torment." Under the word "rack," the *Oxford English Dictionary* includes "racken 'torquere, tendere, tormentis, experime.' See also . . . racken, to vex, torture (Grimm)." There are thus clear associations between the word "torture" and the rack.[34] In contrast to Leibniz and Hamel, Johann Wolfgang von Goethe complained that under scientific investigation "nature falls silent on the rack," and he urged that "phenomena must once and for all be removed from their gloomy empirical-mechanical-dogmatic torture chamber."[35]

Later philosophers also associated the torture of nature with Francis Bacon. In 1878 Thomas Fowler wrote that Bacon "insisted, both by example and precept, on the importance of experiment as well as observation. Nature like a witness, when put to the torture, would reveal her secrets."[36] In 1953 Ernst

[31]Francis Bacon, *The Advancement of Learning* (1605) (*Works*, Vol. 3, p. 331); and Bacon, *De Dignitate et Augmentis Scientiarum* (*Works*, Vol. 4, p. 296).

[32]Quotations and citations have been compiled by Peter Pesic in "Wrestling with Proteus" (cit. n. 16), p. 82; and Pesic, "Nature on the Rack" (cit. n. 16), pp. 195 n 29, 197 nn 34, 35. I have added to and elaborated on them in the notes that follow.

[33]Gottfried Wilhelm Leibniz, *Philosophical Papers and Letters*, ed. Leroy E. Loemker (Chicago: Univ. Chicago Press, 1956), Vol. 2, p. 758; and Jean Baptiste du Hamel, *Regiae scientiarum academiae historia*, 2nd ed. (Paris: Delespine, 1701), p. 16: "sic natura arcana longe facilius deprehendimus, cum per ignem aut alia artis adminicula varie torquetur,

quam ubi itinere quodam suo progreditur." Du Hamel is cited and translated in S. Beasley Linnard Penrose, Jr., "The Reputation and Influence of Francis Bacon" (Ph.D. diss., Columbia Univ., 1934), pp. 97–98. Interestingly, Penrose adds (but without a citation): "Bacon said that nature must be tortured upon the rack to make her give up her secrets. The similarity of expression is striking."

[34]For the Latin "*torquere*" see Sir William Smith, *A Smaller Latin–English Dictionary*, rev. J. R. Lockwood (New York: Barnes & Noble, 1960), p. 759; for "*torqueo*" see *Cassell's New Latin Dictionary*, rev. D. P. Simpson (New York: Funk & Wagnalls, 1959), pp. 607–608. For the definition of "rack" see *Oxford English Dictionary*, compact ed., 2 vols. (Oxford: Oxford Univ. Press, 1971), Vol. 2, p. 2401.

[35]J. W. v. Goethe, *Maximen und Reflexionen: Nach den Handschriften des Goethe- und Schiller-Archivs herausgegeben von Max Hecker* (Weimar: Goethe-Gesellschaft, 1907), maxim 115, p. 21: "Die Natur verstummt auf der Folter; ihre treue Antwort auf redliche Frage ist: Ja! ja! Nein! nein! Alles Übrige ist vom übel." Goethe, *Sämtliche Werke: Jubiläums-Ausgabe*, ed. Eduard von der Hellen, 40 vols. (Stuttgart/Berlin: Cotta, 1902–1912), Vol. 39, maxim 430, p. 64: "Die Phänomene müssen ein für allemal aus der düstern empirisch-mechanischdogmatischen Marterkammer vor die Jurn des gemeinen Menschen-verstandes gebracht werden." For the English see Goethe, *Maxims and Reflections*, trans. Elisabeth Stopp, ed. Peter Hutchinson (London: Penguin, 1998), maxim 115, p. 14: "Nature grows dumb when subjected to torture; the true answer to honest questioning is yes! yes! no! no! All else is idle and basically evil"; and maxim 430, p. 55: "Phenomena must once and for all be removed from their gloomy empirical-mechanical-dogmatic torture chamber and submitted to the jury of plain common sense." See also Erich Heller, *The Disinherited Mind: Essays in Modern German Literature and Thought* (Cambridge: Bowes & Bowes, 1952), p. 18: "Goethe regards it as his own scientific mission to 'liberate the phenomena once and for all from the gloom of the empirico-mechanico-dogmatic torture chamber'"; this is taken from Goethe, *Sämtliche Werke: Jubiläums-Ausgabe*, ed. von der Hellen, Vol. 34, p. 64.

[36]Thomas Fowler, *Bacon's Novum Organum* (Oxford: Clarendon, 1878), p. 124; in the second edition (1889) see p. 127, as noted in Martha [Ornstein] Bronfenbreener, *The Role of Scientific Societies in the Seventeenth Century* (New York: Arno, 1975), p. 40. In his 1990 film *Mindwalk* (directed by Bernt Amadeus Capra), Fritjof Capra used the torture chamber to illustrate the torture of nature under mechanistic science.

Cassirer noted that Bacon's approach to science was to treat nature as if it were a witness on the rack. Cassirer wrote:

> The very style of Bacon's writing evinces everywhere this spirit. Bacon sits as a judge over reality, questioning it as one examines the accused. Not infrequently he says that one must resort to force to obtain the answer desired, that nature must be "put to the rack." His procedure is not simply observational but strictly inquisitorial. The witnesses are heard and brought face to face; the negative instances confront the affirmative ones, just as the witnesses for the defence confront those for the prosecution. After all the available bits of evidence have been gathered together and evaluated, then it is a matter of obtaining the confession which finally decides the issue. But such a confession is not obtainable without resorting to coercive measures. [As Bacon states,] "For like as a man's disposition is never well known or proved till he be crossed . . . so nature exhibits herself more clearly under the trials and vexations of art than when left to herself."[37]

And in 1975, writing in *The Great Instauration: Science, Medicine, and Reform, 1626–1660,* the historian Charles Webster concurred: "By 'interrogation' applied with extreme determination and cunning, nature would be 'tortured' into revealing her secrets; she would then submit to voluntary 'subjugation.'"[38]

Was Bacon's method of interrogating nature to put it on the rack? These philosophers certainly interpreted him that way. To them, the rack exemplified the constraint of nature in a closed, controlled system, responding to questions posed by an inquisitor before witnesses—the very core of experimentation itself. Through metaphor and imagery, Bacon struggled to define experimentation as a new way of learning nature's truths.

A related controversy arises over Bacon's use of the terms "hound," "vex," and the "vexation" of nature. Again, Soble objects to harsh readings of Bacon's usage. "Even though Bacon's use of 'vex' is occasionally strong," he writes, "'vex' does not always or usually carry a pernicious connotation but is meant, innocuously, along the lines of his 'hound' and my 'pester.'" In the passage from *De Dignitate et Augmentis Scientiarum* (Table 1, col. 2), Bacon writes: "For you have but to follow and as it were hound nature in her wanderings, and you will be able when you like to lead and drive her afterward to the same place again." The *Oxford English Dictionary* gives the following definition of the word "hound": "to pursue, chase, or track like a hound, or, as if with hound; esp. to pursue harassingly, to drive as in the chase"; it quotes the phrase from Bacon's 1605 *Advancement of Learning* that I cited earlier (Table 1, col. 1) as the first example. Other definitions of "hound" are equally violent: "to set (a hound, etc.) at a quarry; to incite or urge on to attack or chase anything" and "to incite or set (a person) at or on another; to incite or urge on." Such meanings are reminiscent of the English foxhunt (outlawed by the British Parliament in 2005 for its excessive cruelty to the hounded and tortured foxes). Nature for Bacon, as Soble himself puts it, must be "out-foxed." But, contrary to Soble's desire to read Bacon's rhetoric innocuously, merely "pestering" nature would not produce the results Bacon desired of his new method—extracting the secrets of nature.[39]

Bacon also used the term "vex" to refer to the interrogation of nature under constraint: "The vexations of art are certainly as the bonds and handcuffs of Proteus, which betray the ultimate struggles and efforts of matter." Art in this context meant *techne* or the technologies used to "vex" nature. The term "vex," meaning "to shake, agitate, disturb," likewise carried connotations of violence, including to "harass aggressively," to "physically distress," to "twist," "press," and "strain," and to "subject to violence."[40] All these

---

[37]Ernst Cassirer, *The Platonic Renaissance in England*, trans. James P. Pettegrove (Austin: Univ. Texas Press, 1953), pp. 47–48; he is citing Bacon, *De Dignitate et Augment is Scientiarum* (*Works*, Vol. 4, p. 298). For Cassirer's use of the phrase "nature must be 'put to the rack'" see also Pesic, "Wrestling with Proteus" (cit. n. 16), p. 82 n 4.

[38]Charles Webster, *The Great Instauration: Science, Medicine, and Reform, 1626–1660* (London: Duckworth, 1975), p. 338. See also Pesic, "Wrestling with Proteus," p. 82 n 4.

[39]Soble, "In Defense of Bacon" (cit. n. 16), p. 205; Bacon, *De Dignitate et Augmentis Scientiarum* (*Works*, Vol. 4, p. 296); and *Oxford English Dictionary*, compact ed. (cit. n. 34), Vol. 1, p. 1338.

[40]"The meanings of "vex" included "(1) to trouble, affect, or harass (a person, etc.) by aggression, encroachment, or other interference with peace and quiet. (2) of diseases, etc.: to afflict or distress physically, to afflict with pain or suffering. . . . (6) to disturb by causing physical movement, commotion, or alteration; to agitate, toss about, work, belabour, or tear up; b. to disturb by handling; to twist; c. to press, strain, or urge." Similarly "vexation" was "(1) the act of troubling or harassing by agitation or interference; (2) the action of troubling, disturbing, or irritating by physical means; . . . (5) the action of subjecting to violence or force." *Oxford English Dictionary*, compact ed., Vol. 2, p. 3621.

meanings convey force in ways that range from irritation to inflicting physical pain through intentional violence. All precisely describe much of the early experimentation done on animals and human beings, as I discuss in Part III.

"Vex" and "torture" were closely associated in Bacon's cultural milieu. The French historian Pierre Hadot, in *Le voile d'Isis: Essai sur l'histoire de l'idée de Nature*, quotes a recent French translation of the *Novum Organum* that renders the English phrase "the vexations of art" in French as "la torture des arts [mécaniques]"—that is, "the torture of the [mechanical] arts." A possible French translation of the English word "vex" is in fact "tormenter."[41]

Soble suggests that Bacon's association of the vexations of art with Proteus do not pertain to nature as female because Proteus was a "guy." Yet Bacon himself compares Proteus to nature in the female gender, as was common in the period (translations notwithstanding): "For like as a man's disposition is never well known or proved till he be crossed, nor Proteus ever changed shapes till he was straitened and held fast, so nature exhibits herself more clearly under the trials and vexations of art than when left to herself." The verb "straiten" in the seventeenth century meant "to tighten a knot, cord, or bonds—an act that would hold a body fast as on the rollers and levers of the rack." As John C. Briggs, in his discussion of Bacon's use of the Proteus myth, states: "Still the lesson that Bacon draws from the myth turns upon the wise man's power to chain Proteus to the rack so as to force matter 'to extremities, as if with the purpose of reducing it to nothing.'"[42]

For Bacon, the myth of Proteus was a stand-in for the interrogation of nature under constraint. Proteus was a Greek sea god (the prophetic "old man of the sea"), the son of Neptune and herder of Poseidon's seals. He had the gifts both of prophecy and of changing his shape at will. He would not share his knowledge of the future and changed his shape to avoid doing so unless held fast. He would reveal the future only to someone who could capture and constrain him. Capturing and constraining was the very method used to extract confessions and secrets from witches. Bacon's use of the terms "straiten," "held fast," and "vexed" all indicate violence toward nature; and, I would still argue, casting nature in the female gender (both then and now) legitimates the treatment of nature in ethically questionable ways (that Proteus is a "guy" notwithstanding).[43]

III

Bacon's words and work influenced the growth of scientific societies and experimentation in the early modern period (even if he himself did not anticipate their development). Although condemned by individuals such as John Locke, Samuel Butler, John Wesley, and Leibniz, experiments done on animals during the seventeenth and eighteenth centuries and continuing even to the present can indeed be described as torture. Such experimentation was legitimated by the mechanical philosophy of nature that viewed animals as automata. Both René Descartes and Thomas Hobbes conceptualized the bodies of humans and other animals as machines. Descartes denied thought to animals, although he admitted that they have life and sensation. In his *Meditations on*

---

[41]Francis Bacon, "Paraseve ad Historiam Naturalem et Experimentalem" or "Preparative Towards a Natural and Experimental History" (1620) (*Works*, Vol. 4, p. 257); and Pierre Hadot, *Le voile d'Isis: Essai sur l'histoire de l'idée de Nature* (Paris: Gallimard, 2004), p. 133: "De même, en effet, que, dans la vie publique, le naturel d'un individu et la disposition cachée de son esprit et de ses passions se découvrent, lorsqu'il est plongé dans le trouble, mieux qu'à un autre moment, de même les secrets (*occulta*) de la nature se découvrent mieux sous la torture des arts [mécaniques] que dans son cours naturel." Hadot is citing Bacon, *Novum Organum*, ed. and trans. Michel Malherbe and Jean-Marie Pousseur (Paris: Presses Univ. France, 1986), p. 165. For the French translation of "vex" see E. Clifton and J. McLaughlin, *A New Dictionary of the French and English Languages*, new rev. ed. (New York: McKay, 1904), p. 630.

[42]Soble, "In Defense of Bacon" (cit. n. 16), p. 205; and Bacon, *De Dignitate et Augmentis Scientiarum* (*Works*, Vol. 4, p. 298). In the 1605 English edition of *The Advancement of Learning* (*Works*, Vol. 3, p. 333) the passage reads: "For like as a man's disposition is never well known till he be crossed, nor Proteus ever changed shapes till he was straitened and held fast; so the passages and variations of nature cannot appear so fully in the liberty of nature, as in the trials and vexations of art." The 1623 Latin edition, *De Dignitate et Augmentis Scientiarum* (*Works*, Vol. 1, p. 500) reads: "Quemadmodum enim ingenium alicujus

haud bene noris aut probaris, nisi eum irritaveris; neque Proteus se in varias rerum facies vertere solitus est, nisi manicis arcte comprehensus; similiter etiam nature arte irritata et vexata se clarius prodit, quam cum sibi libera permittitur." *Oxford English Dictionary*, compact ed. (cit. n. 34), Vol. 2, p. 3080 ("straiten"); and John C. Briggs, *Francis Bacon and the Rhetoric of Nature* (Cambridge, Mass.: Harvard Univ. Press, 1989), p. 35.

[43]Briggs, *Francis Bacon and the Rhetoric of Nature*, pp. 32–38. Such descriptions are particularly relevant to biotechnology today. The name "Proteus" comes from the Greek word "*protos*" (also the root of "protein"), meaning "mutable," "changeable," "versatile," and "capable of assuming many forms." The biotechnology company Proteus describes itself as a modern-day Proteus: "Proteus discovers and develops biomolecules of primary importance and turns them to any form that meets the needs of the near future. It is the leading provider of wireless applications and carrier connectivity. . . . Some of the popular programming brands that have been extended to a mobile audience through Proteus's services include HBO's 'The Sopranos' and 'Sex and the City' and ABC's 'The View: His & Her Body Test'": "Proteus," http://proteus.com/hom.jsp.

the First Philosophy (1641) he wrote, "If the body of man be considered as a kind of machine, so made up and composed of bones, nerves, muscles, veins, blood, and skin, that although there were in it no mind, it would still exhibit the same motions which it at present manifests involuntarily." In his introduction to Leviathan ten years later (1651) Hobbes stated, "For what is the heart, but a spring; and the nerves, but so many strings; and the joints, but so many wheels, giving motion to the whole body, such as was intended by the artificer." If animals or even human bodies were thought of as machines, experimentation could be done with impunity.[44]

In England, the Cambridge Platonist Thomas More objected to Descartes's idea of animal automata, writing in 1648: "I recognize in you not only subtle keenness, but also, as it were, the sharp and cruel blade which in one blow, so to speak, dared to despoil of life and sense practically the whole race of animals, metamorphosing them into marble statues and machines." In his response Descartes continued to deny a soul to animals, writing,

> I speak of cogitation, not of life or sense; for to no animal do I deny life, inasmuch as that I attribute solely to the heat of the heart; not do I deny sense in so far as it depends upon the bodily organism. And thus my opinion is not so much cruel to wild beasts as favourable to men, whom it absolves . . . of any suspicion of crime, however often they may eat or kill animals.

Although objections to the concept of the "beast-machine" were voiced in England, the idea nevertheless lent credence to the notion of animal experimentation.[45]

In his History of the Royal Society (1667), Thomas Sprat reported on experiments done on animals under constraint, experiments that could be considered torture. "Experiments of keeping creatures many hours alive, by blowing into the lungs with bellows, after that all the thorax, and abdomen were open'ed and cut away, and all the Intrials save heart, and lungs remov'd: of reviving chickens, after they have been strangled, by blowing into their lungs: to try how long a man can live, by expiring, and inspiring again the same air." Sprat describes the fatal effects of keeping animals in rarified air and of investigations into the amount of air necessary for a breathing animal to survive. Experiments were made on living animals kept in a bell jar with candles to see which would expire first. Vipers, frogs, fish, and insects were subjected both to the removal of air and to increased air pressure.[46]

Experimentation moved from animals to humans. In 1656 Christopher Wren injected a liquid infusion into a dog's veins, with other members of the Royal Society, including Robert Boyle and John Wilkins, as witnesses. Animals were "purg'd, vomited, intoxicated, kill'd, or reviv'd, according to the quality of the liquor injected." A dog was injected with opium, then whipped and beaten to keep it alive. Other dogs and drugs were tested. The experiments soon led to blood transfusions, first on animals and then on humans. Wren used a quill to inject the blood of one animal into another, and Richard Lower described his animal-to-animal transfusions in 1665 and 1666. In 1667 the blood of a sheep was injected into the veins of a spaniel. In France, Jean Baptiste Denis transferred blood between two dogs and experimented with introducing calves' blood into dogs. He then injected lamb's blood into a young woman and, later, blood from a sheep's artery into a human "lunatic," who at first improved but died following a subsequent transfusion. After charges of poisoning were brought by the man's wife, human transfusions were prohibited.[47]

[44]Scott, History of Torture throughout the Ages (cit. n. 25), p. 138; René Descartes, "Animals Are Machines," in Environmental Ethics: Divergence and Convergence, ed. S. J. Armstrong and R. G. Botzler (New York: McGraw-Hill, 1993), pp. 281–285, esp. p. 285; Descartes, "The Meditations," in Meditations and Selections from the Principles of Philosophy (La Salle, Ill.: Open Court, 1952), p. 98; and Thomas Hobbes, Leviathan, in English Works, ed. William Molesworth, 11 vols., rpt. ed. (Aalen, Germany: Scientia, 1966), Vol. 3, p. ix.
[45]Leonora D. Cohen, "Descartes and Henry More on the Beast-Machine: A Translation of Their Correspondence Pertaining to Animal Automatism," Annals of Science, 1936, 1:48–61, on pp. 50, 53. Objections to the concept of animals as machines were voiced by Thomas Willis, John Locke, John Keill, John Ray, David Hartley, and David Hume. See also Albert G. A. Balz, "Cartesian Doctrine and the Animal Soul: An Incident in the Formation of the Modern Philosophical Tradition," in Studies in the History of Ideas, ed. Columbia Department of Philosophy (New York: Columbia Univ. Press, 1935), Vol. 3, pp. 117–177.

[46]Thomas Sprat, History of the Royal Society (1667), ed. Jackson I. Cope and Harold Whitmore Jones (St. Louis: Washington Univ. Press, 1958), pp. 218–219, on p. 218.
[47]Dorothy Stimson, Scientists and Amateurs: A History of the Royal Society (New York: Greenwood, 1968), pp. 84–86; Sprat, History of the Royal Society, p. 317; and Richard Lower, Tractatus de corde (1665). See also Lower [attributed], "The Method Observed in Transfusing the Blood out of One Animal into Another," Philosophical Transactions of the Royal Society of London, Dec. 1666, and Lower, "Extrait du Journal d'Angleterre, contenant la manière de faire passer le sang d'un animal dans un autre," Journal des Sçavans, 31 Jan. 1667, as cited and discussed in Harcourt Brown, Science and the Human Comedy: Natural Philosophy in French Literature from Rabelais to Maupertuis (Toronto: Univ. Toronto Press, 1979), pp. 107–125.

The historian of science Thomas Kuhn noted that Bacon's method of interrogating nature through constraint influenced seventeenth-century experimenters:

> The attitude towards the role and status of experiment is only the first of the novelties which distinguish the new experimental movement from the old. A second is the major emphasis given to experiments which Bacon himself described as "twisting the lion's tail." These were the experiments which constrained nature, exhibiting it under conditions which it could never have attained without the forceful intervention of man. The men who placed grain, fish, mice, and various chemicals *seriatim* in the artificial vacuum of a barometer or an air pump exhibit just this aspect of the new tradition.[48]

Objections to animal torture appeared during the Enlightenment. William Hogarth painted *The Four Stages of Cruelty* in 1751. The series depicted the life and death of the criminal Tom Nero in London. The first stage, the St. Giles Charity Schoolyard, shows acts of cruelty against animals. Nero as a young boy is torturing a dog with an arrow, while other boys are constraining, binding, cutting, goring, hanging, and shooting dogs, cats, and chickens. In the second stage, animal cruelty spreads to the streets and the larger city of London. Nero, now a young man, is beating a horse, while sheep, horses, donkeys, cattle, and humans are tied, beaten, rolled over, and gored. In the third stage, Nero has murdered the woman who carries his child. Finally, Nero himself is publicly dissected in a surgeon's hall. The series was meant to raise consciousness against inhumane methods of treatment, and it ultimately led to the outlawing of vivisection and the formation of the Society against Cruelty to Animals.[49]

Objections to experiments on animals in the bell jar were also mounted. Beginning in 1748, James Ferguson constructed scientific instruments and demonstrated them in lectures around England, writing them up in his 1761 *Lectures on Select Subjects*. Although his lectures included "Experiments with the Air Pump," he warned that

> if a fowl, a cat, rat, mouse or bird be put under the receiver, and the air be exhausted, the animal is at first oppressed as with a great weight, then grows convulsed, and at last expires in all the agonies of the most bitter and cruel death. But as this experiment is too shocking to every spectator who has the least degree of humanity, we substitute a machine called the "lungglass" in place [of] the animal; which by a bladder within it, shows how the lungs of animals are contracted into a small compass when the air is taken out of them.[50]

Perhaps inspired by Ferguson's lectures, but not heeding his admonitions, Joseph Wright of Derby painted *An Experiment on a Bird in the Air Pump* in 1768.[51] In Wright's painting, a pet cockatoo has been removed from a cage (shown in the upper right corner) and placed in a bell jar from which the air is evacuated. The experimenter's hand is placed near the stopcock, and he holds the power to halt the evacuation and return air to the jar to revive the bird. A old man stares at a human skull, contemplating death. A young girl covers her eyes to avoid viewing the impending horror, while a second girl stares anxiously upward and a woman, unable to watch, gazes at the face of another man who views the experiment directly. As Yaakov Garb has pointed out, the men and women have different responses. The women are stereotypically emotional, looking in horror at the bell jar, hiding their eyes, or looking at the men, thereby experiencing the results vicariously. The men, on the other hand, control the outcome via the stopcock, stare directly at the experiment with open curiosity, or contemplate the larger philosophical meaning of death. The men "witness" a scientific truth, the women "experience" a dying bird. The painter has forced social norms about male and female scientific responses to nature onto the audience. The experiment reflects the goals of Francis Bacon's method. A question is asked of nature, a controlled experiment is devised, and the results are witnessed and evaluated for their truth content. Whether a particular experiment reflects the torture of nature (or the mere "pestering" of nature) must be left to the individual to decide.[52]

As I did in *The Death of Nature* in 1980, I would still argue today that Bacon's efforts to define the

[48]Thomas S. Kuhn, "Mathematical vs. Experimental Traditions in the Development of Physical Science," *Journal of Interdisciplinary History*, 1976, 7:1–31, on p. 12.

[49]William Hogarth, *The Four Stages of Cruelty* (1751), http://www.graphicwitness.org/coe/cruel.htm.

[50]James Ferguson, *Lectures on Select Subjects* (1761), cited in http://www.mezzo-mondo.com/arts/mm/wright/wright.html.

[51]Joseph Wright of Derby, *An Experiment on a Bird in the Air Pump* (1768), http://www.nationalgallery.org.uk/cgi-bin/WebObjects.dll/CollectionPublisher.woa/wa/largeImage?workNumber=NG725.

[52]Stephen Daniels, *Joseph Wright* (Princeton, N.J.: Princeton Univ. Press, 1999), p. 40; and Yaakov Garb, personal communication.

experimental method were buttressed by his rhetoric and that the very essence of the experimental method arose out of techniques of human torture transferred onto nature. Such techniques are fundamental to the human domination of nature. Bacon's concept of experiment, along with a mechanistic view of animals as automata, legitimated experiments on living animals—experiments that could be, and later were, considered torture.

Through his use of metaphor, rhetoric, and myth, Francis Bacon developed the idea of the constrained, controlled experiment. Obviously Bacon cannot be held individually responsible for the positive or negative implications or applications of his ideas. He drew on tendencies existing in his culture, and his ideas were augmented by those who followed his direction. Had Bacon lived today he might or might not have supported genetic engineering, factory farming, and biotechnology—rather than stuffing a chicken with snow to see if putrefaction could be halted—as methods of interrogating nature. The development of the scientific method itself was nevertheless strongly influenced by Bacon's rhetoric and his vision of the interrogation and control of nature.

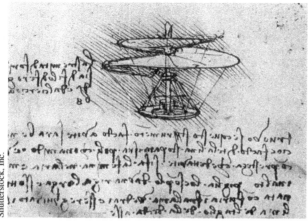

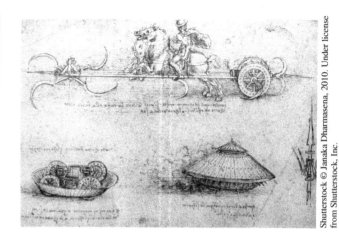

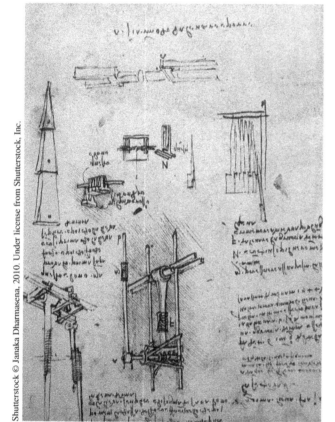

Leonardo Da Vinci—Sketches and Models of His Drawings

Contested Hegemony: The Great War and the Afro-Asian Assault on the Civilizing Mission Ideology

Reading 9

Michael Adas

The civilizing mission has been traditionally seen as an ideology by which late nineteenth century Europeans rationalized their colonial domination of the rest of humankind. Formulations of this ideology varied widely from those of thinkers or colonial administrators who stressed the internal pacification and political order that European colonization extended to "barbaric" and "savage" peoples suffering from incessant warfare and despotic rule, to those of missionaries and reformers who saw religious conversion and education as the keys to European efforts to "uplift" ignorant and backward peoples. But by the late 1800s, most of the fully elaborated variations on the civilizing mission theme were grounded in presuppositions that suggest that it had become a good deal more than a way of salving the consciences of those engaged in the imperialist enterprise. Those who advocated colonial expansion as a way of promoting good government, economic improvement, or Christian proselytization agreed that a vast and ever-widening gap had opened between the level of development achieved by western European societies (and their North American offshoots) and that attained by any of the other peoples of the globe. Variations on the civilizing mission theme became the premier means by which European politicians and colonial officials, as well as popularizers and propagandists, identified the areas of human endeavor in which European superiority had been incontestably established and calibrated the varying degrees to which different non-European societies lagged behind those of western Europe. Those who contributed to the civilizing mission discourse, whether through official policy statements or in novels and other fictional works, also sought to identify the reasons for Europe's superior advance relative to African backwardness or Asian stagnation and the implications of these findings for international relations and colonial policy.

Much of the civilizing mission discourse was obviously self-serving. But the perceived gap between western Europe's material development and that of the rest of the world appeared to validate the pronouncements of the colonial civilizers. Late Victorians were convinced that the standards by which they gauged their superiority and justified their global hegemony were both empirically verifiable and increasingly obvious. Before the outbreak of the Great War in 1914, these measures of human achievement were contested only by dissident (and marginalized) intellectuals, and occasionally by disaffected colonial officials. The overwhelming majority of thinkers and political leaders who concerned themselves with colonial issues had little doubt that the scientific and industrial revolutions—at that point still confined to Europe and North America—*had* elevated Western societies far above all others in the understanding and mastery of the material world. Gauges of superiority and inferiority, such as differences in physical appearance and religious beliefs, that had dominated European thinking in the early centuries of overseas expansion remained important. But by the second half of the nineteenth century, European thinkers, whether they were racists or antiracists, expansionists or anti-imperialists, or on the political left or right,[1] shared the conviction that through their scientific discoveries and inventions Westerners had gained an understanding of the workings of the physical world and an ability to tap its resources that were vastly superior to anything achieved by other peoples, past or present.

Many advocates of the civilizing mission ideology sought to capture the attributes that separated industrialized Western societies from those of the colonized

---

[1] For examples of leftist, anti-imperialist acceptance of these convictions, see Martine Loutfi's discussion of the views of Jean Jaures in *Littérature et colonialisme* (Paris: Mouton, 1971), p. 119; and Raoul Giradet, *L'Idée coloniale en France, 1871–1962* (Paris: La Table Ronde, 1972), pp. 96–98, 104–111.

"Contested Hegemony: The Great War and the Afro-Asian Assault on the Civilizing Mission Ideology" by Michael Adas in *Journal of World History*, Vol. 15, No. 1, 2004, pp. 31–63. Copyright © 2004 by University of Hawai'i Press. Reprinted by permission.

peoples by contrasting Europeans (or Americans) with the dominated "others" with reference to a standard set of binary opposites that had racial, gender, and class dimensions. Europeans were, for example, seen to be scientific, energetic, disciplined, progressive, and punctual, while Africans and Asians were dismissed as superstitious, indolent, reactionary, out of control, and oblivious to time. These dichotomous comparisons were, of course, blatantly essentialist. But the late Victorians were prone to generalizing and stereotyping. They were also determined to classify and categorize all manner of things in the mundane world, and fond of constructing elaborate hypothetical hierarchies of humankind.

For virtually all late Victorian champions of the civilizing mission, the more colonized peoples and cultures were seen to exhibit such traits as fatalism, passivity, and excessive emotionalism, the further down they were placed on imaginary scales of human capacity and evolutionary development, and thus the greater the challenge of civilizing them. For even the best-intentioned Western social theorists and colonial administrators, difference meant inferiority. But there was considerable disagreement between a rather substantial racist majority, who viewed these attributes as innate and permanent (or at least requiring long periods of time for evolutionary remediation), and a minority of colonial reformers, who believed that substantial progress could be made in civilizing stagnant or barbarian peoples such as the Chinese or Indians within a generation, and that even savage peoples such as the Africans or Amerindians could advance over several generations.[2] Those who held to the social evolutionist dogmas interpolated from rather dubious readings of Darwin's writings were convinced that the most benighted of the savage races were doomed to extinction. Some observers, such as the Reverend Frederick Farrar, thought the demise of these lowly peoples who had "not added one iota to the knowledge, the arts, the sciences, the manufactures, the morals of the world,"[3] quite consistent with the workings of nature and God.

Whatever their level of material advancement, "races," such as the Sikhs of India or the bedouin peoples of the African Sahel, that were deemed to be martial—thus, presumably energetic, active, disciplined, in control, expansive, and adaptive—were ranked high in late Victorian hierarchies of human types. The colonizers' valorization of martial peoples underscores the decidedly masculine bias of the desirable attributes associated with the civilizing mission ideology. Colonial administrations, such as the legendary Indian Civil Service, were staffed entirely by males until World War II, when a shortage of manpower pushed at least the British to recruit women into the colonial service for the first time.[4] The club-centric, sports-obsessed, hard-drinking enclave culture of the European colonizers celebrated muscular, self-controlled, direct, and energetic males. Wives and eligible young females were allowed into these masculine bastions. But their behavior was controlled and their activities constricted by the fiercely enforced social conventions and the physical layout of European quarters that metaphorically and literally set the boundaries of European communities in colonized areas. Within the colonizers' enclaves, the logic of the separate spheres for men and women prevailed, under-girded by a set of paired, dichotomous attributes similar to that associated with the civilizing mission ideology. Thus, such lionized colonial proconsuls as Evelyn Baring (the first Earl of Cromer), who ruled Egypt like a monarch for over two decades, saw no contradiction between their efforts to "liberate" Muslim women from the veil and purdah in the colonies and the influential support that they gave to antisuffragist organizations in Great Britain.[5]

As T. B. Macaulay's often-quoted 1840 caricature of the Bengalis as soft, devious, servile, indolent, and effeminate suggests,[6] feminine qualities were often associated in colonial thinking with dominated, inferior races. Some writers stressed the similarities in

---

[2]For representative racist views on these issues, see John Crawfurd, "On the Physical and Mental Characteristics of the European and Asiatic Races," *Transactions of the Ethnological Society of London* 5 (1867); and Robert Knox, *The Races of Men: A Philosophical Enquiry into the Influence of Race over the Destinies of Nations* (London, 1862). For samples of long-term evolutionist thinking, see Gustave Le Bon, *The Psychology of Peoples* (London, 1899); and C. S. Wake, "The Psychological Unity of Mankind," *Memoirs Read before the Anthropological Society of London* 3 (1867–1868). Late nineteenth-century nonracist or anti-racist improvers included Jacques Novicov [*L'Avenir de la race blanche* (Paris, 1897)] and Henry Maine [*Short Essays and Reviews on the Educational Policy of the Government of India from the "Englishman"* (Calcutta, 1866)].

[3]"Aptitudes of the Races," *Transactions of the Ethnological Society of London* 5 (1867), p. 120. For a less celebratory view of this process that was closely tied to evolutionary thinking, see Alfred Russel Wallace, "The Development of Human Races under the Law of Natural Selection," *Anthropological Review* (1865).
[4]Helen Callaway, *Gender, Culture and Empire: European Women in Colonial Nigeria* (Urbana: University of Illinois Press, 1987), esp. pp. 139–145.
[5]Leila Ahmed, *Women and Gender in Islam* (New Haven, Conn.: Yale University Press, 1992), pp. 153–154.
[6]In his essay "On Clive," in Macaulay, *Poetry and Prose*, ed. by G. M. Young (Cambridge, Mass.: Harvard University Press, 1970).

the mental makeup of European women and Africans or other colonized peoples; others argued that key female attributes corresponded to those ascribed to the lower orders of humanity. Again the paired oppositions central to the civilizing mission ideology figured prominently in the comparisons. Though clearly (and necessarily) superior in moral attributes, European women—like the colonized peoples— were intuitive, emotional, passive, bound to tradition, and always late.[7] In addition, the assumption that scientific discovery and invention had been historically monopolized by males (despite the accomplishments of contemporaries such as Marie Curie) was taken as proof that women were temperamentally and intellectually unsuited to pursuits, such as engineering and scientific research, that advocates of the civilizing mission ideology viewed as key indicators of the level of societal development. These views not only served to fix the image and position of the European *memsahib* as passive, domestic, apolitical, and vulnerable, they made it all but impossible for indigenous women in colonized societies to obtain serious education in the sciences or technical training. As Ester Boserup and others have demonstrated, institutions and instruction designed to disseminate Western scientific knowledge or tools and techniques among colonized peoples were directed almost totally toward the male portion of subject populations.[8]

The attributes that the colonizers valorized through the civilizing mission ideology were overwhelmingly bourgeois. Rationality, empiricism, progressivism, systematic (hence scientific) inquiry, industriousness, and adaptability were all hallmarks of the capitalist industrial order. New conceptions of time and space that had made possible and were reinforced by that order informed such key civilizing mission attributes as hard work, discipline, curiosity, punctuality, honest dealing, and taking control—the latter rather distinct from the self-control so valued by aristocrats. Implicit in the valorization of these bourgeois traits was approbation of a wider range of processes, attitudes, and behavior that was not usually explicitly discussed in the tomes and tracts of the colonial proponents of the civilizing mission ideology. Ubiquitous complaints by colonial officials regarding the colonized's lack of

foresight, their penchant for "squandering" earnings on rites of passage ceremonies or religious devotion, and their resistance to work discipline and overtime suggested they lacked proclivities and abilities that were essential to the mastery of the industrial, capitalist order of the West. Implicitly then, and occasionally explicitly, advocates of the civilizing mission ideology identified the accumulation and reinvestment of wealth, the capacity to anticipate and forecast future trends, and the drive for unbounded productivity and the provision of material abundance[9] as key attributes of the "energetic, reliable, improving"[10] Western bourgeoisie that had been mainly responsible for the scientific and industrial revolutions and European global hegemony.

In the decades before the Great War, white European males reached the pinnacle of their power and global influence. The civilizing mission ideology both celebrated their ascendancy and set the agenda they intended to pursue for dominated peoples throughout the world. The attributes that male European colonizers ascribed to themselves and sought—to widely varying degrees in different colonial settings and at different social levels—to inculcate in their African or Asian subjects were informed by the underlying scientific and technological gauges of human capacity and social development that were central to the civilizing mission ideology. Both the attributes and the ideology of the dominant in turn shaped European perceptions of and interaction with the colonized peoples of Africa and Asia in a variety of ways. Many apologists for colonial expansion, for example, argued that it was the duty of the more inventive and inquisitive Europeans to conquer and develop the lands of backward or primitive peoples who did not have the knowledge or the tools to exploit the vast resources that surrounded them.[11] Having achieved political control, it was incumbent upon the Western colonizers to replace corrupt and wasteful indigenous regimes with honest and efficient bureaucracies, to reorganize the societies of subjugated

[7]For examples of these comparisons, see Arthur de Gobineau, *Essai sur l'inégalité des races humaines* (Paris, 1853), vol. 1, pp. 150–152; James Hunt, "On the Negro's Place in Nature," *Memoirs Read before the Anthropological Society of London* (1863–1864), p. 10; and Le Bon, *Psychology of Peoples*, pp. 35–36.

[8]*Women's Role in Economic Development* (New York: St. Martin's Press, 1979), chapter 3.

[9]A stimulating and contentious exploration of these connections can be found in Thomas Haskell's essays on "Capitalism and the Origins of the Humanitarian Sensibility," *American Historical Review* 90, nos. 3 and 4 (1985).

[10]William Greg as quoted in John C. Greene, *Science, Ideology and World View* (Berkeley: University of California Press, 1981), p. 108.

[11]See, for examples, Benjamin Kidd, *The Control of the Tropics* (London: Macmillan, 1898), pp. 14, 39, 52–55, 58, 83–84, 88–90; H. H. Johnson, "British West Africa and the Trade of the Interior," *Proceedings of the Royal Colonial Institute* 20 (1888–1889), p. 91; and Arthur Girault, *Principes de colonisation et de legislation coloniale* (Paris: Larose, 1895), p. 31.

peoples in ways the Europeans deemed more rational and more nurturing of individual initiative and enterprise, and to restructure the physical environment of colonized lands in order to bring them into line with European conceptions of time and space.

The Europeans' superior inventiveness and understanding of the natural world also justified the allotment of tasks in the global economy envisioned by proponents of the civilizing mission. Industrialized Western nations would provide monetary and machine capital and entrepreneurial and managerial skills, while formally colonized and informally dominated overseas territories would supply the primary products, cheap labor, and abundant land that could be developed by Western machines, techniques, and enterprise. Apologists for imperialism argued that Western peoples were entrusted with a mission to civilize because they were active, energetic, and committed to efficiency and progress. It was therefore their duty to put indolent, tradition-bound, and fatalistic peoples to work, to discipline them (whether they be laborers, soldiers, domestic servants, or clerks), and to inculcate within them (insofar as their innate capacities permitted) the rationality, precision, and foresight that were seen as vital sources of Europe's rise to global hegemony. But efforts to fully convert the colonized to the virtues celebrated by the civilizing mission ideology were normally reserved for the Western-educated classes. Through state-supported and missionary education, Western colonizers sought to propagate epistemologies, values, and modes of behavior that had originally served to justify their dominance and continued to be valorized in their rhetoric of governance.

The elite-to-elite emphasis of the transmission of the civilizing mission ideology meant that it was hegemonic in a rather different sense than that envisioned by Gramsci's original formulation of the concept.[12] To begin with, it was inculcated across cultures by colonizer elites onto the bourgeois and petty-bourgeois classes that Western education and collaboration had brought into being. In addition, the proponents of the civilizing mission viewed it only marginally as an ideology that might be employed to achieve cultural hegemony over the mass of colonized peoples. Few of the latter had anything but the most rudimentary appreciation of the scientific and technological breakthroughs that were vital to Western dominance—as manifested in the colonizers' military power, transportation systems,

and machines for extracting mineral and agrarian resources. Only the Western-educated classes among the colonized were exposed to the history of Europe's unprecedented political, economic, and social transformations, and only these groups were expected by their colonial overlords to emulate them by internalizing the tenets of the civilizing mission ideology.

In the pre-World War era, the great majority of Western-educated collaborateur and comprador classes in the colonies readily conceded the West's scientific, technological, and overall material superiority. Spokesmen for these classes—often even those who had already begun to agitate for an end to colonial rule[13]—clamored for more Western education and an acceleration of the process of diffusion of Western science and technology in colonized societies. In Bengal in eastern India in the 1860s, for example, a gathering of Indian notables heartily applauded K. M. Banerjea's call for the British to increase opportunities for Indians to receive advanced instruction in the Western sciences. Banerjea dismissed those who defended "Oriental" learning by asking which of them would trust the work of a doctor, engineer, or architect who knew only the mathematics and mechanics of the Sanskrit sutras.[14] What is noteworthy here is not only Banerjea's confusion of Buddhist (hence Pali) sutras and Sanskrit *shastras*, but his internalization of the Western Orientalists' essentialist conception of Asian thinking and learning as a single "Oriental" whole that had stagnated and fallen behind the West in science and mathematics. Just over two decades later, the prominent Bengali reformer and educator Keshub Chunder Sen acknowledged that the diffusion of Western science that had accompanied the British colonization of India had made it possible for the Indians to overcome "ignorance and error" and share the Europeans' quest to explore "the deepest mysteries of the physical world."[15]

Thus, despite the Hindu renaissance that was centered in these decades in Bengal, as S. K. Saha has observed, the presidency's capital, Calcutta, had been reduced to an intellectual outpost of Europe.[16] But

[12]*Selections from Prison Notebooks*, ed. and trans. Q. Hoare and G. Smith (New York, 1971), pp. 12–14, 55–63, 275–276.

[13]See, for example, Ira Klein, "Indian Nationalism and Anti-Industrialization: The Roots of Gandhian Economics," South *Asia* 3 (1973); and Bade Onimode, A *Political Economy of the African Crisis* (London: Zed, 1988), pp. 14–22 and chapters 6, 9, and 11.

[14]"The Proper Place of Oriental Literature in Indian Collegiate Education," *Proceedings of the Bethune Society* (February 1868), pp. 149, 154.

[15]"Asia's Message to Europe," in *Keshub Sen's Chunder Lectures in India* (London: Cassell, 1901) vol. 2, pp. 51, 61.

[16]"Social Contest of Bengal Renaissance," in David Kopf and Joaarder Safiuddin, eds., *Reflections on the Bengal Renaissance* (Dacca: Institute of Bangladesh Studies, 1977), p. 140.

perhaps a majority of English-educated Indians did not just revere Western scientific and technological achievements, but they accepted their colonial masters' assumption that responsible, cultivated individuals privileged rationality, empiricism, punctuality, progress, and the other attributes deemed virtuous by proponents of the civilizing mission ideology. Just how widely these values had been propagated in the Indian middle classes is suggested by anthropological research carried out among Indian merchant communities in central Africa in the 1960s. Responses to questions relating to the Indians' attitudes toward the African majority in the countries in which they resided revealed that the migrant merchants considered their hosts "illiterate and incomprehensible savages," who were lazy and without foresight, childlike in their thinking (and thus incapable of logical deductions), and self-indulgent and morally reprobate.[17]

As the recollections of one of Zimbabwe's Western-educated, nationalist leaders, Ndabaningi Sithole, make clear, the colonized of sub-Saharan Africa were even more impressed by the Europeans' mastery of the material world than were their Indian counterparts. Because many African peoples had often been relatively isolated before the abrupt arrival of European explorers, missionaries, and conquerors in the last decades of the nineteenth century, early encounters with these agents of expansive, industrial societies were deeply disorienting and demoralizing:

> The first time he ever came into contact with the white man the African was overwhelmed, overawed, puzzled, perplexed, mystified, and dazzled. . . . Motor cars, motor cycles, bicycles, gramophones, telegraphy, the telephone, glittering Western clothes, new ways of ploughing and planting, added to the African's sense of curiosity and novelty. Never before had the African seen such things. They were beyond his comprehension; they were outside the realm of his experience. He saw. He wondered. He mused. Here then the African came into contact with two-legged gods who chose to dwell among people instead of in the distant mountains.[18]

In part because European observers took these responses by (what they perceived to be) materially

impoverished African peoples as evidence of the latter's racial incapacity for rational thought, discipline, scientific investigation, and technological innovation, there were few opportunities before World War I for colonized Africans to pursue serious training in the sciences, medicine, or engineering, especially at the post-secondary level. Technological diffusion was also limited, and the technical training of Africans was confined largely to the operation and maintenance of the most elementary machines.[19] Nonetheless, the prescriptions offered by French- and English-educated Africans for the revival of a continent shattered by centuries of the slave trade shared the assumption of the European colonizers that extensive Western assistance would be essential for Africa's uplift. The Abbé Boilat, for example, a mulatto missionary and educator, worked for the establishment of a secondary school at St. Louis in Senegal, where Western mathematics and sciences would be taught to the sons of the local elite. Boilat also dreamed of an African college that would train indigenous doctors, magistrates, and engineers who would assist the French in extending their empire in the interior of the continent.[20]

Although the Edinburgh-educated surgeon J. A. Horton was less sanguine than Boilat about the aptitude of his fellow Africans for higher education in the Western sciences, he was equally convinced that European tutelage was essential if Africa was to rescued from chaos and barbarism. Horton viewed "metallurgy and other useful arts" as the key to civilized development, and argued that if they wished to advance, Africans must acquire the learning and techniques of more advanced peoples such as the Europeans.[21] Even the Caribbean-born Edward Blyden, one of the staunchest defenders of African culture and historical achievements in the prewar decades, conceded that Africa's recovery from the ravages of the slave trade depended upon assistance from nations "now foremost in civilization and science" and the return of educated blacks from the United States and Latin America. Blyden charged that if Africa had been integrated into the world market

---

[17]Floyd and Lillian Dotson, *The Indian Minority of Zambia, Rhodesia, and Malawi* (New Haven, Conn.: Yale University Press, 1968), pp. 262–268, 320.

[18]In *African Nationalism* (London, Oxford University Press, 1969), p. 157.

---

[19]On British and French educational policies in nineteenth-century Africa and their racist underpinnings, see Michael Adas, *Machines as the Measure of Men: Science, Technology and Ideologies of Western Dominance* (Ithaca, N.Y.: 1989), chapter 5.

[20]*Esquisses sénégalaises* (Paris: Bertrand, 1853), pp. 9–13, 478; and André Villard, *Histoire du Sénégal* (Dakar, 1943), p. 98.

[21]*West African Countries and Peoples* (London: W. J. Johnson, 1868), pp. 1–4; and *Letters on the Political Condition of the Gold Coast* (London: W. J. Johnson, 1870), pp. i–iii.

system through regular commerce rather than the slave trade, it would have developed the sort of agriculture and manufacturing and imported steam engines, printing presses, and other machines by which the "comfort, progress, and usefulness of mankind are secured."[22]

* * *

There were those who contested the self-satisfied, ethnocentric, and frequently arrogant presuppositions that informed the civilizing mission ideology in the decades before World War I. The emergence of Japan as an industrial power undermined the widely held conviction that the Europeans' scientific and technological attainments were uniquely Western or dependent on the innate capacities of the white or Caucasian races. Conversely, the modernists' "discovery" of "primitive" art and the well-publicized conversion of a number of rather prominent European intellectuals to Hinduism, Buddhism, and other Asian religions suggested the possibility of viable alternatives to European epistemologies, modes of behavior, and ways of organizing societies and the natural world. Some European thinkers, perhaps most famously Paul Valéry and Herman Hesse, actually questioned Western values themselves. They asked whether the obsessive drive for increased productivity and profits and the excessive consumerism that they saw as the hallmarks of Western civilization were leading humanity in directions that were conducive to social well-being and spiritual fulfillment.[23]

Before the outbreak of the war in 1914, these critiques and alternative visions were largely marginalized, dismissed by mainstream politicians and the educated public as the rantings of gloomy radicals and eccentric mystics. But the coming of the Great War and the appalling casualties that resulted from the trench stalemate on the Western Front made a mockery of the European conceit that discovery and invention were necessarily progressive and beneficial to humanity. The mechanized slaughter and the conditions under which the youth of Europe fought the war generated profound challenges to the ideals and assumptions upon which the Europeans had for over a century based their sense of racial superiority and from which they had fashioned that ideological

testament to their unmatched hubris, the civilizing mission. Years of carnage in the very heartlands of European civilization demonstrated that Europeans were at least as susceptible to instinctual, irrational responses and primeval drives as the peoples they colonized. The savagery that the war unleashed within Europe, Sigmund Freud observed, should caution the Europeans against assuming that their "fellow-citizens" of the world had "sunk so low" as they had once believed, because the conflict had made it clear that the Europeans themselves had "never risen as high."[24]

Remarkably (or so it seemed to many at the time), the crisis passed, the empire survived, and the British and French emerged victorious from the war. In fact, in the years following the end of the conflict in 1918, the empires of both powers expanded considerably as Germany's colonies and Turkey's territories in the Levant were divided between them.[25] Although recruiting British youths into the Indian Civil Service and its African counterparts became more difficult and such influential proponents of French expansionism as Henri Massis conceded that the Europeans' prestige as civilizers had fallen sharply among the colonized peoples,[26] serious efforts were made to revive the badly battered civilizing mission ideology. Colonial apologists, such as Étienne Richet and Albert Bayet, employed new, less obviously hegemonic slogans that emphasized the need for "mutual cooperation" between colonizers and colonized and programs for "development" based on "free exchanges of views" and "mutual respect." But the central tenets of the colonizers' ideology remained the same: European domination of African and Asian peoples was justified by the diffusion of the superior science, technology, epistemologies, and modes of organization that it facilitated. Though the engineer

---

[22]"Hope for Africa: A Discourse," *Colonization Journal* (August 1861), pp. 7–8; and "The Negro in Ancient History," in *The People of Africa* (New York, 1871), pp. 23–24, 34.

[23]For a discussion of these divergent challenges to the civilizing mission's underlying assumptions, see Michael Adas, *Machines as the Measure of Men: Science, Technology and Ideologies of Western Dominance* (Ithaca, N.Y.: 1989), pp. 345–365.

[24]"Reflections on War and Death," (1915), reprinted in Philip Reiff, ed., *Character and Culture* (New York: Collier, 1963), p. 118. For a fuller discussion of these themes, see *Adas, Machines as the Measure of Men*, chapter 6.

[25]On the postwar expansion of the French colonial empire, see Christopher M. Andrew and A. S. Kanya-Forstner, *France Overseas: The Great War and the Climax of French Imperialism* (London: Thames and Hudson, 1981); and for British expansion in the Middle East, see John Darwin, *The British in the Middle East* 1918–1922 (London: Macmillan, 1981).

[26]Hugh Tinker, "Structure of the British Imperial Heritage," in Ralph Braibanti, ed., *Asian Bureaucratic Systems Emergent from the British Imperial Tradition* (Durham: Duke University Press, 1966), pp. 61–63; and Henri Massis, *Defense of the West* (London: Faber, 1927), pp. 6, 9, 134. For an insightful discussion of the demoralized state of colonial officialdom in Africa in the post-World War I years, see Robert Delavignette, *Freedom and Authority in West Africa* (London, 1965), pp. 149–150.

and the businessman may have replaced the district officer and the missionary as the chief agents of the mission to civilize, it continued to be envisioned as an unequal exchange between the advanced, rational, industrious, efficient, and mature societies of the West and the backward, ignorant, indolent, and childlike peoples of Africa, Asia, and the Pacific.[27]

For many European intellectuals and a handful of maverick politicians, however, postwar efforts to restore credibility to the civilizing mission ideology were exercises in futility. These critics argued that the war had destroyed any pretense the Europeans might have of moral superiority or their conceit that they were innately more rational than non-Western peoples. They charged that the years of massive and purposeless slaughter in the trenches had made a shambles of proofs of Western superiority based on claims to higher levels of scientific understanding and technological advancement. Though the literature in which this discourse unfolded is substantial,[28] I would like to focus here on the writings of Georges Duhamel, and in particular an incident that he relates in his first novel about the war and returns to repeatedly in his later essays, an incident that provides the focal point for his extensive critique of the civilizing mission ideology. That critique in turn fed the growing doubts about European civilization and its global influence that African and Asian intellectuals and political leaders had begun to voice in the years before 1914. Though African and Asian writers rarely cited European authors for support in their assaults on the civilizing mission ideology and colonialism more generally, both metropolitan and colonial intellectuals were engaged in a common discourse in the decades after World War I, a discourse that proved deeply subversive of the colonizers' hegemonic rhetoric and thus a critical force in the liberation struggles of colonized peoples.

As its title, *Civilisation 1914–1918*, suggests, Duhamel's autobiographical novel about a sergeant in the French medical corps on the Western Front is an exercise in irony.[29] Like so many of the millions of young European males who were funneled into the trenches/tombs of the Western Front and lived long enough to tell about it, Duhamel was profoundly disoriented and disillusioned by his wartime experiences. They seemed to contradict all that he thought he knew or believed about Western civilization. Nothing was as it appeared to be or *ought* to be. As in a Max Ernst painting, reality was grotesquely deformed. Everything was bewilderingly inverted. The massive, mechanized, and increasingly senseless slaughter of young men that resulted from the trench stalemate transformed machines from objects of pride and symbols of advancement to barbarous instruments of shame and horror. From the masters of machines, European men had become their slaves, "bent under the burden of tedious or sorrowful work." Even the scientific breakthroughs that Duhamel, a highly trained surgeon and former laboratory technician, had once thought the most unique, noble, and exalted of Europe's achievements had been enlisted by the forces of hate and destruction to sustain the obscene and irrational combat that was destroying Europe from within.[30] The poet Paul Valéry, who shared Duhamel's assessment of the war, lamented the fact that the Europeans' greatest discoveries had been perverted by the need for so "much science to kill so many men, waste so many possessions, and annihilate so many towns in so little time."[31]

The conditions under which the soldiers and Duhamel's protagonist-surgeon lived and the wasteland their combat wrought in northern France made a mockery of the European conviction that their unprecedented mastery of nature was proof of their superiority over all contemporary peoples and past civilizations. The filthy, lice-ridden bodies of the youth of Europe, exposed for weeks on end to the cold and mud of winter in Flanders or the valley of the Somme, fighting with huge rats for their miserable rations or their very limbs, belied the prewar conviction that superior science had given Western man dominion over nature. For Duhamel the flies that swarmed about the open latrines, garbage heaps, and dismembered corpses and carcasses of the unclaimed

[27]For discussions of efforts to revive the civilizing mission ideology in the postwar era, see Raoul Girardet, *L'Idée coloniale en France, 1871–1962* (Paris: La Table Ronde, 1972), pp. 117–132 and chapter 5; and Thomas August, *The Selling of Empire: British and French Imperialist Propaganda, 1890–1940* (Westport: Greenwood, 1985), pp. 126–140. For a thorough exploration of shifts in colonial policy in the 1920s and 1930s, see Frederick Cooper, *Decolonization and African Society: The Labor Question in British and French Africa* (Cambridge, 1996).
[28]For a fuller discussion of these themes, see Michael Adas, *Machines as the Measure of Men*, chapter 6.
[29]The novel was originally published in 1917. The first English translation appeared in 1919 (London: Century Company) with the same title as the original. Direct quotations in this essay are taken from the 1919 translation, and so numbered; otherwise page numbers refer to the 1922 Paris edition, published by Mercure de France.
[30]Duhamel, *Civilisation*, pp. 257–258, 268–269; *Entretien sur l'esprit européen* (Paris: Mercure de France, 1928), pp. 17–18, 36–39; *La Pensée des âmes*, (Paris: Mercure de France, 1949); and *La possession du monde* (Paris: Mercure de France, 1919), pp. 18–19, 140, 242–245.
[31]"Letters from France I: The Spiritual Crisis," *The Athenaeum* (11 April 1919), p. 182.

dead in the no-man's-land moonscape provided an ever-present reminder of the Europeans' reversion to a state of savagery, where they were continually buffeted by the forces of nature. Duhamel's surgeon-protagonist in *Civilisation 1914–1918* is repelled by, but utterly incapable of fending off, the multitude of flies that suck the pus and blood of his patients and the larvae that multiply rapidly in their festering wounds.[32] But the soldiers' vulnerability to the forces of nature represents only one of the inversions Duhamel and other chroniclers of the trench trauma associated with the colossal misuse of science and technology in the war. Duhamel concluded that the Western obsession with inventing new tools and discovering new ways to force nature to support material advancement for its own sake had inevitably led to the trench wasteland where "man had achieved this sad miracle of denaturing nature, of rendering it ignoble and criminal."[33]

The climax of the surgeon's ordeal in Duhamel's *Civilisation 1914–1918* comes in his first encounter with the Ambulance Chirurgicale Automobile (ACA), "the most perfect thing in the line of an ambulance that has been invented . . . the last word in science; it follows the armies with motors, steam-engines, microscopes, laboratories . . . ,"[34] The sergeant is assigned to minister to the casualties delivered to "the first great repair-shop the wounded man encounters." The wounded in question are fittingly cuirassiers—traditionally cavalrymen with shining breastplates and plumed helmets—fighting without their horses and sans plumes, since both had proved positively lethal, given the firepower of the opposing armies, in the first months of the war. The sergeant relates that these once "strong, magnificent creatures," have been shattered and wait "like broken statues" for admission to the ACA. In the midst of the mechanized trench battleground, the cuirassiers are anachronisms, pitiful vestiges of a lost chivalric ethos. They chatter "like well-trained children" about their wounds and fear of anesthesia. In contrast to the active, self-controlled, take-charge European male ideal of the prewar era, the cuirassiers have,

as Sandra Gilbert argues for male combatants more generally,[35] been transformed into "passive, dependent, immanent medical object[s]."

In Gilbert's rendering wounded males are opposed to European women who as nurses and ambulance drivers have become "active, autonomous, and transcendent." But Duhamel recounts an inversion that must have been even more unsettling for his French and British readers in the postwar decades. The cuirassiers are carried into the ACA by African stretcher-bearers, whom Duhamel initially depicts in some of the stock images of the dominant colonizers. With their "thin black necks, encircled by the [stretcher-bearers'] yokes" and their "shrivelled fingers," the "little" Malagasies remind him of "sacred monkeys, trained to carry idols." The sergeant finds the Malagasies "timid," "docile," and "obedient," and compares them (curiously) to "black and serious embryos." But after the Malagasies place the wounded cuirassiers on the operating tables, a revelatory encounter occurs:

> At this moment my glance met that of one of the blacks and I had a sensation of sickness. It was a calm, profound gaze like that of a child or a young dog. The savage was turning his head gently from right to left and looking at the extraordinary beings and objects that surrounded him. His dark pupils lingered lightly over all the marvelous details of the workshop for repairing the human machine. And these eyes, which betrayed no thought, were none the less disquieting. For one moment I was stupid enough to think, "How astonished he must be!" But this silly thought left me, and I no longer felt anything but an insurmountable shame.[36]

The surgeon begins by depicting the African in terms—"child," "young dog," "savage"—that were standard epithets for racists and colonizers alike. But his complacent sense of superiority is shattered by his realization that rather than being impressed by the advanced science and technology that have been packed into the ACA, the "primitive" Malagasy must be appalled or at the very least bewildered by the desperate and costly efforts of the Europeans to repair the devastation wrought by their own civilization's

---

[32]*Civilisation*, pp. 11–12.

[33]*Possession du monde*, p. 99. One is reminded here of Celine's obsessive fear of trees, "since [he] had known them to conceal an enemy. Every tree meant a dead man," (*Journey to the End of Night* [New York: New Directions, 1960], p. 53); or Remarque's tortured account of wounded horses who died in no-man's-land "wild with anguish, filled with terror, and groaning." (*All Quiet on the Western Front* [New York: Fawcett Crest, 1975], p. 61.

[34]Quoted passages in the following are taken from the English translation of *Civilization 1914–1918*, chapter 16, unless otherwise noted.

[35]"Soldier's Heart: Literary Men, Literary Women, and the Great War," *Signs* 8, no. 3 (1983): 435.

[36]*Civilisation 1914–1918*, pp. 282–283.

suicidal war. The savage has the exalted doctor and the frenzied activity of the ACA in his "calm, profound gaze"; an exact reversal of the only permissible relationship between Europeans and "savage" or subordinate peoples according to postmodernist readings of European travel literature and colonial memoirs.[37] The surgeon is embarrassed and angered by his realization that the Malagasy is a witness to the Europeans' irrational, but very destructive, tribal war.

The reversion to barbarism and savagery that Duhamel associates with trench warfare is a pervasive theme in participants' accounts of the conflict. Combatants describe themselves as "wild beasts," "primitives," "bushmen," "ape-men," and "mere brutes." Soldiers at the front compare their mud-caked existence to that of prehistoric men who lived in caves or crude holes dug into the earth.[38] In the trenches or behind the lines, the refinements of civilization receded. Decorum was associated with death; modesty became irrelevant to soldiers who used crudely fabricated latrines as places to congregate, gossip, and curse their leaders. In battle, primal instincts— "the furtive cunning of a stoat or weasel"—were the key to survival. Europeans fought, as Frederic Manning observed in perfect social evolutionist tropes, like peoples at a "more primitive stage in their development, and . . . [became] nocturnal beasts of prey, hunting each other in packs."[39] Infantrymen were forced to listen rather than look for incoming shells, which were often fired from miles away and could not be seen until it was too late. Soldiers who lived long enough to become trench veterans did so by developing an acute sensitivity to the sounds of different sizes and sorts of projectiles and gauging by sound how close they would hit to where the soldiers were dug in. Thus, a refined sense of hearing, which the Europeans had associated with savage or primitive peoples since at least the eighteenth century, superseded sight, which had long been regarded as the most developed sense of civilized peoples like the Europeans.[40]

As these examples suggest, the reversion to savagery that the youth of Europe experienced was mainly of the degraded rather than the noble variety of primitivism that European artists and writers had been trying to sort out for centuries. In a moment that borders on black comedy, Duhamel's surgeon-protagonist fantasizes about escaping the horrors of the trench stalemate by fleeing to the mountains to live among the "savage" blacks. Envisioning, like the impulsive Ernest Psichari who fled to Africa from the Europe of "large stomachs and vain speeches" just before the outbreak of the war,[41] a land where people still lived in a "state of nature," free from the mechanical outrages inflicted continually on those at the front, the sergeant is shocked to encounter Africans riding bicycles at Soissons, and later clamoring for war decorations. Despondent, he concludes that there are no "real black people" left and no place on earth that has not been contaminated by European civilization.[42]

In the many works he published in the decades after the war, Duhamel elaborated and expanded upon the critique of European civilization and of the civilizing mission ideology that had been initially fueled by his experiences on the Western Front.[43] Like many prominent European intellectuals, from Valéry, André Malraux, and René Guenon to Hermann Hesse, Hermann Keyserling, and E. M. Forster, Duhamel concluded that the war was the inevitable outcome of the Europeans' centuries-old obsession with scientific and industrial advance. They had been so captivated by mechanical progress and material increase that they had neglected the needs of the soul and spirit. They had allowed the spiritual ideals and moral dimensions of Western civilization to wither, while subordinating themselves to the machines they had created to serve them. They had confused industry and science with civilization, and become deluded by the conviction that progress, well-being, and goodness could be equated with the ability to go

[37]See, for examples, Mary Louise Pratt, *Imperial Eyes: Travel Writing and Transculturation* (London: Routledge, 1992).

[38]For samples of the use of these metaphors by infantrymen from each of the major combatant nations on the Western Front, see Roland Rorgelès, *Les croix de Bois* (Paris: Albin Michel, 1919), pp. 62, 113; Henri Barbusse, *Under Fire: The Story of a Squad* (London: 1916); Remarque, *All Quiet*, pp. 103–104, 236–237; and Richard Aldington, *The Death of a Hero* (London: 1984), pp. 255, 264, 267.

[39]Quoted portions from Manning's *Middle Parts of Fortune* (New York: St. Martins, 1972), pp. 8, 12, 39–40. Other references to Remarque, *All Quiet*, pp. 12–13, 124–125, 236–237; Aldington, *Death of a Hero*, p. 362; and Ludwig Renn, *War* (London: Antony Mott, 1984), pp. 110–111.

[40]Léon-François Hoffman, *Le nègre romantique* (Paris, 1973); and Eric J. Leed, *No Man's Land: Combat and Identity in World War I* (Cambridge: Cambridge University Press, 1979), pp. 126–127.

[41]Robert Wohl, *The Generation of 1914* (Cambridge, Mass.: Harvard University Press, 1979), pp. 12–13.

[42]*Civilisation 1914–1918*, p. 268; quoted phrases, *Civilization 1914–1918*, p. 272.

[43]The following discussion is based heavily on relevant sections from *La Possession du monde*, esp. pp. 140, 242–246, 254–256, 264–265; and *Entretien sur l'esprit européen*, pp. 17–18, 20–22, 29–37, 40–46, 50.

100 miles per hour. These misunderstandings had led inexorably to Europe's ruin in a war that had devastated its once-prosperous lands, thrown its societies into turmoil, and aroused the colonized peoples to resistance.

Although at times in his later years Duhamel felt compelled to come to Europe's defense in the face of rising challenges from the colonized world,[44] he believed that the Great War had proved decisive in undermining the image of Europeans as "inscrutable masters," "dazzling and terrible demi-gods." In supporting their colonial rulers in the war, Africans and Asians had discovered that the Europeans' claim that they possessed attributes that entitled them to dominate the rest of humankind was false. The "men of color" found that the Europeans inhabited only a small and divided continent, and that their overlords were not gods but "miserable, bleeding animal[s] (the most extreme of inversions from the Western perspective) . . . devoid of hope and pride." The war had taught the colonized peoples that, despite their claims to have mastered the forces of nature, the Europeans submitted to cold and heat, to epidemics, and to innumerable "perils without names." Not surprisingly, Duhamel argues, the colonized felt little pity for the once-proud masters whom they had grown increasingly determined to resist. But he believed that they must not resist European domination alone. They must also resist the spread of the "cruel" and "dangerous" civilization that the Europeans—and, after the war, their American progeny[45]—sought to impose on the rest of humankind. The war had revealed the unprecedented capacity for barbarity of this so-called civilization, as well as the perils of destructiveness and vacuousness that threatened those who sought to emulate its narrowly materialistic achievements. In the decades after the war, a number of Asian and African intellectuals took up Duhamel's call to resist with important consequences for liberation struggles in the colonized world.

* * *

Mounted by Asian and African thinkers and activists who often received little publicity in Europe or the United States, pre–World War I challenges to assumptions of Western superiority enshrined in the civilizing mission ideology were highly essentialist, mainly reactive rather than proactive, and framed by Western gauges of human achievement and worth. The most extensive and trenchant critiques in the case of India were articulated by the Hindu revivalist Swami Vivekananda (Naren Datta), who had won some measure of fame in the West with a brilliant lecture on Vedanta philosophy at the Conference of World Religions held in conjunction with the Chicago World's Fair in 1893. Vivekananda was fond of pitting a highly essentialized spiritual "East" against an equally essentialized materialistic "West." And like the earlier holymen-activists of the Arya Dharm,[46] he claimed that most of the scientific discoveries attributed to Western scientists in the modern era had been pioneered or at least anticipated by the sages of the Vedic age.[47] Vivekananda asserted that after mastering epistemologies devised to explore the mundane world, the ancient Indians (and by inference their modern descendants) had moved on to more exulted, transcendent realms, a line of argument that clearly influenced the thinking of the French philosopher René Guenon in the postwar decades.[48] Vivekananda cautioned his Indian countrymen against the indiscriminate adoption of the values, ways, and material culture of the West, a warning that was powerfully echoed at another level by the writings of Ananda Coomaraswamy, who, like William Morris and his circle in England, called for a concerted effort to preserve and restore the ancient craft skills of the Indian peoples, which he likened to those of Medieval Europe.[49] In what has been seen as a premonition of the coming global conflict, Vivekananda predicted, decades before 1914, that unless the West tempered its obsessive materialistic pursuits by adopting the spiritualism of the East, it would "degenerate and fall to pieces."[50]

---

[44]See, for example, *Les espoirs et les épreuves*, (Paris: Mercure de France, 1953), pp. 135–136, 186–187.

[45]Duhamel visited America in the late 1920s and came away with decidedly negative impressions that are detailed in his *America the Menace: Scenes from the Life of the Future* (London: Allen and Unwin, 1931).

[46]See Mal, *Dayanand* (1962), pp. 66–68, 73, 216; and Dayanda, *Satyarth Prakash*, pp. 292–3.

[47]*Collected Works*, vol. 1, pp. 13, 134; vol. 2, pp. 124, 140–141.

[48]Vivekananda, *Collected Works*, vol. 1, pp. 121, 365. For Guenon's celebration of the higher levels of thinking achieved by Indian philosophers, see *The Crisis of the Modern World* (London, 1924), pp. 24–26, 66–67, 125–126; and *East and West* (London, 1941), pp. 23–26, 36–39, 43–44, 57–62, 68.

[49]Vivekananda, *Collected Works*, vol. 2, pp. 410–411. For Coomaraswamy, see *The Dance of Shiva* (London: Sunrise Turn Press, 1924). For an appreciation of Coomarswamy's message by an influential European thinker, see Romain Rolland's introduction to this edition of the work. Among the many works on Morris and the English arts and crafts revival, two of the best are E. P. Thompson, *William Morris: Romantic to Revolutionary* (London: Lawrence and Wisehart, 1955), and Peter Stansky, *Redesigning the World: William Morris, the 1880s, and the Arts and Crafts* (Princeton, N.J.: Princeton University Press, 1985).

[50]From his *Lectures from Colombo to Almora*, as quoted in V. S. Narvane, *Modern Indian Thought* (Bombay: Asia Publishing House, 1964), p. 106. See also *Collected Works*, vol. 4, pp. 410–411.

Many of Vivekananda's themes had been taken up in the prewar years by two rather different sage-philosophers, Rabindranath Tagore and Aurobindo Ghose, who, like Vivekananda, were both Bengalis who had been extensively exposed to Western learning and culture in their youths. Tagore emerged during the war years as the most eloquent and influential critic of the West, and a gentle advocate of Indian alternatives to remedy the profound distortions and excesses in Western culture that the war had so painfully revealed. But in the prewar decades, knowledge in the West of the concerns of the Hindu revivalists regarding the directions that European civilization was leading the rest of humanity was confined largely to literary and artistic circles, particularly to those, such as the theosophists, that were organized around efforts to acquire and propagate ancient Indian philosophies. Popularists such as Hermann Keyserling had begun in the years before the war to disseminate a rather garbled version of Hinduism to a growing audience in the West. But few Europeans gave credence to the notion that Indian or Chinese learning or values, or those of any other non-Western culture for that matter, might provide meaningful correctives or alternatives to the epistemologies and modes of organization and social interaction dominant in the West. The war changed all of this rather dramatically. Shocked by the self-destructive frenzy that gripped European civilization, Western intellectuals sought answers to what had gone wrong, and some—albeit a small but influential minority—turned to Indian thinkers such as Tagore for tutelage.

In many ways Tagore was the model guru. Born into one of the most intellectually distinguished of modern Bengali families, he was educated privately and consequently allowed to blend Western and Indian learning in his youthful studies. From his father, Devendranath, the founder of the reformist Bramo Samaj, Rabindranath inherited a deep spiritualism and a sense of the social ills that needed to be combated in his colonized homeland. Both concerns were central to his prolific writings that included poems, novels, plays, and essays. Though more of a mystic than an activist, Tagore promoted community development projects on his family estates. And he later founded an experimental school and university at Shantiniketan, his country refuge, which visitors from the West likened to a holyman's ashram. Although Tagore had attracted a number of artistic friends in Europe and America during his travels abroad in the decades before the war, and although his poetry and novels were admired by Yeats, Auden, and other prominent Western authors, he received international recognition only after winning the Nobel Prize for literature in 1913—the first Asian or African author to be so honored.

The timing was fortuitous. When the war broke out in the following year, Tagore was well-positioned to express the dismay and disbelief that so many Western-educated Africans and Asians felt regarding Europe's bitter and seemingly endless intertribal slaughter. He expressed this disenchantment as a loyal subject of King George and the British Empire—which may also help to explain why he received such a careful hearing from educated British, French, and American audiences during and after the war. During the first months of the war, Tagore learned that he had made a bit of money on one of the poems he had sent to his friend William Rothenstein to be published in London. He instructed Rothenstein to use the proceeds to "buy something" for "our" soldiers in France; a gesture he hoped would "remind them of the anxious love of their countrymen in the distant home."[51] But loyalty to the British did not deter Tagore from speaking out against the irrationality and cruelty of the conflict, and using it as the starting point for a wide-ranging critique of the values and institutions of the West. The more perceptive of Tagore's Western readers and the more attentive members of the audiences who attended his well-publicized lectures in Europe, the United States, and Japan could not miss his much more subversive subtexts: Such a civilization was not fit to govern and decide the future of most of the rest of humanity; the colonized peoples must draw on their own cultural resources and take charge of their own destinies.

In his reflections on the meanings of the war Tagore returned again and again to the ways in which it had undermined the civilizing mission ideology that had justified and often determined the course of Western global hegemony. Like Valéry, Hesse, and other critics of the West from within, Tagore explored the ways in which the war had inverted the attributes of the dominant and revealed what the colonizers had trumpeted as unprecedented virtues to be fatal vices. Some of the inversions were incidental, such as Tagore's characterization of the damage to the cathedral town of Rheims as "savage," and others were little more than brief allusions, for example, to science as feminine (the direct antithesis

---

[51]Mary M. Lago, ed., *Imperfect Encounter: Letters of William Rothenstein and Rabindranath Tagore* (New Haven, Conn.: Yale University Press, 1972), pp. 189, 191.

of the masculine metaphors employed in the West[52]) and to Europe as a woman and a child.[53] But many of the inversions were explored in some detail. In a number of his essays and lectures, Tagore scrutinized at some length the colonizers' frequent invocation of material achievement as empirical proof of their racial superiority and fitness to rule less advanced peoples. He charged that the moral and spiritual side of the Europeans' nature had been sapped by their material self-indulgence. As a result, they had lost all sense of restraint (or self-control), as was amply evidenced by the barbaric excesses of trench warfare. Because improvement had come for the Europeans to mean little more than material increase, they could not begin to understand—or teach others—how to lead genuinely fulfilling lives. The much-touted discipline that was thought to be exemplified by their educational systems produced, he averred, little more than dull repetition and stunted minds. The unceasing scramble for profit and material gain that drove Western societies had resulted in a "winning at any cost" mentality that abrogated ethical principles and made a victim of truth, as wartime propaganda had so dramatically demonstrated.

Like Mohandas Gandhi in roughly the same period,[54] Rabindranath Tagore expressed considerable discomfort with railways and other Western devices that advocates of the civilizing mission had celebrated as the key agents of the Europeans' victory over time and space.[55] Forced to rush his meal at a railway restaurant and bewildered by the fast pace at which cinema images flickered across the screen, Tagore concluded that the accelerated pace of living made possible by Western machines contributed to disorientation and constant frustration, to individuals and societies out of sync with the rhythms of nature, each other, and their own bodies. He reversed the familiar, environmental determinists' notion that the fast thinking and acting—hence decisive and aggressive—peoples of the colder northern regions were superior to the languid, congenitally unpunctual peoples of the south. The former, Tagore averred, had lost the capacity for aesthetic appreciation, contemplation, and self-reflection. Without these, they were not fit to shape the future course of human development, much less rule the rest of humankind.

In two allegorical plays written in 1922, Tagore built a more general critique of the science- and industry-dominated societies of the West. The first, titled *Muktadhara*, was translated into French as *La Machine* and published in 1929 with a lengthy introduction, filled with anti-industrial polemic, by Marc Elmer. The second, *Raketh Karabi*, was translated into English as *Red Oleanders*. Both plays detail the sorry plight of small kingdoms that come to be dominated by machines. In each case, the misery and oppression they cause spark revolts aimed at destroying the machines and the evil ministers who direct their operations. Like Vivekananda before him, Tagore warned that science and technology alone were not capable of sustaining civilized life. Like Vivekananda, he cautioned his Indian countrymen against an uncritical adoption of all that was Western, and insisted that the West needed to learn patience and self-restraint from India, to acquire the spirituality that India had historically nurtured and shared with all humankind. With the other major holymen-activists of the Hindu revival, Tagore pitted the oneness and cosmopolitanism of Indian civilization against the arrogance and chauvinism of European nationalism. He argued that the nationalist mode of political organization that the Europeans had long seen as one of the key sources of their global dominion had proved to be the tragic flaw that had sealed their descent into war. Unlike Gandhi, Tagore did not reject the industrial civilization of Europe and North America per se, but concluded that if it was to endure, the West must draw on the learning of the "East," which had so much to share. He urged his countrymen to give generously and to recognize the homage that the Europeans paid to India by turning to it for succor in a time of great crisis.

In sharp contrast to Tagore, Aurobindo Ghose felt no obligation to support the British in the Great War. Educated in the best English-language schools in India and later at St. Paul's School and Cambridge University in Britain, Ghose's life had veered from brilliant student and a stint as a petty bureaucrat in one of India's princely states, to a meteoric career as a revolutionary nationalist that ended with a two-year prison sentence, and finally to an ashram in (French-controlled) Pondicherry in southeast India. Finding refuge in the latter, he began his lifelong quest for

---

[52]As Carolyn Merchant has convincingly demonstrated in *The Death of Nature* (New York: Harper and Row, 1983).

[53]This discussion of Tagore's responses to the war is based upon the following sources: *Diary of a Westward Voyage* (Bombay: Asia Publishing House, 1962), pp. 68–69, 71–74, 96–97; *Letters from Abroad* (Madras: Ganesan, 1924), pp. 18, 56, 66, 83–85, 130; *Personality* (London: Macmillan, 1917), pp. 50, 52, 169–175, 181–182; and *Nationalism* (London, Macmillan, 1917), pp. 33, 37, 44–45, 77, 91–92. Only additional references and quoted portions of Tagore's writings will be individually cited below.

[54]See Gandhi's writings in *Young India* during the war years.

[55]Adas, *Machines as the Measure of Men*, pp. 221–236.

realization and soon established himself as one of India's most prolific philosophers and revered holymen. Aurobindo was convinced that the war would bring an end to European political domination and cultural hegemony throughout Asia.[56] In his view, the conflict had laid bare, for all humanity to see, the moral and intellectual bankruptcy of the West. Fixing on the trope of disease, he depicted Europe as "weak," "dissolute," "delirious," "impotent," and "broken." He believed that the war had dealt a "death blow" to Europe's moral authority, but that its physical capacity to dominate had not yet dissipated. With the alternative for humanity represented by the militarist, materialist West discredited, Aurobindo reasoned, a new world was waiting to be born. And India—with its rich and ancient spiritual legacy—would play a pivotal role in bringing that world into being.

Of all of the Indian critics of the West, Aurobindo was the only one to probe explicitly the capitalist underpinnings of its insatiable drive for power and wealth and the contradictions that had brought on the war and ensuing global crisis. Aurobindo mocked Woodrow Wilson's version of a new world order with its betrayal of wartime promises of self-determination for the colonized peoples. Though he felt that the Bolshevik revolution had the potential to correct some of the worse abuses of capitalism, Aurobindo concluded that socialism alone could not bring about the process of regeneration that humanity needed to escape the *kali yuga* or age of decline and destruction in which it was ensnared. Only Indian spiritualism and a "resurgent Asia" could check socialism's tendency to increase the "mechanical burden of humanity" and usher in a new age of international peace and social harmony.[57]

Although he was soon to become the pivotal leader of India's drive for independence, Mohandas Gandhi was not a major contributor to the cross-cultural discourse on the meanings of World War I for European global dominance. Despite his emergence in the decade before the war as major protest leader in the civil disobedience struggles against the pass laws in South Africa, Gandhi, like Tagore, felt that he must do "his bit" to support the imperial war effort. He served for some months as an ambulance driver, and later sought to assist British efforts to recruit Indians into the military. When the contradiction between his

support of the war and his advocacy of nonviolent resistance was pointed out, Gandhi simply replied that he could not expect to enjoy the benefits of being a citizen of the British Empire without coming to its defense in a time of crisis.[58] But he clearly saw that the war had brutally revealed the limits of Western civilization as a model for the rest of humanity. Even before the war, particularly in a 1909 pamphlet titled "Hind Swaraj," he had begun to dismiss Western industrial civilization in the absolute terms that were characteristic of his youthful thinking on these issues. Like the holymen-activists who had come before him, such as Vivekananda and Aurobindo, and drawing on prominent critics of industrialism and materialism, such as Tolstoy and Thoreau,[59] Gandhi concluded that it was folly to confuse material advance with social or personal progress. But he went beyond his predecessors in detailing alternative modes of production, social organization, and approaches to nature that might replace those associated with the dominant West. The war strengthened his resolve to resist the spread of industrialization in India, and turned him into a staunch advocate of handicraft revival and village-focused community development. Though often neglected in works that focus on his remarkable impact on India's drive for independence, these commitments—fed by his witness of the catastrophic Great War—were central to Gandhi's own sense of mission. As he made clear in an article in *Young India* in 1926, freedom would be illusory if the Indian people merely drove away their British rulers and adopted their fervently nationalistic, industrial civilization wholesale. He urged his countrymen to see that

> India's destiny lies not along the bloody way of the West . . . , but along the bloodless way of peace that comes from a simple and godly life. India is in danger of losing her soul. She cannot lose it and live. She must not, therefore, lazily and helplessly say: "I cannot escape the onrush from the West." She must be strong enough to resist it for her own sake and that of the world.[60]

* * *

---

[56]Aurobindo's responses to the war are set forth in the most detail in his essays on *War and Self-Determination* [Calcutta: Sarojini Chose, n.d. (c. 1924)]; and *After the War* (Pondicherry: Shri Aurobindo Ashram, 1949).

[57]Quoted portions from "After the War," pp. 10, 13.

[58]For Gandhi's activities during the war and justifications of his support for the British, see *The Story of My Experiments with Truth: The Autobiography of Mahatma Gandhi* (Boston: Beacon, 1957 ed.), pp. 346–348; and Louis Fischer, *The Life of Mahatma Gandhi* (New York: Collier, 1962), pp. 133, 164–165, 180, 288–290.

[59]See, for example, his address to the YMCA at Colombo in Ceylon (Sri Lanka), which was reprinted in *Young India*, December 8, 1927.

[60]*Young India*, October 7, 1926.

Because most of sub-Saharan Africa had come under European colonial rule only a matter of decades before 1914, the continent's Western-educated classes were a good deal smaller than their counterparts in India. With important exceptions, such as the Sengalese of the *Quatre Communes*,[61] African professionals and intellectuals tended to have fewer avenues of access to institutions of higher learning in Europe and fewer opportunities for artistic and literary collaboration with their British, French, or German counterparts than the Indians. For these reasons, and because the new Western-educated classes of Africa were fragmented like the patchwork of colonial preserves that the continent had become by the end of the Europeans' late nineteenth-century scramble for territory, African responses to the Great War were initially less focused and forceful than those of Indian thinkers such as Tagore and Aurobindo. Only well over a decade after the conflict had ended did they coalesce in a sustained and cogent interrogation of the imperialist apologetics of the civilizing mission ideology. But the delay in the African response cannot be attributed to an absence of popular discontent or disillusioned intellectuals in either the British or French colonies. In the years following the war, anthropologists serving as colonial administrators and European journalists warned of a "most alarming" loss of confidence in their European overlords on the part of the Africans. They reported widespread bitterness over the post-Versailles denial of promises made to the colonized peoples under the duress of war and a general sense that the mad spectacle of the conflict had disabused the Africans of their prewar assumption that the Europeans were more rational and in control—hence more civilized.[62]

These frustrations and a bitter satire of the Europeans' pretensions to superior civilization were evident in René Maran's novel *Batouala*, which was published in 1921 and was the first novel by an author of African descent to win the prestigious Prix Goncourt in the following year. An *évolué* from Martinique, Maran had been educated from childhood in French schools and had served for decades in the French colonial service. His account of the lives of the people of Ubangui-Shari, the locale in central Africa where the novel takes place, tends to vacillate between highly romanticized vignettes of the lives of African villagers and essentialized depictions of the "natives" as lazy, promiscuous, and fatalistic that are worthy of a European *colon*. But Maran's skillful exposé of the empty promises of civilizing colonizers added an influential African voice to the chorus of dissent that began to drown out Europeans' trumpeting of the global mission in the postwar years.

Although Maran's protagonist, Batouala, admits to an "admiring terror" of the Europeans' technology—including their bicycles and false teeth—he clearly regards them as flawed humans rather than demigods with supernatural powers. In a series of daring inversions, Maran's characters compare their superior bodily hygiene to the sweaty, smelly bodies of the colonizers; their affinity with their natural surroundings to the Europeans "worry about everything which lives, crawls, or moves around [them]"; and their "white" lies to the exploitative falsehoods of the colonizers:

> The "boundjous" (white people) are worth nothing. They don't like us. They came to our land just to suppress us. They treat us like liars! Our lies don't hurt anybody. Yes, at times we elaborate on the truth; that's because truth almost always needs to be embellished; it is because cassava without salt doesn't have any taste.
>
> Them, they lie for nothing. They lie as one breathes, with method and memory. And by their lies they establish their superiority over us.[63]

In the rest of the tale that Maran relates, the vaunted colonizers' mission to civilize is revealed as little more than a string of conscious deceptions and broken promises. In exchange for corvée labor and increasingly heavy taxes, Batouala and his people have been promised "roads, bridges and machines which move by fire on iron rails." But the people of Ubangui-Shari have seen none of these improvements; taxes, Batouala grumbles, have gone only to fill the "pockets of our commandants." The colonizers have done little more than exploit the Africans, whom they contemptuously regard as slaves or beasts of burden. In their arrogant efforts to suppress the exuberant celebrations and sensual pleasures enjoyed by Batouala and his fellow villagers, the Europeans

---

[61]See, Michael Crowder, "Senegal: A Study in French Assimilationist Policy" (London: Methuen, 1967), chapters one and two.

[62]See, for examples, John A. Harris, "The 'New Attitude' of the African," *Fortnightly Review* 108 (1920), pp. 953–960; and G. St. John Orde-Browne, *The Vanishing Tribes of Kenya* (London: Seeley, Service, 1925), p. 271.

[63]*Batouala*, trans. Barbara Beck and Alexandre Mboukou (London: Heinemann, 1973), p. 74.

are destroying the paradisiacal existence the African villagers had once enjoyed.[64]

Maran's essentialized treatment of Africa and Africans is more or less a twentieth-century rendition of the noble savage trope that had long been employed by European travelers and intellectuals. In many ways a testament to the thoroughness of his assimilation to French culture, Maran's depiction of the "natives" of Ubangui-Shari might have been written by a compassionate colonial official who had dabbled in ethnology during his tour of duty. In fact, it is probable that he was influenced by the work of anthropologist colleagues in the colonial civil service, and the pioneering studies of the Sierra Leonean James Africanus Horton and his West Indian-born countryman, Edward Blyden. He would certainly have been familiar with the West African ethnologies compiled by the French anthropologist Maurice Delafosse. Delafosse's works in particular had done much to force a rethinking of Western (and Western-educated African) attitudes toward Africa in the decades before and after the First World War.[65] The revision of earlier assessments of African achievement was also powerfully influenced by the "discovery" of African art in the prewar decades by avant-garde European artists of the stature of Derain, Braque, Matisse, and Picasso. The powerful impact of African masks and sculpture on cubism, abstract expressionism, and other modernist artistic movements bolstered once-despairing African intellectuals in their efforts to fight the racist dismissals of African culture and achievement that had been commonplace in nineteenth-century accounts of the "Dark Continent." The accolades of the European arbiters of high culture energized the delegates who journeyed to Paris in 1919 from all the lands of the slave diaspora and Africa itself for the Second Pan-African Congress, convened by W. E. B. DuBois in 1919.[66] Though most of those attending from colonized areas urged a conciliatory and decidedly moderate approach to the postwar settlement,[67] many took up DuBois's call to combat racism and linked that struggle to the need to remake the image of Africa

that had long been dominant in the West. With its explicit challenges to the assumptions of the civilizing mission ideology and its acclaim by the French literary establishment, Maran's *Batouala* proved a pivotal, if somewhat eccentric, work.

The extent of Maran's influence on the progenitors of the Négritude movement that dominated the thinking of African intellectuals in French-speaking colonies from the late 1930s onward has been a matter of some dispute.[68] But both Maran's efforts to reconstruct pre-colonial life and culture and his challenges to the colonizers' arguments for continuing their domination in Africa, which were grounded in the civilizing mission ideology, figure importantly in the work of the most influential of the Négritude poets. Maran's background as an *évolué* and a scion of the slave diaspora also reflected the convergence of transcontinental influences, energy, and creativity that converged in the Pan-African Congresses in the 1920s and in the Négritude movement in the following decade.

As Léopold Senghor fondly recalls in his reflections on his intellectual development and philosophical concerns,[69] the circle of Négritude writers began to coalesce in Paris in the early 1930s. He credits Aimé Césaire, a poet from Martinique, for the name of the movement, and sees its genesis in the contributions to the short-lived journal *L'Étudiant Noir* and the lively exchanges among the expatriate students and intellectuals drawn to the great universities of Paris from throughout the empire in the interwar decades. Most of the poems that articulated the major themes of Négritude were published after World War II, beginning with the seminal 1948 *Anthologie de la nouvelle poésie négre et malgache de langue française*. But a number of works that were privately circulated in the late 1930s and Aimé Césaire's *Cahier d'un retour au pays natal*, first published in fragments in 1938, suggest the continuing power of recollections of the trauma of the First World War in the African awakening.

In Senghor's evocative "Neige sur Paris,"[70] the poet awakes to find the city covered with newly fallen snow. Though encouraged by the thought that the pure white snow might help to soften the deep divisions that threaten to plunge Europe once again

[64]Ibid., pp. 29–31, 47–50, 75–76.

[65]See, for example, his monumental survey *The Negroes of Africa*, first published in 1921. On his impact, see Gérard Leclerc, *Anthropologie et colonialisme: Essai sur L'histoire de l'africainisme* (Paris, 1972), pp. 43–52; and Girardet, *L'Idée coloniale*, pp. 158–164.

[66]On these connections, see S. Okechukwu Megu, *Léopold Sedar Senghor et la defense et illustration de la civilisation noire* (Paris, 1968), especially pp. 32–36.

[67]Crowder, *Colonial West Africa*, pp. 408–412.

[68]Dorothy S. Blair, *African Literature in French* (Cambridge: Cambridge University Press, 1976), pp. 18–20.

[69]See *Ce que je crois* (Paris: 1988), pp. 136–152.

[70]Quoted portions are taken from the superb translation of "Snow upon Paris," by John Reed and Clive Wake in Senghor, *Selected Poems* (Oxford: 1964).

into war and heal the wounds of a Spain already "torn apart" by civil war, Senghor conjures up the "white hands" that conquered Africa, enslaved its peoples, and cut down its forests for "railway sleepers." He mocks the mission of the colonizers as indifferent to the destruction of the great forests as they are to the suffering they have inflicted on the African people:

> They cut down the forests of Africa to
> save Civilization, for there was a shortage
> of human raw-material.

And he laments the betrayal of his people by those posing as peacemakers, suggesting the ignoble machinations of the Western leaders at Versailles:

> Lord, I know I will not bring out my store
> of hatred against the diplomats who flash
> their long teeth
>   And tomorrow will barter black flesh.

In "For Koras and Balafong," which he dedicated to René Maran, Senghor flees from the factory chimneys and violent conflict of Europe to the refuge of his childhood home, the land of the Serer, south of Dakar along the coast of Senegal. Throughout the poem he celebrates the music and dance, the sensuality and beauty of his people and their communion with the natural world—all central themes in the corpus of Négritude writings. But like Maran, he turns these into inversions of the European societies from which he has fled and that have been defiled by the violence of the Great War. His journey to the land of his ancestors is

> . . . guided through thorns and signs by
> Verdun, yes Verdun the dog that kept
> guard over the innocence of Europe.

In his travels, Senghor passes the Somme, the Seine, the Rhine, and the "savage Slav rivers" all "red under the Archangel's sword." Amid the rhythmic sounds of African celebration, he hears:

> Like the summons to judgment, the burst
> of the trumpet over the snowy graveyards
> of Europe.

He implores the earth of his desert land to wash him clean "from all contagions of civilized man," and prays to the black African night to deliver him from

> . . . arguments and sophistries of salons,
> from pirouetting pretexts, from calculated
> hatred and humane butchery.

These final passages recall the powerful inversions that provide some of the most memorable passages in the verse of Senghor's collaborators and cofounders of the Négritude movement in the 1930s. There is Léon Damas's iconoclastic rejection of the costume of his assimilated self:

> I feel ridiculous
> in their shoes
> in their evening suits,
> in their starched shirts,
> in their hard collars
> in their monocles
> in their bowler hats.[71]

And in Aimé Césaire's *Cahier d'un retour au pays natal*, perhaps the most stirring of the Negritude writers' defiant mockeries of the standards by which the Europeans had for centuries disparaged their people and justified their dominance over them:

> Heia [praise] for those who have never
>   invented anything
> those who never explored anything
> those who never tamed anything
> those who give themselves up to the
>   essence of all things
> ignorant of surfaces but struck by the
>   movement of all things
> free of the desire to tame but familiar with
>   the play of the world.[72]

* * *

The discourse centered on the meanings of the Great War for the future of the science- and technology-oriented civilization pioneered in the West was, I believe, the first genuinely global intellectual exchange. Though the African slave trade had prompted intellectual responses from throughout the Atlantic basin, the post–World War I discourse was the product of the interchange between thinkers from the Americas, Europe, Africa, and Asia. At one level, the postwar discourse became a site for the contestation of the presuppositions of the civilizing mission ideology that had undergirded the West's global hegemony. At another, it raised fundamental questions about the effects of industrialization in the West itself as well as the ways in which that process was being transferred to colonized areas in Asia and Africa. For nearly two decades, philosophers, social commentators, and political activists scrutinized the

---

[71]From "Solde," published in *Pigments* (Paris: 1962) and quoted in Abiola Irele, "Négritude or Black Cultural Nationalism," *The journal of Modern African Studies* 3, no. 3 (1965): 503.

[72]From *Cahier d'un retour au pays natal*, translated as *Return to My Native Land* by John Berger and Anna Bostock (Harmondsworth, 1969), p. 75.

ends to which scientific learning and technological innovation had been put since the industrial watershed. Their profound doubts about the long-term effects of the process itself on human development would not be matched until the rise of the global environmentalist discourse that began in the 1960s and continues to the present.

Although unprecedented in its global dimensions, in the colonized areas of Africa and Asia postwar challenges to the industrial order and the civilizing mission ideology were confined largely to the Western-educated elite. Colonized intellectuals, with such notable (and partial) exceptions as Tagore and Aurobindo, critiqued the hegemonic assumptions of the West in European languages for audiences that consisted largely of Western-educated professionals, politicians, and academics.[73] Even those who wrote in Asian or African languages were also compelled to publish and speak in "strong"[74] languages such as English or French if they wished to participate in the postwar discourse. And as Ngugi wa Thiong'o reminds us, the cage of language set the limits and had much to do with fixing the agenda of that interchange.[75] Not only did Indian and African intellectuals draw on the arguments of Western thinkers such as Tolstoy, Bergson, Thoreau, and Valéry, but the issues they addressed were largely defined by European and, to a lesser extent, American participants in the global discourse. In this sense, the postwar Indian and African assault on the civilizing mission was as reactive as Antenor Firmin's nineteenth-century refutations of "scientific" proofs for African racial inferiority or Edward Blyden's defense of African culture. Even the essentialized stress on the spirituality of Indian civilization or the naturalness of African culture was grounded in tropes employed for centuries by European travelers, novelists, and Orientalists. As the reception of Maran and Tagore (or Vivekananda before them and Senghor afterward) also suggests, Robert Hughes's "cultural cringe"[76] was very much in evidence. European approbation had much to do with the hearing that Asian or African thinkers received not only in the West, but among the Western-educated, elite circles they addressed in colonial settings.

Although the terms of the discourse between colonizer and colonized remained the same in many respects, the Great War had done much to alter its tone and meaning for Indian and African participants. The crisis of the West and the appalling flaws in Western civilization that it revealed did much to break the psychological bondage of the colonized elite, which, as Ashis Nandy has argued,[77] was at once the most insidious and demoralizing of the colonizers' hegemonic devices. World War I provided myriad openings for the reassertion—often in the guise of reinvention—of colonized cultures that were dramatically manifested in the inversions in the postwar writings of Indian and African thinkers of the attributes valorized by the prewar champions of the civilizing mission. The crisis of the Great War gave credence to Gandhi's contention that the path for humanity cleared by the industrial West was neither morally or socially enabling nor ultimately sustainable. And though the circle in which the postwar discourse unfolded was initially small, in the following decades it contributed much to the counterhegemonic ideas of the Western-educated intellectuals of Asia and Africa, ideas that were taken up by the peasants and urban laborers who joined them in the revolt against the European colonial order.

---

[73]These patterns are stressed by Paul Sorum in his treatment of the origins of Négritude in *Intellectuals and Decolonization in France* (Chapel Hill, N.C. University of North Carolina Press, 1977), pp. 213–214.

[74]For a discussion of this useful concept, see Talal Asad, "Two European Images of Non-European Rule," in Asad, *Anthropology and the Colonial Encounter* (New York: Humanities Press, 1973), pp. 103–118.

[75]*Decolonising the Mind: The Politics of Language in African Literature* (London: Heinemann, 1986).

[76]"The Decline of the City of Mahagonny," *The New Republic* (28 June 1990), pp. 27–28.

[77]*The Intimate Enemy: Loss and Recovery of Self under Colonialism* (Delhi: Oxford University Press, 1983), esp. pp. xi–xiii.

Why in America?

Nathan Rosenberg

Let me try to clarify just what question it is I am trying to answer. I am not going to discuss the high overall rate of economic growth in the United States after 1800, that is, the rate of increase of per capita income. In fact, the rate of growth was not particularly high by comparison with later experience, although many scholars believe that it accelerated during the 1830s. If we extend our time horizon well back into the colonial period, we find that the rate of growth of per capita income between 1710 and 1840 did not exceed one-half of 1 percent per year.[1] Nor are we considering the overall rate of technological innovativeness which characterized the American economy, although that rate was doubtless very high. In spite of continued obeisance to the idea of Yankee ingenuity, it cannot be overstressed that America in the first half of the nineteenth century was still primarily a borrower of European technology. Although the rate of technical change was indeed high, most of the new technologies were not American inventions. Americans were rapid adopters of foreign technologies when it suited their economic needs, and they were also skillful in modifying someone else's technology to make it more suitable to their needs. They made abundant albeit selective use of European innovations in power generation, transportation, and metallurgy.[2] It is worth noting, moreover, that high rates of technological improvement were not always a part of the American experience. Such improvements seem to have played an insignificant role in the American economy in the eighteenth century. The latest scholarship on the colonial period

suggests that the rate of inventive activity was very low, and that the slow but steady growth in productivity was dependent upon a cumulatively powerful combination of forces centering upon improvements in the organization of industries and the more effective functioning of markets and ancillary institutions.[3]

The question I propose to address, then, is not the high rate of technological change in America when that rate indeed became high in the nineteenth century. Rather, the question is one of direction and character. By the time of the Crystal Palace Exhibition in London in 1851, there was something so distinctive about many American goods that the British coined the phrase "the American system of manufactures." The expression was used to describe goods which were (1) produced by specialized machines, (2) highly standardized, and (3) made up of interchangeable component parts.[4] I propose to address myself to the question of causality. Why did this system emerge first in America? By midcentury it was apparent that there was an important class of goods which was being produced in a distinctive way in America.[5]

Originally published in Otto Mayr and Robert C. Post (eds.), *Yankee Enterprise, the Rise, of the American System of Manufactures*, Smithsonian Institution Press, Washington (DC), 1981. The focus here on demand- and supply-side factors to explain the particular direction of American technological development complements the discussion of technological change in the energy and forest products sectors in chapters 9 and 12, respectively.

[1] Robert Gallman, "The Record of American Economic Growth," in Lance Davis, Richard A. Easterlin, and William N. Parker (eds.), *American Economic Growth,* Harper & Row, New York, 1972, p. 22.
[2] Nathan Rosenberg, *Technology and American Economic Growth,* Harper & Row, New York, 1972, chapter 3.

[3] See James Shepherd and Gary Walton, *Shipping, Maritime Trade and the Economic Development of Colonial North America*, Cambridge University Press, Cambridge, 1972.
[4] The degree or ease of interchangeability varied between products and over time. The degree of precision required for interchangeability varied considerably from one product to another. As one nineteenth-century commentator observed: "The more prevalent modern idea of the interchangeable in mechanism supposes a super-refinement of accuracy of outline and general proportions that is not always necessary or even desirable. It would be a criminal waste of time and substance to fit a harrow tooth with mathematical accuracy, but yet any harrow tooth should have a practical interchangeable relation to all harrows for which it is designed. The instructive rewards of folly would certainly overtake him who should attempt to make ploughshares and coulters with radical exactness; nevertheless, these essential parts of ploughs should be interchangeable among all ploughs to which they are adapted." W. F. Durfee, "The History of Interchangeable Construction," *American Society of Mechanical Engineers Transactions*, 14 (1893), p. 1228.
[5] The role of the manufacturing sector had increased very sharply during the decade of the 1840s. According to Gallman's estimates the contribution of manufacturing to national commodity output grew from 17 percent to 30 percent during the decade 1839–1849. Robert Gallman, "Commodity Output, 1839–1899," in the Conference on Income and Wealth's *Trends in the American Economy in the 19th Century*, Princeton University Press, 1960, p. 26.

It is important, however, that the distinctiveness should not be overstated, nor should it be thought that America had also assumed a role of broad technological leadership across the whole manufacturing sector. That was far from the case. Indeed, even in gunmaking, commonly regarded as the main triumph of the American system of manufactures, American barrelmaking may still have lagged well behind Britain at midcentury.[6]

Now, if you call upon an economist to try to explain a particular phenomenon, such as why the technology which we have come to call the American system of manufactures first arose in the United States and not elsewhere, you are not entitled to express surprise over an explanation couched in terms of supply-and-demand analysis. That is what I propose to give, and I do so without apology. Presenting an historical explanation in terms of supply-and-demand forces does not *necessarily* mean that I am committed to some form of economic determinism. Supply and demand are nothing more than convenient conceptual categories which enable us to think about an event or process in a more systematic way and to organize our analysis in a way which lends itself more readily to an understanding of cause-effect relationships. The forces *underlying* these economic categories may be social, geographic, technological, or ideological, but such forces have economic content and effects, and I am particularly interested in focusing upon the economic implications or consequences of these forces. There is a difference, which I trust that I need not belabor, between *translating* certain phenomena into their economic consequences and insisting that economic variables are all powerful.

I have already suggested a distinction between the rate and direction of technological progress. However, these two factors are not as easily separable as may be thought at first glance. The point is that differences in the resource endowment and demand conditions of an economy go a long way toward determining what kinds of inventions—with what kinds of product characteristics and factor-saving biases—it will be profitable to develop and exploit. To the extent that technological progress is responsive to economic forces—and I would agree that it is highly responsive and needs to be understood in these terms—the inventions actually brought forth in a particular country will tend to be compatible with those special needs. We need

to distinguish here between invention and adoption. It is obvious that only those inventions which are compatible with a country's needs will be widely adopted. I am making here the stronger assertion that a high proportion of the *inventions made* will also reflect the peculiar needs of the economic environment in which they are developed.

If, at any given time and state of technical knowledge, certain *kinds* of inventions are easier to create, for whatever reason, economies where such inventions are also economically appropriate would be expected (*ceteris paribus*, as economists are fond of saying) to find it easier to generate new and improved technologies. If economic forces tend to push economies with different factor endowments in different directions in the attainment of technological progress, and if the technical problems which have to be overcome are more difficult in some directions than others, then the rate of technological progress may not be independent of the direction in which the economy is being pushed.

I have developed this argument because I believe it has merit. The American experience in the nineteenth century involved an economy with resource endowment and demand conditions that pushed it in a direction somewhat different from the one that prevailed at the time for the economies of western Europe. This turned out to be a direction in which the inventive payoff was rather high. You may, if you wish, take this as a confirmation of the belief that a benevolent Providence watched over the affairs of America in the nineteenth century. Or you may simply regard it as luck or as the consequence of a particular resource endowment.

Whatever the label, I want to argue that what we call the American system of manufactures was a part of a larger process of economic adaptation, or even, if you prefer, economic evolution. American economic and social conditions in the nineteenth century pushed the search for new technologies in a specific direction. It was a direction which, both for social and economic reasons, was less appropriate in Europe. And, as it turned out, it was a direction in which there happened to exist a rich layer of inventive possibilities. In this sense, I will argue, the American system of manufactures was a species of a larger genus and, in this sense also, the high *rate* of technical progress was not entirely independent of the *direction* which that progress took.

The notion that the American system of manufactures requires explanation is reinforced by the facts that (1) the United States was, in the first half of the nineteenth century, still largely a technical borrower

---

[6]Paul Uselding, "Henry Burden and the Question of Anglo-American Technological Transfer in the Nineteenth Century," *Journal of Economic History*, 30 (1970), pp. 312–337.

rather than a pioneer, and (2) technical innovations incorporating some of the elements of the system had been introduced earlier in Britain, mainly in the Portsmouth naval dockyards, but were not widely adopted elsewhere in that country.

Demand

What were the factors on the demand side that were so conducive in early-nineteenth-century America to the emergence of the American system of manufactures? There was a very rapid rate of growth in the aggregate demand for certain classes of commodities, which encouraged entrepreneurs to devise or adopt new production processes whose profitability required very long production runs. Adam Smith long ago taught that "the division of labor is limited by the extent of the market." The proposition applies to the use of specialized machinery as well as to specialized men. The fixed costs embodied in special-purpose machinery are warranted only if they can be distributed over a large volume of output. Therefore the confident expectation of rapidly growing markets is an extremely powerful inducement. Furthermore, the use of highly specialized machinery, as opposed to machinery which has a greater general-purpose capability, is contingent upon expectations concerning the *composition* of demand—specifically, that there will be strict and well-defined limits to permissible variations. Both of these conditions—rapid growth in demand and circumstances conducive to a high degree of product standardization—were amply fulfilled in early-nineteenth-century America.

What forces were responsible for these conditions? Probably the most pervasive force of all was the extremely rapid rate of population growth, primarily from natural increase but with immigration assuming a role of some significance in the 1840s. Between 1790 and 1860, the American population grew at a rate of nearly 3 percent per year (immigration included), a rate more than twice as high as that achieved by any European country even for much shorter periods. American fertility levels at the beginning of the nineteenth century were close to fifty per thousand—again, a rate far higher than European levels—and crude death rates were somewhere around twenty per thousand. The factors underlying this rapid expansion in population are many, but most important was probably the abundant supply of high-quality land in a society which was still predominantly agricultural. The abundance of good land, and the concomitant optimism about the

future, were conducive to high fertility levels as well as an inducement to immigration.[7]

Although scholars differ concerning the long-term trend in the rate of growth of per capita income during the nineteenth century, there can be no doubt that the strong rate of population growth resulted in an exceedingly rapid rate of market growth. As early as 1840 the net national product of the United States was two-thirds or more of the size of Great Britain's.[8] In subsequent years America's net national product grew far more rapidly than Britain's.

Rapid population growth resulted in a very high rate of new household formation and therefore a rapid rate of growth in the demand for a wide range of manufactured commodities. This growth in market size was reinforced by other developments. Improvements in transportation, especially after the beginning of the canal-building period in 1815 and the beginning of railroad construction in the 1830s, served to link industrial centers with remote potential markets that previously had been largely self-sufficient. Thus, in many cases industrial producers and consumers were in a position to establish economically feasible market relationships with one another for the first time.

Many of the distinctive characteristics of the American market can be properly appreciated only in the context of its predominantly rural and agricultural character. Over 80 percent of the American labor force was in agriculture in 1810 and this remained true for well over 60 percent in 1840.[9] The predominance of the agricultural sector in the nineteenth century, and the abundant endowment of natural resources more generally, served to shape the American environment in numerous ways which made it more receptive to the American system of manufactures. First of all, we have already noted how the rich and abundant supply of agricultural land contributed to high fertility levels and therefore to a rapid growth in the market. In addition, the abundance of land conferred another advantage upon the American economy, for it meant that food prices were relatively low. As a result, for

---

[7]For an excellent, concise treatment of long-term demographic trends in America, see Richard A. Easterlin, "The American Population," chapter 5 in Davis, Easterlin, and Parker (eds.), *American Economic Growth*.
[8]Gallman, "Commodity Output, 1839–1899," p. 33.
[9]Lebergott's estimates of the percentage of the labor force in agriculture are as follows: 1810—80.9%; 1820—78.8%; 1830—68.8%; 1840—63.1%; 1850—54.8%; 1860—52.9%; 1870—52.5%; 1880—51.3%. Stanley Lebergott, "Labor Force and Employment, 1800–1960," in *Output, Employment and Productivity in the U.S. after 1800*, ed. Dorothy Brady, Studies in Income and Wealth, vol. XXX, National Bureau of Economic Research, New York, 1966, p. 119.

any given income level or family size, there was a larger margin left over for the purchase of nonfood products, including those produced under the American system of manufactures. Thus the structure of relative prices in the United States, as a result of its resource endowment, was distinctly more favorable than in Europe for the emergence of manufactured goods. Both American farmers and urban industrial workers spent less of their income on food than did their European counterparts. Empirical evidence in support of this hypothesis has been carefully marshalled and analyzed by Albert Fishlow, relying upon data collected for the period 1888–91.[10] Since American food prices were, relatively, lower earlier in the nineteenth century, the case for the more favorable American conditions is even stronger.[11]

Yet another advantage conferred by America's natural-resource abundance was the easy accessibility to land ownership and the relatively egalitarian social structure which consequently emerged—with the major exception, of course, of the southern plantation system. Outside of the south there had emerged in colonial times and later a rural society which was far less hierarchical, and with property ownership far less concentrated, than in Europe. American society included a large component of middle-class farmers and craftsmen. This was in sharp contrast with European societies in which poor peasants and farm laborers, with very little income beyond subsistence needs, constituted a large fraction of the total population, while at the top of the social pyramid was a small group of large landowners whose expenditure patterns made little contribution to the emergence of a sector producing standardized manufactured goods.

In America, to a degree which is now largely lost from the collective memory of a highly urbanized population, markets in the first half of the nineteenth century were dominated by the tastes and require-

ments of middle-class rural households.[12] In such households, often living under frontier or at least isolated conditions, there was little concern—indeed there was little opportunity—for ostentation or conspicuous consumption. Out of these social conditions emerged a relative simplicity of taste and a stress upon functionalism in design and structure. Rural isolation also strongly favored reliability of performance and ease of repair in case of breakdown. Such characteristics were particularly important for agricultural machinery and firearms. Simplicity of design and uniformity of components meant that a farmer could repair a broken plough in the field himself without having to depend upon a skilled repairman or distant repair shop. Thus, out of the social and geographic conditions of land-abundant America emerged a set of tastes and preferences highly congenial to a technology capable of producing large quantities of standardized, low-priced goods. These circumstances even left their indelible imprint on the American automobile in the early years of the twentieth century. The Ford Model T was designed in a manner which strongly resembled the horse and buggy, and the primary buyers were farmers for whom a cheap car offered a unique opportunity for overcoming rural isolation.

Supply

The point has already been made that in the early nineteenth century America was a large-scale borrower of European technology. But the point also has been made that America's factor proportions, especially its rich abundance of natural resources, differed substantially from Europe's. These two propositions have an interesting implication that provides a useful way of entering into an examination of supply-side influences upon the emergence of the American system of manufactures. For, if factor proportions between the two continents were sufficiently disparate, it would follow that the technology devised to accommodate European needs did not always

---

[10]Albert Fishlow, "Comparative Consumption Patterns, the Extent of the Market, and Alternative Development Strategies," in *Micro-Aspects of Development*, ed. Eliezer Ayal, Praeger Publishers, New York, 1973.
[11]"The very fact that food prices should have been relatively lower earlier would have given them more leverage at precisely the right moment—when per capita incomes were smaller and the percentages allocated to food higher. What is significant is that they did not rise rapidly thereafter and impede the extension of the market. American food prices in the 1830s and 1840s were not only lower relative to those of a nonfood composite, but also compared to British foodstuffs. That is, British food prices fell more rapidly from 1831–50 to 1866–90 than did American, and we have already seen that American foods were absolutely cheaper at that later date. Hence, the comparative advantage the United States enjoyed was greater in the earlier period." *Ibid.*, pp. 77–78.

[12]"The share in aggregate expenditures of different groups in the population in the nineteenth century and earlier can be estimated only roughly, for want of statistical information representative of all sections of the country. Even a crude statistical picture, however, can indicate how important rural households were in the market for manufactured goods and services. In the 1830s, about 66 percent of all households were on farms, 23 percent in villages, and 11 percent in cities. On the supposition that farm families on the average were able to spend $100, village families $200, and city families $300 over and above their outlays for housing and fuel, rural families (farm and village) would have accounted for 76 percent of all such expenditures." D. Brady in Davis, Easterlin, and Parker (eds.), *American Economic Growth*, p. 62.

constitute, in present-day terminology, an "appropriate technology" in America. We now need to consider, therefore, how differences in the supply of available resources pushed Americans in a direction that helps to account for the country's unique technological contributions.

An abundance of natural resources means, in economic terms, that it is rational to employ methods of production which are resource intensive. Much of American inventive activity in the first half of the nineteenth century aimed at substituting abundant natural resources for scarcer labor and capital—although this aim is often not apparent from a present-day examination of the hardware devised for the purpose. America's early world leadership in the development of specialized woodworking machinery—machines for sawing, planing, mortising, tenoning, shaping, and boring—was a consequence of an immense abundance of forest products. Although these machines were wasteful of wood, that was of little consequence in a country where wood was very cheap. The substitution of abundant wood for scarce labor was, in fact, highly rational.[13]

Similarly, the immense available supply of potential farmland was not conducive to inventive activity that would maximize output per acre (as would be sensible in a land-scarce economy like Japan's), but rather to activity that would maximize the amount of land that could be cultivated by a single worker. This is precisely the thrust of nineteenth-century American mechanical innovation in agriculture, based upon animal power which supplied the traction for the steel plough (as well as Jethro Wood's earlier plough which already had replaceable cast-iron parts); the cultivator which replaced the hand-operated hoe in the corn and cotton fields; and the awesome reaper which swept away what once had been a basic constraint upon grain cultivation—fluctuations in labor requirements that reached a sharp peak during the brief harvesting season. Later, output per worker was farther increased by binders and threshing machines and, eventually, by combine harvesters. In corn cultivation, the corn shelter and the corn picker were particularly valuable for their laborsaving characteristics.

In the gunmaking trade, usually regarded as the *locus classicus* of the distinctly American mass-production technology, some of the most original contributions were in the development and elaboration of a set of lathes for shaping the gunstock with a minimum amount of labor. The fact that these laborsaving machines were initially highly wasteful of wood—as compared to the shaping of a gunstock by a Birmingham craftsman—was a trivial consideration in a wood-abundant economy. Blanchard lathes were widely adopted in America.

What is the relevance of these woodworking and agricultural machines to the emergence, by midcentury, of the American system of manufactures? The essential point is that resource abundance provided an incentive in America to explore the possibilities of certain new machine technologies earlier and more deeply than in Europe.[14] The fact that these machines, particularly in their early stages of development, were not only laborsaving but also resource intensive, was not the economic deterrent that it was in Britain or France. Thus, in a wide range of manufacturing activities, Americans had a strong incentive to develop and adopt new technologies which, in effect, traded off abundant natural-resource inputs for labor, as well as machines which were wasteful of natural resources yet could be constructed more cheaply. But the pressures generated by America's unique resource endowment led to exploratory activities and to eventual learning experiences the outcome of which cannot be adequately summarized merely in terms of factor-saving of factor-using biases. For they also led to new patterns of specialization and division of labor between firms—especially between the producers and users of capital goods—as a result of which the American economy developed a degree of technological dynamism and creativity greater than existed in other industrial economies in the second half of the nineteenth century.

I believe that this technological dynamism was due in large measure to the unique role played by the capital-goods industries in the American industrialization process and the especially favorable conditions under which they operated. For these capital-goods industries—I refer here primarily to those involved in the forming and shaping of metals—became learning centers where metalworking skills were acquired and developed, and from which such skills were eventually transferred to the production of a sequence of new standardized products—interchangeable firearms, clocks and watches,

---

[13]For a more detailed treatment, see Nathan Rosenberg, *Perspectives on Technology*, Cambridge, 1976, chapter 2, "America's Rise to Woodworking Leadership."

[14]The rest of this paragraph draws upon Nathan Rosenberg, "American Technology: Imported or Indigenous?" *American Economic Review Papers and Proceedings*, Feb. 1977.

agricultural machinery, hardware, sewing machines, typewriters, office machinery, bicycles, automobiles. A key feature of the story of American industrialization is that it involved the application of certain basically similar production techniques to a growing range of manufactures. Moreover, the technological knowledge and competence that gradually accumulated in this sector were directly applicable to generating cost reduction in the production of capital goods themselves. A newly designed turret lathe or universal milling machine, or a new steel alloy permitting a lathe to remove metal at higher speeds—each of these innovations not only resulted in better machines but also reduced the cost of producing the machines in the first place.

Thus, although the initial shift to the capital-using end of the spectrum was generated by the unique pattern of American resource scarcities, it is by no means obvious that the final outcome of this process was in terms of factor biases. For the capital-using path was also a path which, *eventually*, generated a much-increased capacity for capital-saving innovations. The attempt to deal with labor scarcity in a regime of natural-resource abundance pushed us quickly in a direction in which there turned out to be rich inventive possibilities. In turn, the skills acquired in a more capital-abundant society with an effectively organized capital-goods sector provided the basis—in terms of knowledge and engineering skills and expertise—for innovations which were capital saving as well as labor saving. Indeed, most new products, after their technical characteristics became sufficiently stabilized, have passed through such a cost-reducing stage during which capital-goods producers accommodated themselves more efficiently to the large-quantity production of the new product. American industry seems to have particularly excelled at these activities.

Aside from the highly visible major inventions, capital-intensive technologies have routinely offered extensive opportunities for improvements in productivity which seem to have had no equivalent at the labor-intensive end of the spectrum. Knowledge of mechanical engineering, metallurgy, and, perhaps most important of all, the kind of knowledge which comes from day-to-day contact with machine technology, provide innumerable opportunities for small improvements—minor modifications, adaptation to some special-purpose use, design alterations, substitution of a superior or cheaper material—the cumulative effects of which have, historically, been very great.

It is important that these developments be seen in their actual historical sequence. America began the growth of her capital-goods sector not only with a strong preoccupation with standardization and interchangeability, but also with some early experience with techniques as well as a market which readily accepted standardized products. This shaped the eventual outcome of the industrialization process in some decisive ways. The acceptance of standardization and interchangeability vastly simplified the production problems confronting the makers of machinery and provided the technical basis for cost reductions in machinemaking. At the same time it provided the conditions which encouraged the emergence of highly specialized machine producers as well as the transfer of specialized technical skills from one industrial application to another. Indeed, America's most significant contributions to machine-tool design and operation and related processes—profile lathes, turret lathes, milling machines, die-forging techniques, drilling and filing jigs, taps and gauges—were associated with specialized, high-speed machinery devoted to the production of standardized components of complex products.

Thus, as a result of certain initial conditions, America's industrialization in the early nineteenth century proceeded along one out of a variety of possible paths. It was a path dictated by peculiar resource conditions and also a path rich in inventive possibilities. Movement along this path generated a dynamic interaction of forces, a dialectical process in which cause and effect become exceedingly difficult to disentangle. While greater homogeneity of tastes was originally conducive to the introduction of goods produced according to the American system of manufactures, it is also true that, once this technology began to spread, it in turn shaped and influenced tastes in the direction of simplicity and functionality.[15] Furthermore, although American factor endowment pushed in the direction of mechanization, the experience with mechanization *in itself* brought about an improvement in inventive ability and its more rapid diffusion. This was an interactive process, whereby the successful application of mechanical skills in one sector improved the likelihood of their successful application in other sectors.

What, finally, can we add by way of explaining why Britain—which was far more advanced industrially than the United States in the early

---

[15]Dorothy Brady, "Relative Prices in the Nineteenth Century," *Journal of Economic History*, 24 (1964), pp. 147–148.

nineteenth century—did not go further than it did in developing the distinctive features of the American system of manufactures? Much of the answer is already implicit in the preceding discussion of Anglo-American differences in conditions of demand-and-resource endowment. There is, however, an additional factor which deserves mention, namely, the persistence and continuing powerful influence of certain traditional values and attitudes which had their roots in a preindustrial craft society. Such values and attitudes had, of course, played a much more significant role in shaping the outlook of workers and engineers in Britain than in the United States. Strong craft traditions, with their emphasis upon pride in workmanship, individuality, and high standards of product quality, were often inimical to standardization and the alterations in final product design which were essential to low-cost, high-volume production. To some extent the persistence of traditional craft attitudes in Britain reflected the demand-side phenomena to which we have already referred, as well as other market peculiarities. But beyond that there is evidence from many industries—firearms, automobiles, locomotives, a wide range of production machinery, clocks and watches—suggesting a preoccupation with technical perfection beyond what could be justified by narrowly utilitarian considerations.[16] And perhaps this observation suggests what should be an appropriate final comment upon why this particular system first emerged in America. For the American system of manufactures was, above all, a totally unsentimental approach to the productive process in industry, one in which purely commercial considerations prevailed. The British long consoled themselves with the belief that standardization and mass-production techniques inevitably resulted in an inferior product. Unhappily, from their point of view, they persisted in this belief long after it ceased to contain even an element of truth.

_____

[16]S.B. Saul, "The Market and the Development of the Mechanical Engineering Industries in Britain, 1860–1914," _Economic History Review_, 41 (1967); S.B. Saul, "The Engineering Industry," chapter 7 in _The Development of British Industry and Foreign Competition, 1875–1914_, ed. D. Aldcroft, University of Toronto Press, Toronto, 1968; S.B. Saul, "The Motor Industry in Britain to 1914," _Business History_, 5 (1962); R.A. Church, "Nineteenth Century Clock Technology in Britain, the U.S., and Switzerland," _Economic History Review_, 49 (1975); and Nathan Rosenberg, introduction to _The American System of Manufactures_, Edinburgh University Press, Edinburgh, 1969.

The System Must Be First

Thomas P. Hughes

[W]e reflect too little about the influences and patterns of a world organized into great technological systems. Usually we mistakenly associate modern technology not with systems but with such objects as the electric light, radio and television, the airplane, the automobile, the computer, and nuclear missiles. To associate modern technology solely with individual machines and devices is to overlook deeper currents of modern technology that gathered strength and direction during the half-century after Thomas Edison established his invention factory at Menlo Park. Today machines such as the automobile and the airplane are omnipresent. Because they are mechanical and physical, they are not too difficult to comprehend. Machines like these, however, are usually merely components in highly organized and controlled technological systems. Such systems are difficult to comprehend, because they also include complex components, such as people and organizations, and because they often consist of physical components, such as the chemical and electrical, other than the mechanical. Large systems—energy, production, communication, and transportation—compose the essence of modern technology. . . .

. . . In seeking the creators of modern industrial America, we must consider the system builders as well as the independent inventors and the industrial scientists.

Henry Ford's production system remains the best known of the large technological systems maturing in the interwar years. Contemporaries then usually perceived it as a mechanical production system with machine tools and assembly lines. But Ford's system also included blast furnaces to make iron, railroads to convey raw materials, mines from which these came, highly organized factories functioning as if they were a single machine, and highly developed financial, managerial, labor, and sales organizations. . . .

Electrical light-and-power systems, such as those managed and financed by the system-building utility magnate Samuel Insull of Chicago, incorporated not only dynamos, incandescent lamps, and transmission lines, but hydroelectric dams, control or load-dispatching centers, utility companies, consulting-engineering firms, and brokerage houses, as well. When Ford placed a mechanical assembly line in motion, the public was greatly impressed, but electrical systems transmitted their production units too rapidly to perceive: 186,000 miles per second, the speed of light. . . .

Of the American system builders, none took on a more difficult and controversial task than Frederick Winslow Taylor. Ford directed his ordering-and-controlling drive primarily to production machines; Insull focused his on ensuring the large and steady flow of electrical power; Taylor tried to systematize workers as if they were components of machines. Ford's image was of a factory functioning as a machine; Insull envisaged a network or circuit of interacting electrical and organizational components; and Taylor imagined a machine in which the mechanical and human parts were virtually indistinguishable. Idealistic, even eccentric, in his commitment to the proposition that efficiency would benefit all Americans, Taylor proved naïve in his judgments about complex human values and motives. In the history of Taylorism we find an early and highly significant case of people reacting against the system builders and their production systems, a reaction widespread today among those who fear being co-opted by "the system." . . .

Taylor was not the first to advocate a so-called scientific approach to management, but the enthusiasm and dedication, bordering on obsession, with which he gave himself to spreading his views on management, his forceful personality, and his highly unusual and erratic career filled with failure as well as success have left a strong, indelible impression on his contemporaries and succeeding generations. More than a half-century after his death, many persons in Europe, the Soviet Union, and the United States continue to label scientific management "Taylorism." Labor-union leaders and radicals then and now find Taylor convenient to attack as a symbol of a despised system of labor organization and control. In the early decades of this century, Europeans and Russians adopted "Taylorism" as the catchword for the much-admired and -imitated American system of industrial management and mass production. The publication in 1911 of Taylor's *Principles of Scientific Management* remains a landmark in the history of management-labor relations. Within two years of publication, it had been translated into French, German, Dutch, Swedish, Russian, Italian, Spanish, and Japanese. In his novel *The Big Money* (1936), John Dos Passos gave a sketch of Taylor, along with ones of Edison, Ford, Insull, and a few others, because he believed that they expressed the spirit of their era. Dos Passos noted that Taylor never smoked or drank tea, coffee, or liquor, but found comparable stimulation in solving problems of efficiency and production. For him, production was the end-all, whether it be armor plate for battleships, or needles, ball bearings or lightning rods.

Taylor's fundamental concept and guiding principle was to design a system of production involving both men and machines that would be as efficient as a well-designed, well-oiled machine. He said, "in the past, the man has been first; in the future the system must be first," a remark that did not sit well then with workers and their trade-union leaders and that today still rankles those who feel oppressed by technology. He asked managers to do for the production system as a whole what inventors and engineers had done in the nineteenth century for machines and processes. Highly efficient machines required highly efficient functionally related labor. When several Taylor disciples, including later U.S. Supreme Court Justice Louis D. Brandeis, sought a name for Taylor's management system, they considered "Functional Management," before deciding on "Scientific Management." Taylor and his followers unfeelingly compared an inefficient worker to a poorly designed machine member.

Taylor developed his principles of management during his work as a machinist and then as a foreman in the Midvale Steel Company of Philadelphia. . . .

Worker soldiering, variously called "stalling," "quota restriction," "goldbricking" "by Americans, *Bremsen* by Germans, and "hanging it out" or "Ca' canny" by the English and Scots, greatly offended Taylor's sense of efficiency. Having concluded that workers, especially the skilled machinists, were the major industrially inefficient enclave remaining after the great wave of nineteenth-century mechanization, Taylor proposed to eliminate "soldiering." He later wrote that "the greater part of systematic soldiering . . . is done by the men with the deliberate object of keeping their employers ignorant of how fast work can be done." The machinists at Midvale, for example, were on a "piecework" schedule, so they were determined that the owners not learn that more pieces could be turned out per hour and therefore demand an increase in the number of pieces required. They did not trust the owners to maintain the piece rate and allow the workers, if they exerted themselves, to take home more pay. The workers believed that the increased effort would become the norm for the owners. We can only conjecture about the natural rhythm and reasonableness of the pace that the workers maintained over the long duration; Taylor believed that they were soldiering. Nevertheless, he also showed that he was determined that the diligent worker be rewarded with a share of the income from more efficient and increased production. To his consternation, he later found that management and the owners also soldiered when it came time to share the increased income. Taylor was no close student of human nature; his approach was, as he described it, scientific.

After being put in charge of the machinists working at the lathes, Taylor set out to end soldiering among them. His friends began to fear for his safety. As Taylor recalled, the men came to him and said, "Now, Fred, you are not going to be a damn piecework hog, are you?" To which he replied, "If you fellows mean you are afraid I am going to try to get a larger output from these lathes," then "Yes; I do propose to get more work out." The piecework fight was on, lasting for three years at Midvale. Friends begged Taylor to stop walking home alone late at night through deserted streets, but he said that they could shoot and be damned and that, if attacked, he would not stick to the rules, but resort to biting, gouging, and brickbats. At congressional hearings in 1912—about thirty years later—he insisted:

> I want to call your attention, gentlemen, to the bitterness that was stirred up in this fight before the men finally gave in, to the meaness of it. . . . I did not have any bitterness against any particular man or men. Practically all of those men were my friends, and many of them are still my friends. . . . My sympathies were with workmen, and my duty lay to the people by whom I was employed.

In his search for the one best way of working, of deciding how and how fast a lathe operator should work, he used a method that he considered scientific. He believed values and opinions of neither workers nor managers influenced his objective, scientific approach. Beginning in 1882, first he, then an assistant began using a stopwatch to do time studies of worker's motions. Timing was not a new practice, but Taylor did not simply time the way the men worked: he broke down complex sequences of motions into what he believed to be the elementary ones and then timed these as performed by workers whom he considered efficient in their movements. Having done this analysis, he synthesized the efficiently executed component motions into a new set of complex sequences that he insisted must become the norm. He added time for unavoidable delays, minor accidents, inexperience, and rest. The result was a detailed set of instructions for the worker and a determination of time required for the work to be efficiently performed. This determined the piecework rate; bonuses were to be paid for faster work, penalties for slower. He thus denied the individual worker the freedom to use his body and his tools as he chose.

Taylor stressed that the time studies, with their accompanying analysis and synthesis, did not alone constitute scientific management. He realized and insisted that, for the work to be efficiently performed, the conditions of work had to be reorganized. He called for better-designed tools and became known for his near-fixation about the design of shovels. He ordered the planning and careful management of materials handling so that workers would have the materials at hand where and when needed. Often, he found, men and machines stood idle because of bottlenecks in complex manufacturing processes. Taylor even attended to lighting, heating, and toilet facilities. Seeing inanimate machines and men together as a single machine, he also looked for ways in which the inanimate ones failed. Believing that machine tools could also be driven faster, he invented a new chromium-tungsten steel for cutting tools that greatly increased their speed. As we would expect, he did not leave decisions about even the cutting speed of the machine tool or the depth of the cut to the subjective judgment of the machinist. In his book *On the Art of Cutting Metals* (New York, 1907), he described his thousands of experiments that extended over twenty-six years.

As a system builder seeking control and order, Taylor was not content to redesign machines, men, and their relationship; he was set upon the reorganization of the entire workplace or factory as a machine for production. Stimulated by his example,

individuals with special education, training, and skill contributed to the establishment of "the new factory system." To understand this achievement, we need to consider the way in which the work process was carried out in many machine shops, engineering works, and factories before Taylor's reforms. After the concern received an order, copies specifying the product and quantity to be made were sent to the foremen. They carried most of the responsibility for the production process. Once draftsmen had prepared detailed drawings, foremen in the machine shop, the foundry, pattern-making shop, and forge determined the various component parts needed, ordered the raw materials, and wrote out job cards for the machinists. The machinists then collected drawings, raw materials, and tools, and planned the way in which the job for a particular component part would be done. When the machinists had completed the particular job, they reported to the foreman for another. The foremen had overall supervision, but there was little scheduling and, therefore, little planned coordination of the various jobs. Components sometimes reached the assembly point, or erecting shop, haphazardly. Because of lack of planning, scheduling, and close monitoring of the progress of work, raw materials were often not on hand. How the workmen might then use their time is not clear, but proponents of Taylorism leave the impression that they were idle.

Taylor found the disorder and lack of control unbearably inefficient and declared war on traditional methods responsible for these. His reform specified that an engineering division take away from the foremen overall responsibility for the preparation of drawings, the specification of components, and the ordering of raw materials. Upwardly mobile young graduates from the rising engineering schools were soon displacing their "fathers," the foremen. The planning department in the engineering division coordinated deliveries of materials, and the sequence in which component parts would be made. The planning department prepared detailed instructions about which machines would be used, the way in which machinists, pattern makers, and other workmen would make each part, and how long the job should take. Careful records were kept of the progress being made in the manufacture of each part, including materials used and time consumed. Unskilled workers moved materials and parts around shops so that they would be on hand where and when needed. By an elaborate set of instruction cards and reports, the planning department had an overall picture of the flow of parts throughout the shops, a flow that prevented the congestion of the work at particular machines and the idleness of other machines and workmen. The reports of worker time and materials consumed greatly facilitated cost accounting.

The complexity and holism of Taylor's approach was often ignored because of the widespread publicity given to some of his simplest and most easily reported and understood successes. Taylor often referred to the "story of Schmidt," who worked with the pig-iron gang at the Bethlehem Steel Corporation in Pennsylvania. When Taylor and his associates came to Bethlehem in 1897 to introduce their management techniques and piecework, they found the pig-iron gang moving on the average about twelve and a half tons per day. Each man had repeatedly to lift ninety-two pounds of iron and carry it up an inclined plant onto a railroad car. After careful inquiry into the character, habits, and ambition of each of the gang of seventy-five men, Taylor singled out a "little Pennsylvanian Dutchman who had been observed to trot backhome for a mile or so after his work in the evening about as fresh as he was when he came trotting down to work in the morning." After work he was building a little house for himself on a small plot of ground he had "succeeded" in buying. Taylor also found out that the "Dutchman" Henry Noll, whom Taylor identified as Schmidt, was exceedingly "close," or one who placed "very high value on the dollar." The Taylorites had found their man.

Taylor recalled the way he and Schmidt talked, a story that tells us more of Taylor's attitudes than of what actually transpired:

> "Schmidt, are you a high-priced man? . . . What I want to find out is whether you want to earn $1.85 a day or whether you are satisfied with $1.15, just the same as all those cheap fellows are getting?"
>
> "Did I vant $1.85 a day? Vas dot a high-priced man? Vell, yes, I vas a high-priced man."
>
> ". . . Well, if you are a high-priced man, you will load that pig iron on that car tomorrow for $1.85. You will do exactly as this man tells you tomorrow, from morning till night. When he tells you to pick up a pig, and walk, you pick it up and you walk, and when he tells you to sit down and rest, you sit down. . . . And what's more, no back talk."

Taylor found it prudent to add:

> This seems to be rather rough talk. And indeed it would be if applied to an educated mechanic or even intelligent laborer. With a man of the mentally sluggish type of Schmidt it is appropriate and not unkind, since it is effective in fixing his attention on high wages which he wants. . . .

Perhaps Taylor, the upper-middle-class Philadelphian, forgot that the Pennsylvania Dutchman was not so mentally sluggish that he could not save for land and build a house. Schmidt moved the forty-seven tons of pig that the Taylorites had decided should be the norm, instead of the former twelve and a half tons, and soon all the gang was moving the same and receiving sixty percent more pay than other workmen around them. We are not told whether Schmidt was still able to trot home and work on his house.

Numerous other examples of Taylor's methods increasing worker output and production abound, but there is also abundant evidence of failures. Ultimately his efforts at Bethlehem Steel exhausted him, and the head of the company summarily dismissed him. Taylor had come to the steel company with the full support of Joseph Wharton, a wealthy Philadelphian into whose hands the company had passed. Wharton wanted a piecework system installed in the six-thousand-man enterprise. Taylor warned that his system would be strongly opposed by all of the workmen, most of the foremen, and even a majority of the superintendents. Bold and determined, he forged relentlessly ahead, introducing a planning department and new administrative roles for the foremen. Instructions for routine were codified with time cards, work sheets, order slips, and so on. As worker resistance stiffened over several years, Taylor became rigid, even arbitrary, in dealing with labor and management. His achievements were impressive, but "as time went on, he exhibited a fighting spirit of an intensity almost pathological," an admirer wrote. Taylor's communications to the Bethlehem president were tactless and peremptory (he believed Wharton would shelter him). He complained of poor health and nervous strain. He thought that some of the major stockholders opposed him because he was cutting the labor force, and they were losing rents on the workers' houses. The curt note dismissing him came in April 1901.

Many workers were unwilling, especially the skilled ones, to give control of their bodies and their tools to the scientific managers, or, in short, to become components in a well-planned system. An increase in pay often did not compensate for their feeling of loss of autonomy. Taylor's scientific analysis did not take into account worker independence and pride in artful craftsmanship—even artful soldiering. Perhaps this was because Taylor, despite his years of experience on the shop floor, did not come from a blue-collar worker culture.

Samuel Gompers, a labor leader, said of Taylorism and similar philosophies of management:

> So, there you are, wage-workers in general, mere machines—considered industrially, of course. Hence, why should you not be standardized and your motion-power brought up to the highest possible perfection in all respects, including speeds? Not only your length, breadth, and thickness as a machine, but your grade of hardness, malleability, tractability, and general serviceability, can be ascertained, registered, and then employed as desirable. Science would thus get the most out of you before you are sent to the junkpile.

One of the most publicized setbacks for Taylorism took place at the Watertown Arsenal when Carl G. Barth, a prominent Taylor follower and a consultant on scientific management, tried to introduce the Taylor system. Serious trouble started in the foundry when one of Barth's associates began stopwatch-timing the men's work procedures. The skilled workers in the shop discovered that the man carrying out the study knew little about foundry practice. The foundrymen secretly made their own time study of the same work process and complained that the time specified by the "expert" was uninformed and represented an unrealistic speedup. The Watertown project was also flawed because Taylor's practice was to reorganize and standardize a shop before doing time-and-motion studies, and this had not been carried out at Watertown Arsenal. On the evening after the initiation of the stopwatch studies, the workers met informally and in a petition to the commanding officer of the arsenal they stated:

> The very unsatisfactory conditions which have prevailed in the foundry among the molders for the past week or more reached an acute stage this afternoon when a man was seen to use a stop watch on one of the molders. This we believe to be the limit of our endurance. It is humiliating to us, who have always tried to give to the government the best that was in us. This method is un-American in principle, and we most respectfully request that you have it discontinued at once.

When stopwatch timing continued, the molders walked out on 11 August 1911.

Promised an investigation of the "unsatisfactory conditions," the molders returned to work after a week, but the publicity given a strike against the U.S. government intensified, fermenting union opposition to scientific management, specifically Taylorism, at Watertown and at another U.S. arsenal, at Rock, Illinois. August brought the formation of a special congressional committee of three to investigate scientific management in government establishments. The committee took extensive testimony from Taylor, among others. He became so exercised by hostile questions that his remarks had to be removed from the record. The report of the committee did not immediately call for any legislation. In 1914, however, Congress attached to appropriations bills the proviso that no time studies or related incentive payments should be carried out in government establishments, a prohibition that survived for over thirty years. Yet Taylorism involved, as we have seen, more than time studies and incentive payments, so work processes in government establishments continued to be systematically studied, analyzed, and changed in ways believed by management experts to be scientific.

Worn down by the never-ending opposition and conflict, Taylor moved in 1902 to a handsome house in me Chestnut Hill area of Philadelphia. He no longer accepted employment or consulting fees but announced that he was ready to advise freely those interested in Taylorism. . . .

Free from the confrontations in the workplace, Taylor dedicated himself to showing that his philosophy of management would ultimately promote harmony between management and labor. He argued that increasing production would increase wages and raise the national standard of living. His principles of scientific management struck responsive chords in a nation intent on ensuring economic democracy, or mass consumption, through mass production and also on conserving its natural resources. Taylor wrote that maximum prosperity could exist only as a result of maximum productivity. He believed that the elimination of wasted time and energy among workers would do more than socialism to diminish poverty and alleviate suffering.

Because of his firm belief that his method was objective, or scientific, he never fully comprehended the hostile opposition of aggressive, collective-bargaining labor-union leaders. He found the unions mostly standing "for war, for enmity," in contrast to scientific management, which stood for "peace and friendship." Nor could he countenance unenlightened and "hoggish" employers who either found his approach and his college-educated young followers unrealistic or were unwilling to share wholeheartedly with the workers the increased profits arising from scientific management. He considered the National Manufacturers Association a "fighting association," so he urged his friends in scientific management to cut all connections with it and its aggressive attitudes toward labor unions. Firmly persuaded that conflict and interest-group confrontations were unnatural, he awaited, not too patiently, the day when management and labor would realize, as he, that where the goal was increased productivity there were discoverable and applicable scientific laws governing work and workplace. Scientific managers were the experts who would apply the laws. He wrote:

> I cannot agree with you that there is a conflict in the interests of capital and labor. I firmly believe that their interests are strictly mutual, and that it is practicable to settle by careful scientific investigation the proper award that labor should receive for the work it renders.

Their interest was not only a mutual but a national one—production and democracy, production and Democracy. Taylor's times were not ones of affluence for workers, so his means to the end of mass production, thereby raising living standards of the glasses, seemed in accord with democratic principle. Within a few years, Vladimir Lenin argued that Taylor principles accorded with socialism, as well.

Taylor became nationally known when [Louis] Brandeis, the Boston "people's lawyer," argued in 1911 that scientific management especially Taylorism, could save the nation's railroads so much money that the increased rates that the railroads were requesting from the Interstate Commerce Commission would not be needed. Since the rate hearings were well publicized, writers from newspapers and magazines descended on Taylor to find out about his

system and then, at his suggestion, visited Philadelphia plants to see firsthand Taylorism in practice. The favorable publicity induced Taylor to write that "the interest now taken in scientific management is almost comparable to that which was aroused in the conservation of our natural resources by Roosevelt."

Taylor rightly associated his scientific management with the broader conservation movement that had attracted national interest and support during Theodore Roosevelt's terms as president, 1901–08. This progressive program for conservation focused on the preservation and efficient utilization of lands and resources. Like scientific management, it advocated that decisions about conservation be made scientifically by experts. Like Taylor, the progressive conservationists did not countenance as inevitable conflict of interests among ranchers, farmers, lumbermen, utilities, manufacturers, and others. To the contrary, they believed that such conflict was regressive, that it must be displaced by a scientific approach expected to bring harmonious and rational compromises in the general interest. This approach expressed a technological spirit spread by engineers, professional managers, and appliers of science, a belief that there was one best way. College-educated foresters, hydraulic engineers, agronomists should be, the progressives argued, the decision makers about resources; professional managers about the workplace.

Taylor and the growing number of his followers wrote books, published articles, gave lectures, and acted as consultants. He authorized C. G. Barth, H. K. Hathaway, Morris L. Cooke, and Henry L. Gantt to teach his system of management: "All others were operating on their own." Frank Gilbreth, among those who operated "on their own," became well known for his *A Primer of Scientific Management* (1914) and for the use he and his wife, Lillian Gilbreth, made of the motion-picture camera to prepare time-and-motion studies. Her contribution to scientific management has yet to be generally acknowledged. She, not her husband, had a Ph.D. degree in psychology (Brown University, 1915). Perhaps because of her study of psychology, she sensitively took into account complex worker characteristics. The Gilbreths' articles on scientific management show the influence of her concern that the worker should not be seen simply as a component in a Taylor system. After her husband's death, she continued her consulting work and served as a professor of industrial management at Purdue University.

Mass Production

David A. Hounshell

Mass production became the Great American art.
—*Paul Mazur, American Prosperity (1920)*

Since the 1920s the term "mass production" has become so deeply ingrained in our vocabulary and our thought that we seldom stop to ask how it arose and what lay behind its appearance. The purpose of this brief essay is to provide an overview of the development of mass production in America as a means of getting at these questions. In the first half of the nineteenth century, manufacturing in the United States developed along such distinct lines that by the 1850s English observers came to speak of an "American system" of manufactures. Subsequently, the American system grew and changed in character so much that by the 1920s, the United States possessed the most prolific production technology the world has ever known. This was "mass production."

In 1925, the American editor of the *Encyclopaedia Britannica* wrote to Henry Ford asking him to submit an article on "Mass Production" for the three-volume supplement to the *Britannica*, the so-called 13th Edition. Apparently Ford's office, if not Ford himself, responded favorably and promptly set Ford's spokesman, William J. Cameron, to work on the article. Cameron consulted the company's chief production planner about how the "general reader" might comprehend the principles of mass production. When Cameron completed the article, he placed Henry Ford's name beneath it and sent it to the *Britannica's* New York office.

Although Cameron would later say that he "should be very much surprised to learn that [Henry Ford] read it," this article played a fundamental role in giving the phrase "mass production" a place in the English vocabulary. Even before the article appeared in the Britannica, the *New York Times* published it as a full-page, feature article in a Sunday edition. Under the banner, "Henry Ford Expounds Mass Production: Calls It the Focusing of the Principles of Power, Economy,

Continuity and Speed," the article attracted the attention of a wide segment of the American population, especially since it also went through the wire service. While one can certainly wonder what led the *Britannica* editor to choose the term, mass production, there is little doubt that the ghost-written Ford article led to the widespread use of the term and its identification with the assembly line manufacturing techniques that were the hallmark of automobile production. Immediately, the article proved interesting enough to provoke a *Times* editorial. The term, which had not commonly appeared in reference works such as the *Reader's Guide to Periodic Literature* prior to the *Britannica Times* article, soon passed into general use in both popular and scholarly literature. After the appearance of "Mass Production," the previously popular expression, "Fordism," soon disappeared. The Ford article endowed mass production with a certain universality despite its ambiguity and its status as poor grammar.

Much more important than the story of how mass production entered the English vocabulary are the developments that lay behind the manufacturing system described in the article. Commenting in 1940 on Henry Ford and the *Britannica* article in his *Engines of Democracy*, Roger Burlingame raised the essential questions:

> With [Ford's] great one-man show moving toward a dictatorship of which any totalitarian leader might well be proud he was ready for what he calls [Mass] Production. [Mass] Production, Ford believes, had never existed in the world before. With the magnificent contempt of men immune to history, he disregards all predecessors: Whitney, Evans, Colt, Singer, McCormick, the whole chain of

"Mass Production" by David A. Hounshell as appeared in "Mass Production in American History, 1800–1932" from *Polhem 2* (1984): 1–28. © David A. Hounshell. Reprinted by permission of the author.

patient, laborious workers who wrought his assembly lines and all the ramifications of his processes out of the void of handicrafts. In a colossal blurb printed in the *Encyclopaedia Britannica* under the guise of an article on mass production, he writes: "In origin, mass production is American and recent; its notable appearance falls within the first decade of the 20th century," and devotes the remainder of the article and two full pages of half-tone plates to the Ford factory.

Burlingame was obviously contemptuous of the claims that mass production was a creation of the Ford Motor Company. Eli Whitney, Oliver Evans, Samuel Colt, Isaac Singer, and Cyrus McCormick, among others, he implied, provided essential building blocks for development at Ford. Burlingame was even more pointed when he later asked rhetorically, "What are those production methods in use today in every large automobile plant with scarcely any variation? They are simply the methods of Eli Whitney and Samuel Colt, improved, coordinated and applied with intelligent economy—economy in time, space, men, motion, money and material."

Since the establishment of the history of technology as an academic discipline in the United States, the assertions contained both in Ford's encyclopaedia article and in Burlingame's popular work have come under close study by a number of investigators. Indeed, the so-called "American system of manufactures," which describes the methods of Whitney, Colt, and the rest, has become one of the most productive areas of American scholarship in the history of technology, and there now exists a rich body of literature on this historical phenomenon. Portions of that new scholarship . . . indicate that the Ford article came much nearer the truth than did Burlingame and his followers. "[I]n origin," as the Ford piece suggested, "mass production is American and recent"—what Whitney et al., did in the nineteenth century was not true mass production. . . . [M]ass production differed in kind as well as in scale from the techniques referred to in the antebellum period as the American system of manufactures. This can be seen most clearly by first considering the American system itself.

Two decades of research on this topic have yielded a number of conclusions, particularly concerning a basic aspect of modern manufacturing, the interchangeability of parts. The symbolic kingpin of interchangeable parts production fell in 1960 when Robert S. Woodbury published his essay, "The Legend of Eli Whitney," in the first volume of *Technology and Culture*. Woodbury convincingly argued that the parts of Whitney's guns were not in fact constructed with interchangeable parts. In 1966, the artifactual research of Edwin A. Battison solidly confirmed Woodbury's more traditional, document-based research findings. Eugene S. Ferguson later wrote of Woodbury's pioneering article, "Except for Whitney's ability to sell an undeveloped idea, little remains of his title as father of mass production."

With Eli Whitney reinterpreted as a *promoter* rather than as a *pioneer* of machine-made interchangeable parts manufacture, it remained for Merritt Roe Smith to identify conclusively the personnel and the circumstances of this fundamental step in the development of mass production. Smith demonstrated that the United States Ordinance Department was the prime mover in bringing about machine-made interchangeable parts production of small arms. The national armory at Springfield, Massachusetts, played a major role in this process, especially as it tried to coordinate its operations with those of its sister armory at Harpers Ferry and John Hall's experimental rifle factory, also located in Harpers Ferry. While these federally owned arms plants occupied a central place in its efforts, the Ordnance Department also used contracts with private arms makers to further its aims. By specifying interchangeability in its contracts and by giving contractors access to techniques used in the national armories, the Ordinance Department contributed significantly to the growing sophistication of metalworking and woodworking (in the case of gunstock production) in the United States by the 1850s. British observers found these techniques sufficiently different from their own and alluded to them in expressions such as the "American system," the "American plan," and the "American principle."

Although British visitors to the United States in the 1850s, especially Joseph Whitworth and John Anderson, were impressed with every aspect of American manufacturing, small arms production received their most careful and detailed analysis. Certainly this was Anderson's job, for he had been sent to the United States to find out everything he could about small arms production and to purchase arms-making machinery for the Enfield Arsenal. In his report, Anderson indicated that the federal armory at Springfield had indeed achieved what the Ordnance Department had sought since its inception: true interchangeability of parts. Anderson and his committee went into Springfield's arsenal and randomly selected ten muskets, each made in a different year from 1844 to 1853. A workman then disassembled these muskets, and their parts were mixed together. According to Anderson, the committee then "requested the workman, whose duty it is to 'assemble' the arms, to put them together, which he did—the Committee handing him the parts, taken at hazard—with the use of a turn-screw only, and as quickly as though they had been English muskets whose parts had carefully been kept separate."

What Anderson was not likely to have known was the extraordinary sum of money that the Ordnance Department had expended over a forty or fifty year period, "[i]n order," as an Ordnance Officer wrote in 1819, "to attain this grand object of uniformity of parts." Nor was Anderson necessarily aware that the unit cost of Springfield small arms with interchangeable parts almost certainly was significantly higher than arms produced by more traditional methods. He should, however, have been aware that the Ordnance Department could annually turn out only a relatively small number of Springfield arms manufactured with interchangeable parts. Despite the high costs and limited output, Anderson pointed out that the special techniques used in the Springfield Armory as well as in some private armories could be applied almost universally in metalworking and woodworking establishments. In fact, by the time Anderson reached this conclusion, the application of those techniques in other industries was already under way.

The new manufacturing technology spread first to the production of a new consumer item, the sewing machine, and eventually it diffused into other areas, including consumer durables such as typewriters, bicycles, and eventually automobiles. Nathan Rosenberg has provided economic and technological historians with an excellent analysis of a major way in which this diffusion occurred. Rosenberg identified the American machine tool industry, which grew out of the small arms industry (notably the Colt armory and the firm of Robbins & Lawrence in Windsor, Vermont, and Hartford, Connecticut) as the key agent for introducing arms-making technology into the sewing machine industry, the bicycle industry, and the automobile industry. The makers of machine tools worked with manufacturers in various industries as they encountered and overcame production problems relating to the cutting, planing, boring, and shaping of metal parts. As each problem was solved, new knowledge went back into the machine tool firms, which then could be used for solving production problems in other industries. Rosenberg called

this phenomenon "technological convergence." In many industries that worked with metal, the final products were vastly different in terms of the kinds of markets in which they were sold—the Springfield Armory, for example, "sold" its products to a single customer, the government, while sewing machines producers faced a widely scattered group of individual consumers. Nevertheless, these products had *technological* things in common because their manufacture depended upon similar metalworking technique. These common needs "converged" at the point where the machine tool industry interacted with the firms that bought its machine tools.

Although he did not emphasize the point, Rosenberg recognized that individual mechanics played an equally important role in diffusing know-how as they moved from the firearms industry to sewing machine manufacture to bicycle production and even to automobile manufacture. Examples of such mechanics abound. Henry M. Leland is an obvious example: he worked at Springfield Armory, carried this knowledge to Brown & Sharpe Manufacturing Company when it was making both machine tools and Wilcox & Gibbs sewing machines, next created the Cadillac Motor Car Company and finally the Lincoln Motor Company.

But the process of diffusion was neither as smooth nor as simple as Rosenberg and others would have it. New research suggests that the factories of two of the giants of nineteenth century manufacturing, the Singer Manufacturing Company and the McCormick Harvesting Machine Company, were continually beset with production problems. Previously, many historians attributed the success of these two companies to their advanced production technology. But it now appears that a superior marketing strategy (including advertising and sales techniques and policies) proved to be the decisive factor. . . .

Joseph Woodworth, author of *American Tool Making and Interchangeable Manufacturing,* argued that the "manufacture of the bicycle . . . brought out the capabilities of the American mechanic as nothing else had ever done. It demonstrated to the world that he and his kind were capable of designing and making special machinery, tools, fixtures, and devices for economic manufacturing in a manner truly marvellous; and has led to the installation of the interchangeable system of manufacturing in a thousand and one shops when it was formerly thought to be impractical." Clearly the bicycle industry as a staging ground for the diffusion of armory practice cannot be overemphasized. Rosenberg's idea that the machine tool industry played a leading role in this diffusion applies even more clearly to the bicycle than to the sewing machine. The bicycle boom of the 1890s kept the machine tool industry in relatively good health during the serious depression that began in 1893, and it was accompanied by changes in production techniques.

Entirely new developments occurred in bicycle production—sheet metal stamping and electric resistance welding techniques. These new techniques rivalled in importance the diffusion of older metalworking technologies. During the 1890s, bicycle makers located principally but not exclusively in areas west of New England began to manufacture bicycles with many components (pedals, crank hangers, steering heads, joints, forks, hubs, etc.) made from sheet steel. Punch pressing or stamping operations were combined with the recent invention of electric resistance welding to produce parts at significantly lower costs. This technology would become fundamental to the automobile industry. . . .

Bicycle makers such as [Albert A.] Pope who used traditional armory-type production techniques looked with disdain at those who manufactured bicycles with parts made by the new techniques in pressing and stamping steel. An executive at the Columbia works [Pope's factory] called them cheap and nasty. Despite such views, the one manufacturer that outstripped Pope's production at the peak of the bicycle boom was the Western Wheel Works of Chicago, which made a "first class bicycle out of pressed steel hubs, steering head, sprocket, frame joints, crank hanger, fork, seat, handlebar, and various brackets. Although not quite as expensive as the Columbia, the Western Wheel bicycle ranked high in the top price category among some 200 to 300 manufacturers. Production of this bicycle reached 70,000 in 1896, an output that was significantly less than that of the Ford Model T in 1912, the last full year of its pre-assembly line manufacture.

. . . In terms of production, it is only with the rise of the Ford Motor Company and its Model T that there clearly appears an approach to manufacture capable of handling an output of multi-component durables ranging into the millions each year.

Moreover, the rise of Ford marks an entirely new epoch in the manufacture of consumer durables in America. The Ford enterprise may well have been more responsible for the rise of "mass production," particularly for the attachment of the noun "mass" to the expression, than we have realized. Unlike Singer, McCormick, and Pope, Ford sought to manufacture the lowest priced automobile and to use continuing

price reductions to produce ever greater demand. Ford designed the Model T to be a "car for the masses." Prior to the era of the Model T, the word "masses" had carried a largely negative connotation, but with such a clearly stated goal and his company's ability to achieve it, Ford recognized "the masses" as a legitimate and seemingly unlimited market for the most sophisticated consumer durable product of the early twentieth century. Whether Henry Ford envisioned "the masses" as the populace or 'lower orders' of late nineteenth parlance or merely as a large number of potential customers hardly matters, for the results were the same. Peter Drucker long ago maintained that Ford's work demonstrated for the first time that maximum profit could be achieved by maximizing production while minimizing cost. He added that "the essence of the mass-production process is the reversal of the conditions from which the theory of monopoly was deduced. The new assumptions constitute a veritable economic revolution." For Drucker, mass production was as much an economic doctrine as an approach to manufacture. For this reason if for no other, the work of the national armories, Singer, McCormick, Pope et al., differed substantially from Ford's. But Ford was able to initiate this new "economic revolution" because of advances in production technology, especially the assembly line.

Before the adoption of the revolutionary assembly line in 1913, Ford's production engineers had synthesized the two different approaches to manufacture that had prevailed in the bicycle era. On the other hand, Ford adopted the techniques of armory practice. All of the company's earliest employees recalled how ardently Henry Ford had supported efforts to improve precision in machining. Although he knew little about jig, fixture, and gauge techniques, Ford nevertheless became a champion of interchangeability within the Ford Motor Company, and he hired mechanics who knew what was required to achieve that goal. Certainly by 1913, most of the problems of interchangeable parts manufacture had been solved at Ford. In addition to armory practice, Ford adopted sheet steel punch and press work in an important way. Initially he contracted for stamping work with the John R. Keim Company in Buffalo, New York, which had been a major supplier of bicycle components. Soon after opening his new Highland Park factory in Detroit, however, Ford purchased the Keim plant and promptly moved its presses and other machines to the new factory. More and mote Model T components were stamped out of sheet steel rather than being fabricated with traditional machining methods.

Together, armory practice and sheet steel work equipped Ford with the capability to turn out virtually unlimited numbers of components. It remained for the assembly line to eliminate the remaining bottleneck—how to put these parts together.

The advent of line assembly at Ford Motor Company in 1913 is one of the most confused episodes in American history. . . . First, the assembly line, once it was first tried on April 1, 1913, came swiftly and with great force. Within eighteen months of the first experiments with moving line assembly, assembly lines were used in almost all sub-assemblies and in the most symbolic mass production operation of all, the final chassis assembly. Ford engineers witnessed productivity gains ranging from fifty percent to as much as ten times the output of static assembly methods. Allan Nevins quite correctly called the moving assembly line "a lever to move the world."

Secondly, there can be little doubt that Ford engineers received their inspiration for the moving assembly line from outside the metalworking industries. Henry Ford himself claimed that the idea derived from the "disassembly lines" of meatpackers in Chicago and Cincinnati. William Klann, a Ford deputy who was deeply involved in the innovation, agreed but noted that an equally important source of inspiration was flour milling technology as practiced in Minnesota. . . . Although there may have been a clear connection in the minds of Ford engineers between "flow production" and the moving assembly line, there is little justification for saying that the assembly line came directly from flour milling. The materials and processes of both were simply too different to support such a view.

The origins of the Ford assembly line are less important than its effect. While providing a clear solution to the problems of assembly, the line brought with it serious labor problems. Already, Ford's highly mechanized and subdivided manufacturing operations imposed severe demands on labor. Even more than previous manufacturing technologies, the assembly line implied that men, too, could be mechanized. Consequently during 1913 the Ford company saw its annual labor turnover soar to 380 percent and even higher. Henry Ford moved swiftly to stem this inherently inefficient turn-over rate. On January 5, 1914, he instituted what became known as the "five-dollar day." Although some historians have argued that this was a wage system that more than doubled the wages of "acceptable" workers, most recently the five-dollar day has been interpreted as a profit

sharing plan whereby Ford shared excess profits with employees who were judged to be fit to handle such profits. In any case, the five-dollar day effectively doubled the earnings of Ford workers, and provided a tremendous incentive for workers to stay "on the line." With highly mechanized production, moving line assembly, high wages, and low prices on products, "Fordism" was born.

During the years between the birth of "Fordism" and the wide-spread appearance of the term, "mass production," the Ford Motor Company expanded its annual output of Model T's from 300,000 in 1914 to more than 2,000,000 in 1923. In an era when most prices were rising, those of the Model T dropped significantly—about sixty percent in current dollars. Throughout the Model T's life, Henry Ford opened his factories to technical journalists to write articles, series of articles, and books on the secrets of production at Ford Motor Company. Soon after the appearance of the first articles on the Ford assembly lines, other automobile companies began putting their cars together "on the line." Manufacturers of other consumer durables also followed suit. Ford's five-dollar day forced automakers in the Detroit vicinity to increase their wage scales. Because Ford secured more than fifty percent of the American automobile market by 1921, his actions had a notable impact on American industry.

Ford's work and the emulation of it on the part of other manufacturers led to the establishment of what could be called an "ethos of mass production" in America. The creation of this ethos marks a significant moment in the development of mass production and consumption in America. Certain segments of American society looked at Ford's and the entire automobile industry's ability to produce large quantities of goods at surprisingly low costs. When they did so, they wondered why, for example, housing, furniture, and even agriculture could not be approached in precisely the same manner in which Ford approached the automobile.

Consequently, during the years that the Model T was in production, movements arose within each of these industries to introduce mass production methods. In housing, an industry always looked upon as one of the most staid and pre-industrial of all, pre-fabrication efforts reached heights not achieved by the pioneers of prefabrication. Foster Gunnison, for example, strove to become the "Henry Ford of housing" by establishing a factory to turn out houses on a moving assembly line, and Gunnison was only one among many such entrepreneurs. Furniture production also saw the influence of Ford and the automobile industry. In the 1920's a large number of mechanical engineers in America banded together within the American Society of Mechanical Engineers in an effort to bring the woodworking industry into the twentieth century—into the century of mass production. Consequently, the ASME established in 1925 a Wood Industries Division, which served to focus the supposed great powers of mechanical engineering on all aspects of woodworking technology. In agriculture, Henry Ford himself argued that the problems of American agriculture could all be solved simply by adopting mass production techniques. Ford conducted experiments in this direction, but he was no more successful in agriculture than the mechanical engineers and housing fabricators were in bringing about mass production in their respective industries. One could argue, however, that today such an agricultural product as the hybrid tomato, bred to be picked, sorted, packaged, and transported by machinery, demonstrates that mass production methods have penetrated American agriculture. But in furniture and housing, there seems to be no equivalent to the hybrid tomato. . . .

. . . American furniture manufacturers continued to operate relatively small factories employing around 150 workers, annually turning out between 5,000 and 50,000 units. Beliefs that automotive production technology holds the key to abundance in all areas of consumption persist today. As recently as 1973, Richard Bender observed in his book on industrial building that "much of the problem of industrializing the building industry has grown out of the mistaken image of the automobile industry as a model." In many areas, the panacea of Fordism will continue to appeal to those who see in it solutions to difficult economic and social problems. The ethos of mass production, established largely by Ford, will die a hard death, if it ever disappears completely.

Yet the very timing of the rise of this ethos along with the appearance of the *Encyclopedia Britannica* article, "Mass Production," shows how full of paradox and irony history is. Although automotive America was rapidly growing in its consumption of everything under the sun and although Ford's achievements were known by an, mass production as Ford had made it and defined it was, for all intents and purposes, dead by 1926. Ford and his production experts had driven mass production into a deep cul-de-sac. American buyers had given up on the Ford Model T, and the Ford Motor Company watched its

sales drop precipitously amid caustic criticism of its inability to accept and make changes. In mid-1927, Henry Ford himself finally gave up on the Model T after 15,000,000 of them had been produced. What followed in the changeover to the Model A was one of the most wrenching nightmares in American industrial history. Designing the new model, tooling up for its production, and achieving satisfactory production levels posed an array of unanticipated problems that led to a long delay in the Model A's introduction. In some respect, the Ford Motor Company never recovered from the effects of its first big changeover. Changes in consumers' tastes and gains in their disposable income made the Model T and the Model T idea obsolete. Automobile consumption in the late 1920s called for a new kind of mass production, a system which could accommodate frequent change and which was no longer wedded to the idea of maximum production at minimum cost. General Motors, not Ford, proved to be in tune with changes in American consumption with its explicit policy of "a car for every purpose and every purse," its unwritten policy of annual change, and its encouragement of "trading up" to a more expensive car. Ford learned painfully and at great cost that the times called for a new era, that of "flexible mass production."

The Great Depression dealt additional blows to Ford's version of mass production. With dramatic decreases in sales following the Great Crash, Ford and the entire industry began laying off workers. As a result, Detroit became known as the "beleaguered capital of mass production." Mass production had not prevented mass unemployment or, more properly, unemployment of the masses, but seemed rather to have exacerbated it. While overproduction had always posed problems for industrial economies, the high level of unemployment in the Great Depression made mass production an easy culprit for critics as they saw hundreds of thousands of men out of work in the Detroit area alone. Writing in the *New York Times* in 1931 Paul Mazur stressed that, "mass production has not proved itself to be an unmixed blessing; in the course of its onward march lie overproduction and the disastrous discontinuity of industry that comes as a consequence." Call it Fordism or mass production, it was nonetheless, "an alluring but false doctrine." Moreover, Mazur argued, "it is essential for business to realize that unquestioning devotion to mass production can [only] bring disaster.

Mazur's comments came in the wake of a previous *Times Magazine* article entitled, "Gandhi Dissects the

Ford Idea." The article's author, Harold Callender, pitted Ford's doctrine of mass production against Mahatma Gandhi's notion that handicrafts, not mechanization, offered the solution to global problems of unemployment and hunger. The *Times Magazine* juxtaposed a photograph of an assembly line against one of a group of Indian hand spinners. Captions under the two photographs read as follows: "The Ford Formula for Happiness—A Mass-Production Line" and "The Gandi Formula for Happiness—A Group of Handicraft Spinners." While few would have agreed with the Gandhi, formula, Americans in the depths of the Depression certainly seem to have concluded that developments in mass production had not been matched by the development of "mass consumption." As Mazur put it, "the power of production . . . has been so great that its products have multiplied at geometric rates . . . at the same time the power of consumption—even under the influence of stimuli damned as unsocial and tending toward profligacy [e.g., advertising and built-in obsolescence (frequent style changes)]—has expanded only at a comparatively slow arithmetic rate."

While Americans may have had doubts about the doctrine of mass production, they by no means were willing to scrap it in favor of the Gandhi formula. Already their desire for style and novelty, coupled with increased purchasing power in the 1920s, had forced even Henry Ford to change his system of mass production. When pushed by the Depression, the greater part of Americans looked for solutions in the sphere of "mass consumption." The 1930s witnessed the publication of an extensive amount of literature on the "economics of consumption." As history would have it, the prophets of mass production were proven at least temporarily correct as the United States pulled itself out of the Depression by the mass consumption of war material and, after the war, by the golden age of American consumption in the 1950s and 1960s.

Nazi Science and Nazi Medical Ethics: Some Myths and Misconceptions

Robert N. Proctor*

We often hear that the Nazis destroyed science and abandoned ethics. That was the view of Telford Taylor in his opening statement at the Nuremberg "Doctor's Trial" of 1946–1947, where he stated that the Nazi doctors had turned Germany "into an infernal combination of a lunatic asylum and a charnel house" where "neither science, nor industry, nor the arts could flourish in such a foul medium" [1]. Similar views were expressed by Franz Neumann, author of the 1942 treatise *Behemoth*, the first major analysis of how the Nazis came to power [2]. Neumann predicted "a most profound conflict" between the "magic character" of Nazi propaganda and the "rational" processes of German industry, a conflict the emigré political theorist believed would culminate in an uprising on the part of engineers to combat the irrationalist regime. Such an uprising, needless to say, never materialized.

It would be comforting to believe, of course, that good science tends to travel with good ethics, but the sad truth seems to be that cruelty can coexist fairly easily with "good science." There is a convenience of sorts in the myth: it makes it easy to argue that "Nazi science" was not really science at all, and therefore there is no ethical dilemma. One needn't talk about the ethics of Nazi medicine, since there was no legitimate medicine to speak of. Nazi science in one swift blow is reduced to an oxymoron, a medical non-problem.[1]

Myth-Making and Exculpation

Different groups have participated in this process of myth-making; let me mention four.

First, for *Germans who stayed in Germany*, the myth of suppressed science served as a way for post-war scientists to distance themselves from their Nazi past, to block historical investigators from dredging up potentially embarrassing collaborations. If there was no such thing as Nazi science, what is there to investigate? As an epidemiologist at Germany's National Cancer Center in Heidelberg once confessed: "1945 was a scientific *Stunde Null*, we don't look at what came before." This allowed scientists to argue that the very act of scientific research under Nazi rule was a form of resistance. It was often advertised as such, after the war.

Second, for *Jewish scholars forced from Germany*, there was also the unwillingness to believe that the system that had treated them so shoddily had continued to produce good science. This view was reinforced by the fact that fields in which Jews had been prominent (biochemistry and quantum mechanics, for example, but also medical specialties like dermatology) tended to be the fields most heavily gutted by Nazi policies. Fields with lesser Jewish representation—like veterinary medicine or surgery—generally speaking suffered less and therefore have drawn less critical historical attention because there were fewer emigrés in these fields [4].

Third, for *American military authorities*, the myth of flawed science also served to disguise the fact that even as U.S. officials were denouncing Nazi science, they were also busily trying to recruit Nazi talent for use in U.S. military projects. At least 1,600 German scientists came to the United States under the rubric of "Operation Paperclip"—including not just SS officers like Werner von Braun, but also a number of medical professionals, some of whom had been implicated in abusive human experimentation [5].

*Department of History, Pennsylvania State University, 0311 Weaver Building, University Park, PA 16802.

Email: rnp5@psu.edu or proctor@mpiwg-berlin.mpg.de.

[1]Analyzing the notorious Dachau hypothermia experiments, Robert Berger claimed that "the Nazi data" were not scientific, any ethical debate over whether to use Nazi science was misconceived from the outset; see [3].

"Nazi Science and Nazi Medical Ethics: Some Myths and Misconceptions" by Robert N. Proctor, *Perspectives in Biology & Medicine* 43:3 (2000), 335–346. © 2000 The Johns Hopkins University Press. Reprinted with permission of The Johns Hopkins University Press.

Finally, this myth, as I have identified it, served to reassure *the American public* that abuses like those of the Nazi era could never occur in a liberal democracy. Nazi science was pseudo-science, science out-of-control; American science was genuine science, secure within democratic institutions, obedient to the rule of law. Post-war ethical codes of conduct could even be dismissed as unnecessary—after all, weren't they designed to prevent abuses that could only occur in a totalitarian society? That, apparently, was the thinking of some of those involved in the radiation experiments investigated by the President's Advisory Committee on Human Radiation Experiments: for example, the Vanderbilt nutritionist William J. Darby, who from 1945 through 1947 supervised the feeding of radioactive iron to 829 pregnant women without their consent, when asked whether he recognized the legitimacy of the Nuremberg Code, responded that he did not, because the code applied only to "experiments of a medical nature committed by the Germans . . . a different setting entirely" [6].

Triumphs in Science

Germany was not a technical or medical backwater in the 1930s and 1940s. Anyone who has ever examined a V-2 engine will have little doubts about this, but there are many other examples. Nazi-era scientists and engineers were pioneers of television, jet-propelled aircraft, guided missiles, electronic computers, the electron microscope and ultracentrifuge, atomic fission, new data processing technologies, new pesticides, and the world's first industrial murder factories (including the use of gas chambers disguised as showers)—all of which were either first developed in Nazi Germany or reached their high point at that time. We should recall that the first magnetic tape recording was of a speech by Hitler, that the V-2 emerged from a plan for intercontinental ballistic missiles designed to be able to reach New York City, and that the nerve gases sarin and tabun were Nazi inventions. I have recently published a book arguing that German cancer research at this time was the most advanced in the world: Nazi-era health reformers built on this research base, introducing smoke-free public spaces, bans on carcinogenic food dyes, and new means of controlling dust exposure on factory floors. The period saw extensive work in the area of occupational carcinogenesis, and in 1943, Germany became the first nation to recognize lung cancer and mesothelioma as compensable occupational illnesses caused by asbestos inhalation [see 7, 8].

The story of science under German fascism must therefore be more than a narrative of suppression and survival; it must also tell how Nazi ideology *promoted* certain areas of inquiry, how projects and policies came and went with the movement of political forces. The frightening truth is that the Nazis supported many kinds of science, left politics (as we often think about it) out of most, and transformed but did not abandon ethics. There is an ethics of Nazi medical practice, sometimes explicit, sometimes not; sometimes cruel, sometimes not. This is important to understand if we are not to demonize the Nazi phenomenon as something absolutely alien and otherworldly.

Appreciating Nazi support for science and medicine can help us understand the appeal of Nazism within German intellectual culture. I also want to argue, though, that what went wrong in the Nazi period is not best understood as a subordination of the good of the individual to the good of the whole. Rather, one has to understand who was included within "the whole" and who was banished. It was not, after all, "the individual" in the abstract who suffered but particular kinds of individuals. Public health protections were extended to the "healthy" majority, while so-called "enemies of the people" were first excluded and then exterminated. Medicine was complicit in both ends of this moral scale: in public health reforms that brought the majority of Germans cleaner air and water, and in "health reforms" that involved sterilization and eventually wholesale murder.

The Example of Tobacco Research

If you ask most people when the first good evidence arose that tobacco was a major cause of lung cancer, they will point to a series of epidemiological studies by English and Americans in the early 1950s. If you ask when a medical consensus on this question first arose, they will most likely point to the 1964 Surgeon General's report, which took a strong stand on this question, or a similar report by Britain's Royal College of Physicians two years earlier [9, 10].

I have become convinced, however, that there is an earlier and overlooked consensus during the Nazi period. The Nazis had a powerful anti-tobacco movement, arguably the most powerful in the world at that time. Tobacco was opposed by racial hygienists who feared the corruption of the German germ plasm, by industrial hygienists who feared a reduction of work capacity, and by nurses and midwives who feared harms for the "maternal organism." Tobacco was said to be "a corrupting force in a rotting civilization that has become lazy." The Nazi-era anti-tobacco rhetoric drew from an earlier generation's eugenic rhetoric, combining this with an ethic of bodily purity and performance at work [11]. Tobacco use was attacked as "epidemic," as a "plague," as "dry drunkenness," and as "lung masturbation"; tobacco and alcohol abuse were "diseases of civilization" and "relics of a liberal lifestyle" [12].

Anti-tobacco research also flourished in the Nazi period. Animal experimental work demonstrated that the tar extracted from cigarette smoke could cause cancer, and physical chemists distilled tobacco tars to identify the carcinogenic components. The editor of Germany's *Monatsschrift für Krebsbekämpfung* organized animal experiments to test whether smoking causes lung cancer, putting rats in a "gas chamber" with cigarette smoke pumped in from the top—until the animals suffocated. The description of the dying rats, gasping and falling over one another, is chilling, given subsequent events [13].

Germans also pioneered what we now call experimental tobacco epidemiology, the two most striking papers being a 1939 article by Franz H. Müller of Cologne's City Hospital, and a 1943 paper by Eberhard Schairer and Erich Schöniger at Jena, presenting the most convincing demonstrations up to that time that cigarettes were a major cause of lung cancer [14, 15]. In his paper, Müller analyzed the smoking habits of 83 male lung cancer patients and compared these with the habits of age-standardized "controls" not suffering from lung cancer. His findings were clear-cut and striking: the lung cancer patients were much more likely to be heavy smokers and much less likely to be non-smokers. Sixteen percent of the healthy group were non-smokers, compared with only 3.5 percent for the lung cancer group. The 86 lung cancer patients smoked a total of 2,900 grams of tobacco per day, while the 86 healthy men smoked only 1,250 grams. Müller concluded that tobacco was not just "an important cause" of lung cancer, but also that *"the extraordinary rise in tobacco use"* was *"the single most important cause of the rising incidence of lung cancer"* in recent decades [14, emphasis in original].

Müller's work was taken one step further by Schairer and Schöniger, two physicians working at Jena's Institute for Tobacco Hazards Research. The authors were aware that German lung cancer rates were on the rise, and that many of the non-tobacco explanations of the rise were flawed (the automotive exhaust theory, for example, failed to explain the fact that rural rates were also rising). The authors drew attention to the fact that a heavy smoker could inhale as much as four kilograms of tar over a lifetime, a frightening figure given Angel H. Roffo's demonstration that animals painted with tobacco tars develop high rates of cancer [15].

Following closely the method pioneered by Müller, Schairer and Schöniger sent questionnaires to the relatives of 195 lung cancer victims, inquiring into the smoking habits of the deceased. An additional 555 questionnaires were sent to the families of patients who had died from other kinds of cancer—the presumption being that smokers would be more likely to develop certain kinds of cancer rather than others. Questionnaires were also sent to 700 male residents of Jena to determine the smoking habits among a population apparently free of cancer. The results again were clear: among the 109 lung cancer cases for which useable data were obtained, only three were non-smokers, a far lower proportion than among the population as a whole (about 3 percent, versus 16 percent for the non-cancer controls). The smokers were not necessarily "cancer prone," because when other kinds of cancer were looked at—stomach, for example—smokers were found to be no more likely to develop cancer than non-smokers. Their conclusion: smoking was very likely a cause of lung cancer, but much less likely a cause of other kinds of cancer. The results were of the "highest" statistical significance; a 1994 reevaluation of the study showed that the probability that the results could have come about by chance was less than one in ten million [16].

Questions of Interpretation

How do we interpret such studies? How do we explain the fact that Nazi Germany was home to the world's foremost tobacco-cancer epidemiology, the world's strongest cancer prevention policy, or the world's first recognition that asbestos could cause lung cancer? Do we say that "pockets of innovation" existed in Nazi Germany, resistant to ideological influence?[2] What if we find, on closer inspection, that Germany's anti-tobacco research flourished not just *despite* the Nazis, but in large part *because of* the Nazis? Is it kosher then to cite such studies?

---

[2]On the "pockets of innovation" thesis, see [17, 18].

I ask this last question, partly because the tobacco studies mentioned above have, in fact, been occasionally cited, though rarely with any mention of the social context within which they arose. (A notable exception is [19].) There is never any mention, for example, of the fact that the founding director of Schöniger and Schairer's Institute was Karl Astel, president of the University of Jena and a vicious racial hygienist and SS officer. There is never a mention of the fact that the grant application for the Institute was written by Gauleiter Fritz Sauckel, chief organizer of Germany's system of forced labor, a man hanged after the war for crimes against humanity. (Most leaders of Germany's anti-tobacco movement were silenced in one way or another after Germany's defeat. Müller disappeared in the war, Hans Reiter lost his job, Astel and Leonardo Conti committed suicide, etc.) No mention is ever made of the fact that funding for Astel's institute came from a RM 100,000 personal gift from the Führer, himself an ardent anti-smoking activist. It is clear to anyone who follows the money trail that Schairer and Schöniger's study would not have been undertaken had it not been for Hitler's interest, and the interest of several of his underlings. I might also point out that Hitler at one point even attributed the rise of German fascism to his quitting smoking: the young and struggling artist smoked a couple of packs a day until 1919, at which time he threw his cigarettes into the Danube and never reached for them again. In a 1942 conversation, the Führer said that Germany owed its liberation (i.e., the triumph of Nazism) to his quitting smoking—though I guess one could say, contra Hitler and with a hint of Freud, that sometimes giving up smoking is just giving up smoking.

What do we make of the fact that Nazi ideology in this case (and there are others) appears not just not to have hindered research, but actually to have promoted it?

In drawing attention to such studies, my intention is not to argue that there was "something good" that came from the Nazi era. I have no desire to rescue the honor of this era, or to "balance the historical record" for balance's sake. You cannot balance the heavy weight of genocide against a few flashy epidemiological studies or a glitzy V-2 engine. That is not the point—and I should add that I have little sympathy for those who argue that republishing brutal Nazi experiments may mean, as some have argued, that the victims of such experiments "may not have died in vain." They did die in vain, and a better designed life preserver or Saturn V rocket is no compensation.

My point is, rather, that it does us little good to caricature Nazi medicine as irrational or anti-science *in the abstract*. What we have to look at more carefully is the relationship between science and ideology at this time. I do not believe, for example, that the papers on tobacco epidemiology I have cited were uninfluenced by Nazi ideology. The anti-tobacco program was motivated by Nazi ideals of bodily purity and racial hygiene: there is a kind of "homeopathic paranoia" pervading Nazi ideology that led many to believe that tiny, corrosive poisons were infiltrating the German *Volkskörper*, sapping its strength, causing harm. Appreciating this helps us understand how Nazi medical ideologues could argue that lead, mercury, asbestos, Jews, and tobacco tars were all a menace to the German body. I don't believe that doctors were "doubling" when they sought to cleanse the German population of those elements; nor do I believe that we can identify a sharp boundary between the science and the politics in this process. The two are painfully intertwined.

Why were German doctors such avid fans of fascism? I don't think it was the tirades of Julius Streicher in *Der Stürmer* that caught their eye, but rather the promises of Nazi leaders to solve Germany's problems medically, surgically. The Nazi state was supposed to be a hygienic state; Nazism was supposed to be "applied biology." (Fritz Lenz coined this widely cited phrase in his popular human genetics textbook of 1931 [20].) Hitler was celebrated as the "Robert Koch of politics" and the "great doctor" of German society. The seductive power of National Socialism lay in its promise to cleanse German society of its corrosive elements—not just communism and Jews, but pollutants in the air and water, along with TB, homosexuality, and the "burdensome" mentally ill—maladies which could all be traced, in the Nazi view of the world, to the "false humanitarianism" of previous political regimes. The doctors in this process were not victims but coconspirators, seduced by the powers being offered to them and the promises of an orderly, hygienic state.

Nazi Medical Ethics

We often hear that the Nazis abandoned ethics. A recent Israeli film on the euthanasia operation says that Nazi doctors' overzealous scientific curiosity led them "to abandon all moral sense in the pursuit of medical knowledge" [21]. The image is of the unfettered quest for knowledge, a kind of scientific zealotry reminiscent of the Faustian bargain, of science practiced "without limits" or of an overly "aggressive search for the truth."

The problem with this view is that there were in fact ethical standards at this time. Medical students took courses on medical ethics, and medical textbooks from the time treated medical ethics.[3] There are discussions in German journals of the obligations of physicians to society, to the state, and sometimes even to the individual. Nazi medical philosophers were critical of the ideal of "neutral" or value-free science, which was often equated with apathetic ivory-tower liberal-Jewish "science for its own sake." Science was supposed to serve the German Volk, the healthy and productive white races of Europe. We have to distinguish between no ethics and a lot of bad ethics, between chaos and evil.

Surprisingly, there never has been a systematic study of medical ethics under the Nazis. We know a lot about the postwar rationalizations of physicians facing death at the hands of Allied prosecutors [24, pp. 133–34], but we need to know more about the implicit and explicit norms of the era; we need to know more about bedside manners and student-teacher relations, and what kinds of things were discussed when medical malpractice suits went to court. Such questions are especially intriguing, given that Germany prior to the Nazi era had taken steps to protect patients' rights in the experimental context. The 1900 code promulgated by the Prussian Ministry of Religion, Education, and Medical Affairs, for example, was the world's first official regulation of human experimentation, barring non-therapeutic interventions without voluntary consent, along with experiments on minors and others judged vulnerable or incompetent. Experiments had to be authorized by the director of the institution involved, and records had to be kept in writing [24, p. 127].

A 1931 code issued by the Reich Health Office strengthened sanctions against inappropriate human experiments. The edict was issued in response to a growing public clamor about experimental abuses—most notably the deaths of 75 children in Lübeck in the course of testing an experimental TB vaccine. The 1931 guidelines banned experiments on the dying and on anyone under the age of 18 if it posed a risk. The document has been called the world's most comprehensive code governing human experimentation; in certain respects it is stricter even than the subsequent Nuremberg Code or Helsinki Accord [24, pp. 121–44].

Medical ethical discourse continued throughout the Nazi period. On 24 November 1933, for example, a law for the protection of cruelty to animals was passed, barring experimentation causing pain or injury to animals. The law specifically disallowed experiments involving exposure to cold, heat, or infection. Hermann Göring threatened to toss vivisectionists into a concentration camp, though it is unclear whether this ever took place. Jewish doctors were commonly accused of excessive experimental zeal: Erwin Liek, a kind of spiritual godfather for many Nazi physicians, linked abuses in this sphere with the "Jewish abstract scientific spirit" [25, 26].

There was also a great deal of discussion of physicians' duties and responsibilities—especially their obligations to serve ("unflinchingly") the Volk and the Führer. Doctors were instructed to counsel their patients against tobacco use [27], and were enlisted in the task of safeguarding public and genetic health. The job of maintaining labor productivity fell to physicians, who also were asked to devote increasing attention to neonatal care. There are discussions of who should be allowed to practice medicine—whether natural healers were to be barred from treating cancer patients, for example (they were) [28]—and there are discussions of the limits of medical confidentiality and of medical disclosure. A 1943 article in a leading cancer journal cited the "demands of medical ethics" to inform patients of the severity of their diseases, and in at least one case a physician was prosecuted for failing to inform a woman she had cancer [29].

We also know that doctors did not stand idly by when they thought things were going wrong. There was a culture of complaint in the Nazi period: doctors complained that sterilizations were going too fast or too slow, that too many people were being X-rayed or too few, and so forth. Physicians complained about the use of fancy-sounding obfuscatory Latin to characterize diseases (doctors were supposed to use plain German, so patients could understand the nature of their illness), and thousands of women were said to be needlessly dying from lack of access to preventive care (e.g., regular, cost-free gynecologic exams, proposed by many Nazi doctors) [30, 31].

The Nazi medical profession, in other words, was not without its ethics. Nazi medical ethics was typically sexist and racialist, but there was also stress on cleanliness, punctuality, orderliness, and obedience to legal authority, especially the supreme authority of Adolf Hitler, the beloved Führer. Nazi medical ethics tended to reduce morality to efficiency, economics,

---

[3]An example of Nazi medical ethical discussion is [22]. The most widely cited medical ethics text at this time was [23].

and aesthetics, relegating to the trash heap everything that was seen as ugly and a burden. Preventive medicine was emphasized, as was Nordic (and especially Germanic) supremacy, cost-efficiency, the natural lifestyle, and the superiority of the productive worker over persons deemed inferior or burdensome.

Ethical norms are implicit even in the most horrific experimental practices of the camps: how else does one explain the fact that "healthy" German citizens were never experimented on? Those subjected to human experimental violence were invariably people judged less than fully human in the Nazi scale of values. Jews and Gypsies were considered diseased races, tumors in the German body politic; Russian POWs were vermin deserving enslavement or extermination; the unproductive handicapped were encumbrances on the German Volk, warranting "euthanasia."

It is important not to make a caricature of Nazi ideology. The doctors who exploited prisoners at Buchenwald or Dachau were not morally blind or devoid of the power of moral reflection. Upholding such a view would almost render the guilty parties not responsible for their actions: you cannot hold guilty parties accountable if they are lunatics devoid of morals. Some of the worst were thinking men, acting consistently within a frame of callous and often criminal logic.

Research Integrity versus Research Ethics

The U.S. Office of Research Integrity (ORI) defines "scientific misconduct" to include "fabrication, falsification . . . or other practices that seriously deviate from those that are commonly accepted within the scientific community for proposing, conducting, or reporting research." The NSF similarly defines misconduct as "fabrication, falsification, plagiarism, or other serious deviation from accepted practices in proposing, carrying out, or reporting results from activities funded by the NSF" [32]. What is interesting about these definitions is that they avoid the difficult issue of how one deals with large-scale, deeply ingrained prejudices that may be widely accepted in both the scientific community and the society as a whole. Narrow definitions of scientific misconduct say nothing about whether a given research practice may be abusive, or racist, or sexist—nothing about its larger context or implications.

Some of the worst Nazi research cannot even be considered misconduct according to such definitions.

Most Nazi doctors did not lie, or cheat, or misrepresent their credentials. They did not falsify or fabricate data to an unusual degree, and there is little evidence of plagiarism. The evil must be sought elsewhere.

The primary failing of Nazi medicine, I would argue, was the failure of physicians to challenge the rotten, substantive core of Nazi values. Too many physicians were willing to go with the political flow; too many were unwilling to resist, to "deviate" from "commonly accepted" practices. The ORI definition of "misconduct" misses this larger point entirely. Ian Kershaw reminds us that "the road to Auschwitz was paved with indifference"; I would only add that conformist urges blocked the many side paths leading in other directions. Doctors should have acted up, broken with convention, defended their patients against the long arm of the law.

The Black in White, the White in Black

My point is not to rescue the honor of medicine in its darkest hour, but rather to stress its subtlety and complexity. The history of medicine in this period is a history of both forcible sterilization and herbal remedies; we cannot forget the crimes of a Karl Brandt or a Hermann Voss [33], but we also should not forget that the SS built the world's largest botanical medical garden in Dachau, or that German nutritionists mandated the production of whole-grain bread. Fascist physicians willfully murdered their handicapped patients, but organic farming and species protection were also going concerns. The question is not one of balance, but of the proper understanding of origins, context, continuities, and contradictions. It is part of the horror of this period, that such an "advanced" technological society could fall so far into butchery and barbarism.

I do not believe there is an inherently totalitarian tendency in modern science, but I do think it is important to recognize that, just as the routine practice of science is not incompatible with the routine exercise of cruelty, so the dictatorial and murderous aspirations of fascism were not necessarily at odds with the promotion of cutting edge science and progressive public health—at least for certain elements of the population. The exclusive focus on the heinous aspects of Nazi medical practice makes it easy for us to relegate the events of this era to the monstrous or otherworldly, but there is more to the story than "medicine gone mad." The Nazi campaign against carcinogenic food dyes, the world-class asbestos and tobacco epidemiology, and much else as well, are all in some sense as fascist as the yellow stars and the death camps. There is sometimes white in black, and black in white; appreciating some of these subtler speckles and shadings may open our eyes to new kinds of continuities binding the past to the present. It may also help us better see how fascism triumphed in the first place.[4]

References

1. Taylor, T. Opening statement of the prosecution, December 9, 1946. In *The Nazi Doctors and the Nuremberg Code*, edited by G. J. Annas and M. A. Grodin. New York: Oxford Univ. Press, 1992. 69.

2. Neumann, F. *Behemoth: The Structure and Practice of National Socialism*. New York: Oxford Univ. Press, 1942. 471–72.

3. Berger, R. Nazi science: The Dachau hypothermia experiments. *New Engl. J. M.* 322:1435–40, 1990.

4. Fischer, W., et al., eds. *Exodus von Wissenschaften aus Berlin: Fragestellung, Ergebnisse, Desiderate.* Berlin: de Gruyter, 1994.

5. Hunt, L. *Secret Agenda: The United States Government, Nazi Scientists, and Project Paperclip, 1945–1990.* New York: St. Martin's Press, 1991.

6. Sea, G. Time distance, shielding: Human radiation experiments and the laying of blame. Paper presented to the Conference on Humans in Experiments, Hamburger Institut für Sozialforschung, 28 June 1995. 9. See also his The radiation story no one would touch. *Columbia Journalism Rev.* (March/April) 1994, 37–40.

7. Enterline, P. E. Changing attitudes and opinions regarding asbestos and cancer 1934–1965," *Am. J. Industr. Med.* 20:685–700, 1991.

8. Proctor, R. N. *The Nazi War on Cancer.* Princeton: Princeton Univ. Press, 1999.

9. Doll, R. Uncovering the effects of smoking: Historical perspective. *Stat. Methods Med. Res.* 7:87–117, 1998.

10. Proctor, R. N. The anti-tobacco campaign of the Nazis: A little known aspect of public health in Germany, 1933–1945. *BMJ* 313:1450–53, 1996.

11. Proctor, R. N. *Racial Hygiene: Medicine Under the Nazis.* Cambridge: Harvard Univ. Press, 1988. 228, 239–40.

12. Proctor, R. N. The Nazi war on Tobacco: Ideology, evidence, and public health consequences. *Bull. Hist. Med.* 71:435–88, 1997.

13. Mertens, V. E. Zigarettenrauch eine Ursache des Lungenkrebses? (Eine Anregung). *Zeitschr. Krebsforschung* 32:82–91, 1930.

14. Müller, F. H. Tabakmissbrauch und Lungencarcinom. *Zeitsch. Krebsforschung* 49:78, 1939.

15. Schairer, E., and E. Schöniger. Lungenkrebs und Tabakverbrauch. *Zeitschr. Krebsforschung* 54:261–69, 1943.

16. Smith, G. D., S. A. Ströbele, and M. Egger. Smoking and health promotion in Nazi Germany. *J. Epidem. Comm. Health* 48:220, 1994.

17. Macrakis, K. *Surviving the Swastika: Scientific Research in Nazi Germany.* New York: Oxford Univ. Press, 1993. 110–11.

18. Proctor, R. N. Nazi doctors, racial medicine, and human experimentation. In *The Nazi Doctors and the Nuremberg Code*, edited by G. J. Annas and M. A. Grodin. New York: Oxford Univ. Press, 1992. 17–31.

19. Smith, G. D., S. A. Ströbele, and M. Egger. Smoking and death. *BMJ* 310:396, 1995.

20. Lenz, F. *Menschliche Auslese und Rassenhygiene (Eugenik)*, 3rd ed. Munich: J. F. Lehmann Verlag, 1931. 417.

21. Fishkoff, S. They called it mercy killing. *Jerusalem Post (International Ed.)*, 20 April 1996, p. 17.

22. Ramm, R. *Arztliche Rechts- und Slandeskunde.* Berlin: De Gruyter, 1943.

23. Moll, A. *Ärztliche Ethik: Die Pflichten des Arztes in allen Beziehungen seiner Thätigkeit.* Stuttgart: F. Enke, 1902.

24. Grodin, M. A. Historical origins of the Nuremberg Code. In *The Nazi Doctors and the Nuremberg Code*, edited by G. J. Annas and M. A. Grodin. New York: Oxford Univ. Press, 1992. 133–34.

25. Liek, E. *Der Arzt und seine Sendung.* Munich: J. F. Lehmanns Verlag, 1926.

26. Like, E. *Der Welt des Arztes.* Munich: J. F. Lehmanns Verlag, 1933.

27. Klarner, W *Vom Rauchen: Eine Sucht und ihre Bekämpfung.* Nuremberg: Med. Diss., 1940. 43.

28. Steinwallner, B. Zur Pflicht des Heilpraktikers, bei Krebskranken sofort einen Arzt zuzuziehen. *Monatsschr. Krebsbekämpfung* 7:137–38, 1940.

---

[4]I make a similar argument in [34].

29. Schläger, G. Aufklärung und Verschwiegenheit bei Krebsverdacht. *Monatsschr. Krebsbekämpfung* 10:154, 1943.

30. Niderehe, W. Bericht über die Vorträge betr. Krebsbekämpfung auf dem 3. internat. Kongress f.d. ärztl. Fortbildungswesen in Berlin. 24 Aug. 1937, E 1496 Thür. Landeshauptarchiv Weimar.

31. Hinselmann, H. 14 Jahre Kolposkopie: Ein Rückblick und Ausblick. *Hippokrates* 9:661–67, 1938.

32. Parrish, D. The scientific misconduct definition and falsification of credentials. *Prof. Ethics Report* 9 (Fall):6, 1996.

33. Aly, G., P. Chroust, and C. Pross. *Cleansing the Fatherland: Nazi Medicine and Racial Hygiene.* Baltimore: Johns Hopkins Univ. Press, 1994.

34. Proctor, R. N. Why did the Nazis have the world's most aggressive anti-cancer campaign? *Endeavor* 23: 76–79, 1999. Compare also my *Nazi War on Cancer*, pp. 277–78.

The Atomic Bombings Reconsidered

Barton J. Bernstein

The Questions America Should Ask

Fifty years ago, during a three-day period in August 1945, the United States dropped two atomic bombs on Japan, killing more than 115,000 people and possibly as many as 250,000, and injuring at least another 100,000. In the aftermath of the war, the bombings raised both ethical and historical questions about why and how they were used. Would they have been used on Germany? Why were cities targeted so that so many civilians would be killed? Were there likely alternative ways to end the war speedily and avoid the Allies' scheduled November 1, 1945, invasion of Kyushu?

Such questions often fail to recognize that, before Hiroshima and Nagasaki, the use of the A-bomb did not raise profound moral issues for policymakers. The weapon was conceived in a race with Germany, and it undoubtedly would have been used against Germany had the bomb been ready much sooner. During the war, the target shifted to Japan. And during World War II's brutal course, civilians in cities had already become targets. The grim Axis bombing record is well known. Masses of noncombatants were also intentionally killed in the later stages of the American air war against Germany; that tactic was developed further in 1945 with the firebombing of Japanese cities. Such mass bombing constituted a transformation of morality, repudiating President Franklin D. Roosevelt's prewar pleas that the warring nations avoid bombing cities to spare civilian lives. Thus, by 1945, American leaders were not seeking to avoid the use of the A-bomb on Japan. But the evidence from current archival research shows that by pursuing alternative tactics instead, they probably could still have obviated the dreaded invasion and ended the war by November.

Shifting from Germany to Japan

In 1941, urged by émigré and American scientists, President Roosevelt initiated the atomic bomb project—soon code-named the Manhattan Project—amid what was believed to be a desperate race with Hitler's Germany for the bomb. At the beginning, Roosevelt and his chief aides assumed that the A-bomb was a legitimate weapon that would be used first against Nazi Germany. They also decided that the bomb project should be kept secret from the Soviet Union, even after the Soviets became a wartime ally, because the bomb might well give the United States future leverage against the Soviets.

By mid-1944, the landscape of the war had changed. Roosevelt and his top advisers knew that the likely target would now be Japan, for the war with Germany would undoubtedly end well before the A-bomb was expected to be ready, around the spring of 1945. In a secret September 1944 memorandum at Hyde Park, Roosevelt and British Prime Minister Winston Churchill ratified the shift from Germany to Japan. Their phrasing suggested that, for the moment anyway, they might have had some slight doubts about actually using the bomb, for they agreed that "it might *perhaps*, after mature consideration, be used against the Japanese" (my emphasis).

Four days later, mulling over matters aloud with a visiting British diplomat and chief U.S. science adviser Vannevar Bush, Roosevelt briefly wondered whether the A-bomb should be dropped on Japan or whether it should be demonstrated in America, presumably with Japanese observers, and then used as a threat. His speculative notion seemed so unimportant and so contrary to the project's long-standing operating assumptions that Bush actually forgot about it when he prepared a memo of the meeting. He only recalled the president's remarks a day later and then added a brief paragraph to another memorandum.

Put in context alongside the dominant assumption that the bomb would be used against the enemy, the significance of F.D.R.'s occasional doubts is precisely that they were so occasional—expressed twice in almost four years. All of F.D.R.'s advisers who knew about the bomb always unquestioningly assumed that it would be used. Indeed, their memoranda frequently spoke of "after it is used" or "when it is used," and never "*if* it is used." By about mid-1944, most had comfortably concluded that the target would be Japan.

The bomb's assumed legitimacy as a war weapon was ratified bureaucratically in September 1944 when General Leslie Groves, the director of the Manhattan Project, had the air force create a special group—the 509th Composite Group with 1,750 men—to begin practicing to drop atomic bombs. So dominant was the assumption that the bomb would be used against Japan that only one high-ranking Washington official, Undersecretary of War Robert Patterson, even questioned this notion after V-E Day. He wondered whether the defeat of Germany on May 8, 1945, might alter the plans for dropping the bomb on Japan. It would not.

The Assumption of Use

The Manhattan Project, costing nearly $2 billion, had been kept secret from most cabinet members and nearly all of Congress. Secretary of War Henry L. Stimson, a trusted Republican, and General George C. Marshall, the equally respected army chief of staff, disclosed the project to only a few congressional leaders. They smuggled the necessary appropriations into the War Department budget without the knowledge—much less the scrutiny—of most congressmen, including most members of the key appropriations committees. A conception of the national interest agreed upon by a few men from the executive and legislative branches had revised the normal appropriations process.

In March 1944, when a Democratic senator heading a special investigating committee wanted to pry into this expensive project, Stimson peevishly described him in his diary as "a nuisance and pretty untrustworthy . . . He talks smoothly but acts meanly." That man was Senator Harry S Truman. Marshall persuaded him not to investigate the project, and thus Truman did not learn any more than that it involved a new weapon until he was suddenly thrust into the presidency on April 12, 1945.

In early 1945, James F. Byrnes, then F.D.R.'s "assistant president" for domestic affairs and a savvy Democratic politician, began to suspect that the Manhattan Project was a boondoggle. "If [it] proves a failure," he warned Roosevelt, "it will be subjected to relentless investigation and criticism." Byrnes' doubts were soon overcome by Stimson and Marshall. A secret War Department report, with some hyperbole, summarized the situation: "If the project succeeds, there won't be any investigation. If it doesn't, they won't investigate anything else."

Had Roosevelt lived, such lurking political pressures might have powerfully confirmed his intention to use the weapon on the enemy—an assumption he had already made. How else could he have justified spending roughly $2 billion, diverting scarce materials from other war enterprises that might have been even more useful, and bypassing Congress? In a nation still unprepared to trust scientists, the Manhattan Project could have seemed a gigantic waste if its value were not dramatically demonstrated by the use of the atomic bomb.

Truman, inheriting the project and trusting both Marshall and Stimson, would be even more vulnerable to such political pressures. And, like F.D.R., the new president easily assumed that the bomb should and would be used. Truman never questioned that assumption. Bureaucratic developments set in motion before he entered the White House reinforced his belief. And his aides, many inherited from the Roosevelt administration, shared the same faith.

Picking Targets

Groves, eager to retain control of the atomic project, received Marshall's permission in early spring 1945 to select targets for the new weapon. Groves and his associates had long recognized that they were considering a weapon of a new magnitude, possibly equivalent to the "normal bombs carried by [at least] 2,500 bombers." And they had come to assume that the A-bomb would be "detonated well above ground, relying primarily on blast effect to do material damage, [so that even with] minimum probable efficiency, there will be the maximum number of structures (dwellings and factories) damaged beyond repair."

On April 27, the Target Committee, composed of Groves, army air force men like General Lauris Norstad, and scientists including the great mathematician John Von Neumann, met for the first time to discuss how and where in Japan to drop the bomb. They did not want to risk wasting the precious weapon, and decided that it must be dropped visually and not by radar, despite the poor weather conditions in Japan during the summer, when the bomb would be ready.

Good targets were not plentiful. The air force, they knew, "was systematically bombing out the following cities with the prime purpose . . . of not leaving one stone lying on another: Tokyo, Yokohama, Nagoya, Osaka, Kyoto, Kobe, Yawata, and Nagasaki . . . The air force is operating primarily to laying [sic] waste all the main Japanese cities . . . Their existing procedure is to bomb the hell out of Tokyo."

By early 1945, World War II—especially in the Pacific—had become virtually total war. The fire-bombing of Dresden had helped set a precedent for the U.S. air force, supported by the American people, to intentionally kill mass numbers of Japanese citizens. The earlier moral insistence on noncombatant immunity crumbled during the savage war. In Tokyo, during March 9–10, a U.S. air attack killed about 80,000 Japanese civilians. American B-29S dropped napalm on the city's heavily populated areas to produce uncontrollable firestorms. It may even have been easier to conduct this new warfare outside Europe and against Japan because its people seemed like "yellow subhumans" to many rank-and-file American citizens and many of their leaders.

In this new moral context, with mass killings of an enemy's civilians even seeming desirable, the committee agreed to choose "large urban areas of not less than three miles in diameter existing in the larger populated areas" as A-bomb targets. The April 27 discussion focused on four cities: Hiroshima, which, as "the largest untouched target not on the 21st Bomber Command priority list," warranted serious consideration; Yawata, known for its steel industry; Yokohama; and Tokyo, "a possibility [though] now practically all bombed and burned out and . . . practically rubble with only the palace grounds left standing." They decided that other areas warranted more consideration: Tokyo Bay, Kawasaki, Yokohoma, Nagoya, Osaka, Kobe, Kyoto, Hiroshima, Kure, Yawata, Kokura, Shimonoseki, Yamaguchi, Kumamoto, Fukuoka, Nagasaki, and Sasebo.

The choice of targets would depend partly on how the bomb would do its deadly work—the balance of blast, heat, and radiation. At their second set of meetings, during May 11–12, physicist J. Robert Oppenheimer, director of the Los Alamos laboratory, stressed that the bomb material itself was lethal enough for perhaps a billion deadly doses and that the weapon would give off lethal radioactivity. The bomb, set to explode in the air, would deposit "a large fraction of either the initial active material or the radioactive products in the immediate vicinity of

the target; but the radiation . . . will, of course, have an effect on exposed personnel in the target area." It was unclear, he acknowledged, what would happen to most of the radioactive material: it could stay for hours as a cloud above the place of detonation or, if the bomb exploded during rain or in high humidity and thus caused rain, "most of the active material will be brought down in the vicinity of the target area." Oppenheimer's report left unclear whether a substantial proportion or only a small fraction of the population might die from radiation. So far as the skimpy records reveal, no member of the Target Committee chose to dwell on this matter. They probably assumed that the bomb blast would claim most of its victims before the radiation could do its deadly work.

In considering targets, they discussed the possibility of bombing the emperor's palace in Tokyo and "agreed that we should not recommend it but that any action for this bombing should come from authorities on military policy." They decided to gather information on the effectiveness of using the bomb on the palace.

The Target Committee selected their four top targets: Kyoto, Hiroshima, Yokohama, and Kokura Arsenal, with the implication that Niigata, a city farther away from the air force 509th group's Tinian base, might be held in reserve as a fifth. Kyoto, the ancient former capital and shrine city, with a population of about a million, was the most attractive target to the committee. "From the psychological point of view," the committee minutes note, "there is the advantage that Kyoto is an intellectual center for Japan and [thus] the people there are more apt to appreciate the significance of such a weapon." The implication was that those in Kyoto who survived the A-bombing and saw the horror would be believed elsewhere in Japan.

Of central importance, the group stressed that the bomb should be used as a terror weapon—to produce "the greatest psychological effect against Japan" and to make the world, and the U.S.S.R. in particular, aware that America possessed this new power. The death and destruction would not only intimidate the surviving Japanese into pushing for surrender, but, as a bonus, cow other nations, notably the Soviet Union. In short, America could speed the ending of the war and by the same act help shape the postwar world.

By the committee's third meeting, two weeks later, on May 28, they had pinned down matters. They chose as their targets (in order) Kyoto, Hiroshima, and Niigata, and decided to aim for the center of each city. They agreed that aiming for industrial areas would be a mistake because such targets were small, spread on the cities' fringes, and quite dispersed. They also knew that bombing was imprecise enough that the bomb might easily miss its mark by a fifth of a mile, and they wanted to be sure that the weapon would show its power and not be wasted.

The committee understood that the three target cities would be removed from the air force's regular target list, reserving them for the A-bomb. But, the members were informed, "with the current and prospective rate of . . . bombings, it is expected to complete strategic bombing of Japan by 1 Jan 46 so availability of future [A-bomb] targets will be a problem." In short, Japan was being bombed out.

The Ratification of Terror Bombing

On may 28, 1945, physicist Arthur H. Compton, a Nobel laureate and member of a special scientific panel advising the high-level Interim Committee newly appointed to recommend policy about the bomb, raised profound moral and political questions about how the atomic bomb would be used. "It introduces the question of mass slaughter, really for the first time in history," he wrote. "It carries with it the question of possible radioactive poison over the area bombed. Essentially, the question of the use . . . of the new weapon carries much more serious implications than the introduction of poison gas."

Compton's concern received some independent support from General Marshall, who told Secretary Stimson on May 29 that the A-bomb should first be used not against civilians but against military installations—perhaps a naval base—and then possibly against large manufacturing areas after the civilians had received ample warnings to flee. Marshall feared "the opprobrium which might follow from an ill considered employment of such force." A graduate of Virginia Military Institute and a trained soldier, Marshall struggled to retain the older code of not *intentionally* killing civilians. The concerns of Compton the scientist and Marshall the general, their values so rooted in an earlier conception of war that sought to spare noncombatants, soon gave way to the sense of exigency, the desire to use the bomb on people, and the unwillingness or inability of anyone near the top in Washington to plead forcefully for maintaining this older morality.

On May 31, 1945, the Interim Committee, composed of Stimson, Bush, Harvard President James Conant, physicist and educator Karl T. Compton,

Secretary of State designate James F. Byrnes, and a few other notables, discussed the A-bomb. Opening this meeting, Stimson, the aged secretary of war who had agonized over the recent shift toward mass bombing of civilians, described the atomic bomb as representing "a new relationship of man to the universe. This discovery might be compared to the discoveries of the Copernican theory and the laws of gravity, but far more important than these in its effects on the lives of men."

Meeting, as they were, some six weeks before the first nuclear test at Alamogordo, they were still unsure of the power of this new weapon. Oppenheimer told the group that it would have an explosive force of between 2,000 and 20,000 tons of TNT. Its visual effect would be tremendous. "It would be accompanied by a brilliant luminescence which would rise to a height of 10,000 to 20,000 feet," Oppenheimer reported. "The neutron effect [radiation] would be dangerous to life for a radius of at least two-thirds of a mile." He estimated that 20,000 Japanese would be killed.

According to the committee minutes, the group discussed "various types of targets and the effects to be produced." Stimson "expressed the conclusion, on which there was general agreement, that we could not give the Japanese any warning; that we could not concentrate on a civilian area; but that we should seek to make a profound psychological impression on as many of the inhabitants as possible. At the suggestion of Dr. Conant, the secretary agreed that the most desirable target would be a vital war plant employing a large number of workers and closely surrounded by workers' houses."

Directed by Stimson, the committee was actually endorsing terror bombing—but somewhat uneasily. They would not focus exclusively on a military target (the older morality), as Marshall had recently proposed, nor fully on civilians (the emerging morality). They managed to achieve their purpose—terror bombing—without bluntly acknowledging it to themselves. All knew that families—women, children, and, even in the daytime, during the bomb attack, some workers—dwelled in "workers' houses."

At the committee's morning or afternoon session, or at lunch, or possibly at all three times—different members later presented differing recollections—the notion of a noncombat demonstration of the A-bomb came up. The issue of how to use the bomb was not even on Stimson's agenda, nor was it part of the formal mandate of the Interim Committee, but he may have showed passing interest in the subject of a noncombat demonstration. They soon rejected it. It was deemed too risky for various reasons: the bomb might not work, the Japanese air force might interfere with the bomber, the A-bomb might not adequately impress the Japanese militarists, or the bomb might incinerate any Allied POWs whom the Japanese might place in the area.

The discussion on May 31 had focused substantially on *how* to use the bomb against Japan. At one point some of the members had considered trying several A-bomb strikes at the same time and presumably on the same city. Groves opposed this notion, partly on the grounds that "the effect would not be sufficiently distinct from our regular air force bombing program." Like the others, he was counting on the dramatic effect of a single bomb, delivered by a single plane, killing many thousands. It was not new for the air force to kill so many Japanese, but this method would be new. And the use of the new weapon would carry, as stressed by American proclamations in early August, the likelihood of more nuclear attacks on Japanese cities—a continuing "rain of ruin."

Two weeks after the Interim Committee meeting, on June 16, after émigré physicists James Franck and Leo Szilard and some colleagues from the Manhattan Project's Chicago laboratory raised moral and political questions about the surprise use of the bomb on Japan, a special four-member scientific advisory committee disposed of the matter of a noncombat demonstration. The group was composed of physicists Arthur Compton, J. Robert Oppenheimer, Enrico Fermi, and Ernest O. Lawrence. By one report, Lawrence was the last of the four to give up hope for a noncombat demonstration. Oppenheimer, who spoke on the issue in 1954 and was not then controverted by the other three men, recalled that the subject of a noncombat demonstration was not the most important matter dealt with during the group's busy weekend meeting and thus did not receive much attention. On June 16, the four scientists concluded: "We can propose no technical demonstration likely to bring an end to the war; we see no acceptable alternative to direct military use."

At that time, as some members of the scientific panel later grudgingly acknowledged, they knew little about the situation in Japan, the power of the militarists there, the timid efforts by the peace forces there to move toward a settlement, the date of the likely American invasion of Kyushu, and the power of the still untested A-bomb. "We didn't know beans

about the military situation," Oppenheimer later remarked pungently.

But even different counsel by the scientific advisers probably could not have reversed the course of events. The bomb had been devised to be used, the project cost about $2 billion, and Truman and Byrnes, the president's key political aide, had no desire to avoid its use. Nor did Stimson. They even had additional reasons for wanting to use it: the bomb might *also* intimidate the Soviets and render them tractable in the postwar period.

Stimson emphasized this theme in a secret memorandum to Truman on April 25: "If the problem of the proper use of this weapon can be solved, we should then have the opportunity to bring the world into a pattern in which the peace of the world and our civilization can be saved." Concern about the bomb and its relationship to the Soviet Union dominated Stimson's thinking in the spring and summer of 1945. And Truman and Byrnes, perhaps partly under Stimson's tutelage, came to stress the same hopes for the bomb.

The Agonies of Killing Civilians

During 1945, Stimson found himself presiding, with agony, over an air force that killed hundreds of thousands of Japanese civilians. Usually, he preferred not to face these ugly facts, but sought refuge in the notion that the air force was actually engaged in precision bombing and that somehow this precision bombing was going awry. Caught between an older morality that opposed the intentional killing of noncombatants and a newer one that stressed virtually total war, Stimson could neither fully face the facts nor fully escape them. He was not a hypocrite but a man trapped in ambivalence.

Stimson discussed the problem with Truman on June 6. Stimson stressed that he was worried about the air force's mass bombing, but that it was hard to restrict it. In his diary, Stimson recorded: "I told him I was anxious about this feature of the war for two reasons: first, because I did not want to have the United States get the reputation of outdoing Hitler in atrocities; and second, I was a little fearful that before we could get ready the air force might have Japan so thoroughly bombed out that the new weapon would not have a fair background to show its strength." According to Stimson, Truman "laughed and said he understood."

Unable to reestablish the old morality and wanting the benefits for America of the new, Stimson proved decisive—even obdurate—on a comparatively small matter: removing Kyoto from Groves' target list of cities. It was not that Stimson was trying to save Kyoto's citizens; rather, he was seeking to save its relics, lest the Japanese become embittered and later side with the Soviets. As Stimson explained in his diary entry of July 24: "The bitterness which would be caused by such a wanton act might make it impossible during the long post-war period to reconcile the Japanese to us in that area rather than to the Russians. It might thus . . . be the means of preventing what our policy demanded, namely, a sympathetic Japan to the United States in case there should be any aggression by Russia in Manchuria."

Truman, backing Stimson on this matter, insisted privately that the A-bombs would be used only on military targets. Apparently the president wished not to recognize the inevitable—that a weapon of such great power would necessarily kill many civilians. At Potsdam on July 25, Truman received glowing reports of the vast destruction achieved by the Alamogordo blast and lavishly recorded the details in his diary: a crater of 1,200 feet in diameter, a steel tower destroyed a half mile away, men knocked over six miles away. "We have discovered," he wrote in his diary, "the most terrible bomb in the history of the world. It may be the fire destruction prophesied." But when he approved the final list of A-bomb targets, with Nagasaki and Kokura substituted for Kyoto, he could write in his diary, "I have told Sec. of War . . . Stimson to use it so that military objectives and soldiers and sailors are the target and not women and children. Even if the Japs are savages, ruthless, merciless, and fanatic . . . [t]he target will be a purely military one." Truman may have been engaging in self-deception to make the mass deaths of civilians acceptable.

Neither Hiroshima nor Nagasaki was a "purely military" target, but the official press releases, cast well before the atomic bombings, glided over this matter. Hiroshima, for example, was described simply as "an important Japanese army base." The press releases were drafted by men who knew that those cities had been chosen partly to dramatize the killing of noncombatants.

On August 10, the day after the Nagasaki bombing, when Truman realized the magnitude of the mass killing and the Japanese offered a conditional surrender requiring continuation of the emperor, the president told his cabinet that he did not want to kill any more women and children. Rejecting demands to drop more atomic bombs on Japan, he hoped not to use

them again. After two atomic bombings, the horror of mass death had forcefully hit the president, and he was willing to return partway to the older morality—civilians might be protected from A-bombs. But he continued to sanction the heavy conventional bombing of Japan's cities, with the deadly toll that napalm, incendiaries, and other bombs produced. Between August 10 and August 14—the war's last day, on which about 1,000 American planes bombed Japanese cities, some delivering their deadly cargo after Japan announced its surrender—the United States probably killed more than 15,000 Japanese.

The Roads Not Taken

Before august 10, Truman and his associates had not sought to avoid the use of the atomic bomb. As a result, they had easily dismissed the possibility of a noncombat demonstration. Indeed, the post-Hiroshima pleas of Japan's military leaders for a final glorious battle suggest that such a demonstration probably would not have produced a speedy surrender. And American leaders also did not pursue other alternatives: modifying their unconditional surrender demand by guaranteeing the maintenance of the emperor, awaiting the Soviet entry into the war, or simply pursuing heavy conventional bombing of the cities amid the strangling naval blockade.

Truman and Byrnes did not believe that a modification of the unconditional surrender formula would produce a speedy surrender. They thought that guaranteeing to maintain the emperor would prompt an angry backlash from Americans who regarded Hirohito as a war criminal, and feared that this concession might embolden the Japanese militarists to expect more concessions and thus prolong the war. As a result, the president and his secretary of state easily rejected Stimson's pleas for a guarantee of the emperor.

Similarly, most American leaders did not believe that the Soviet entry into the Pacific war would make a decisive difference and greatly speed Japan's surrender. Generally, they believed that the U.S.S.R.'s entry would help end the war—ideally, before the massive invasion of Kyushu. They anticipated Moscow's intervention in mid-August, but the Soviets moved up their schedule to August 8, probably because of the Hiroshima bombing, and the Soviet entry did play an important role in producing Japan's surrender on August 14. Soviet entry without the A-bomb *might* have produced Japan's surrender before November.

The American aim was to avoid, if possible, the November 1 invasion, which would involve about 767,000 troops, at a possible cost of 31,000 casualties in the first 30 days and a total estimated American death toll of about 25,000. And American leaders certainly wanted to avoid the second part of the invasion plan, an assault on the Tokyo plain, scheduled for around March 1, 1946, with an estimated 15,000–21,000 more Americans dead. In the spring and summer of 1945, no American leader believed—as some later falsely claimed—that they planned to use the A-bomb to save half a million Americans. But, given the patriotic calculus of the time, there was no hesitation about using A-bombs to kill many Japanese in order to save the 25,000–46,000 Americans who might otherwise have died in the invasions. Put bluntly, Japanese life—including civilian life—was cheap, and some American leaders, like many rank-and-file citizens, may well have savored the prospect of punishing the Japanese with the A-bomb.

Truman, Byrnes, and the other leaders did not have to be reminded of the danger of a political backlash in America if they did not use the bomb and the invasions became necessary. Even if they had wished to avoid its use—and they did not—the fear of later public outrage spurred by the weeping parents and loved ones of dead American boys might well have forced American leaders to drop the A-bomb on Japan.

No one in official Washington expected that one or two atomic bombs would end the war quickly. They expected to use at least a third, and probably more. And until the day after Nagasaki, there had never been in their thinking a choice between atomic bombs and conventional bombs, but a selection of both—using mass bombing to compel surrender. Atomic bombs and conventional bombs were viewed as supplements to, not substitutes for, one another. Heavy conventional bombing of Japan's cities would probably have killed hundreds of thousands in the next few months, and might have produced the desired surrender before November 1.

Taken together, some of these alternatives—promising to retain the Japanese monarchy, awaiting the Soviets' entry, and even more conventional bombing—very probably could have ended the war before the dreaded invasion. Still, the evidence—to borrow a phrase from F.D.R.—is somewhat "iffy," and no one who looks at the intransigence of the Japanese militarists should have full confidence in those other strategies. But we may well regret that these alternatives were not pursued and that there was not an

effort to avoid the use of the first A-bomb—and certainly the second.

Whatever one thinks about the necessity of the first A-bomb, the second—dropped on Nagasaki on August 9—was almost certainly unnecessary. It was used because the original order directed the air force to drop bombs "as made ready" and, even after the Hiroshima bombing, no one in Washington anticipated an imminent Japanese surrender. Evidence now available about developments in the Japanese government—most notably the emperor's then-secret decision shortly before the Nagasaki bombing to seek peace—makes it clear that the second bomb could undoubtedly have been avoided. At least 35,000 Japanese and possibly almost twice that number, as well as several thousand Koreans, died unnecessarily in Nagasaki.

Administration leaders did not seek to avoid the use of the A-bomb. They even believed that its military use might produce a powerful bonus: the intimidation of the Soviets, rendering them, as Byrnes said, "more manageable," especially in Eastern Europe. Although that was not the dominant purpose for using the weapon, it certainly was a strong confirming one. Had Truman and his associates, like the dissenting scientists at Chicago, foreseen that the A-bombing of Japan would make the Soviets intransigent rather than tractable, perhaps American leaders would have questioned their decision. But precisely because American leaders expected that the bombings would also compel the Soviet Union to loosen its policy in Eastern Europe, there was no incentive to question their intention to use the atomic bomb. Even if they had, the decision would probably have been the same. In a powerful sense, the atomic bombings represented the implementation of an assumption—one that Truman comfortably inherited from Roosevelt. Hiroshima was an easy decision for Truman.

The Redefinition of Morality

Only years later, as government archives opened, wartime hatreds faded, and sensibilities changed, would Americans begin seriously to question whether the atomic bombings were necessary, desirable, and moral. Building on the postwar memoirs of Admiral William Leahy and General Dwight D. Eisenhower, among others, doubts began to emerge about the use of the atomic bombs against Japan. As the years passed, Americans learned that the bombs, according to high-level American military estimates in June and July 1945, probably could not have saved a half million American lives in the invasions, as Truman sometimes contended after Nagasaki, but would have saved fewer than 50,000. Americans also came slowly to recognize the barbarity of World War II, especially the mass killings by bombing civilians. It was that redefinition of morality that made Hiroshima and Nagasaki possible and ushered in the atomic age in a frightening way.

That redefinition of morality was a product of World War II, which included such barbarities as Germany's systematic murder of six million Jews and Japan's rape of Nanking. While the worst atrocities were perpetrated by the Axis, all the major nation-states sliced away at the moral code—often to the applause of their leaders and citizens alike. By 1945 there were few moral restraints left in what had become virtually a total war. Even F.D.R.'s prewar concern for sparing enemy civilians had fallen by the wayside. In that new moral climate, any nation that had the A-bomb would probably have used it against enemy peoples. British leaders as well as Joseph Stalin endorsed the act. Germany's and Japan's leaders surely would have used it against

cities. America was not morally unique—just technologically exceptional. Only it had the bomb, and so only it used it.

To understand this historical context does not require that American citizens or others should approve of it. But it does require that they recognize that pre- and post-Hiroshima dissent was rare in 1945. Indeed, few then asked why the United States used the atomic bomb on Japan. But had the bomb not been used, many more, including numerous outraged American citizens, would have bitterly asked that question of the Truman administration.

In 1945, most Americans shared the feelings that Truman privately expressed a few days after the Hiroshima and Nagasaki bombings when he justified the weapons' use in a letter to the Federal Council of Churches of Christ. "I was greatly disturbed over the unwarranted attack by the Japanese on Pearl Harbor and their murder of our prisoners of war," the president wrote. "The only language they seem to understand is the one we have been using to bombard them. When you have to deal with a beast you have to treat him as a beast."

British Tank from World War I

WWI Airplane

Soldier in WWI Gas Mask

WWI Machine Gun

WWI Trench Warfare

1954 BAC (Electric) Lightning Supersonic Fighter Aircraft

WWII Radar Station

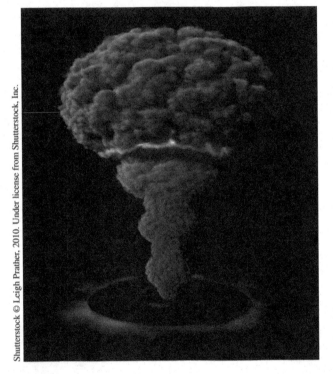

Atomic explosion

Auschwitz-Birkenau camp

Reading 15

Ideologically Correct Science

Michael Gordin, Walter Grunden, Mark Walker, and Zuoyue Wang[1]

Introduction

The historical study of science and ideology is really a twentieth-century phenomenon, for it is only after the First World War that stark differences appeared in political ideology and regime: liberal capitalist democracy, Marxist-Leninist communism, fascism and National Socialism. Furthermore, as the introduction to this volume argues, this historiography has been profoundly influenced by the Cold War, with the result that certain themes have been dominant. This essay will investigate perhaps the most striking examples of science being influenced by ideology, which here will be called "ideologically-correct-science" (ICS). The French Revolution will be included because, as the above-mentioned introduction also describes, it was both the first case study for the interaction of science and ideology, and a surrogate for other Cold War case studies. Not all relevant types of ICS will be, or even could be covered here. Indeed, this article will sacrifice depth in favor of breadth and use the comparative approach in order to provide a suggestive analysis of science under different ideological regimes.

Scholars have generally assumed that a political environment can influence science, but relatively little is known as to how this functioned in particular circumstances or across national boundaries. ICS refers to attempts by the state (or at least some representatives of, or forces within the state) to not only use science, but also to transform it into a more ideologically acceptable form, both with regard to scientific content and institutions. These efforts were often inconsistent, and not always entirely rational, but they existed all the same. Jacobins called for a "democratic," not "aristocratic" science in the French Revolution. Bolsheviks called for a "Marxist," not a "bourgeois" science in the Soviet Union. National Socialists in Germany called for an "Aryan," not a "Jewish" science. Ideologues in Second World War

Japan demanded a nationalistic, "Japanese" science and technology. During the McCarthy era in the U.S.A., politicians and some scientists tried to reshape science to help win the Cold War; sometimes calling for what Jessica Wang has described as an "anti-communist" science.[2] Finally, the Red Guards demanded a "people's science" during Mao Zedong's Cultural Revolution in China.

ICS often followed the same pattern: (1) purge of unacceptable scientists and purge or transformation of unacceptable scientific institutions; (2) the enlistment/recruitment of acceptable scientists; (3) the training of new scientists and creation of new institutions; and (4) the production of ICS. This essay will examine this admittedly ideal pattern in order to shed light on ICS in particular and the interaction of science and ideology in general.

In cases of ICS, the state often rewarded scientists and sciences that were, or appeared to be, politically and ideologically correct, while those who deviated from the prescribed path might receive punishment. This pressure was sometimes overt, as during the Chinese Cultural Revolution, or subtle, as in the McCarthy era of the American national security state. ICS could also be self-imposed. Some scientists, voluntarily or otherwise, sought to apply what they viewed as the official ideology in their scientific work. Finally, ICS could also provide "protective coloration,"[3] whereby a straightforward piece of research was wrapped in the official ideology for self-protection or self-promotion. These attempts to make a science that had merely the forms and trappings of ideology should be distinguished from efforts which were made to influence the content, but which fell short and only had an external effect on the social position of the sciences.

Undoubtedly the classic examples of ICS are the "Aryan Physics" (*Deutsche Physik*) movement during the Third Reich and the "Lysenko Affair" in the Soviet Union. Indeed, these are two of the most-studied examples of ideology influencing science. But Aryan Physics, like its counterparts in other sciences under National Socialism, was neither typical of science in the Third Reich, nor very successful. A search for ICS under Hitler would certainly include "Aryan Science," but must also go beyond it. Similarly, Lysenkoism was not typical of science under Stalin. Although it is debatable whether or not it was "successful," it did not spread from plant breeding and genetics to other sciences, and like Aryan science, it did not go unchallenged.

Furthermore, the history of "ideologically-correct-science" is not merely the story of the perversion or destruction of "good" science. Although this was sometimes the case, it is equally true that there are many examples in which ICS either failed to have a particularly harmful effect, or even produced benefits and positive scientific results. This latter case can be made both for a direct effect, as Loren Graham has argued with regard to dialectical materialism sometimes facilitating scientific progress, and an indirect effect, such as Mark Walker's thesis that the fight against Aryan Physics actually strengthened the hands of some in the German physics community.

ICS is a useful concept, and was very real, but it is also something to be used carefully, for it can also be seen as a straw man. If by ICS one means the total, coordinated, systematic, and intentional implementation of an ideologically determined program in science by the state, then ICS never happened, indeed never came close. But interesting and important things did happen. This article will therefore examine several case studies, comparing and contrasting them, with an eye towards gauging the limits and usefulness of this concept.

France

The varied fortunes of science in the French Revolution, like the other case studies, is a story of complexity and nuance, which can hardly be given adequate justice here. The two most celebrated incidents of the "attack" on science during the Jacobin Terror—the closing of the Academy of Sciences (*Académie des Sciences*), and the execution of the chemist Antoine Lavoisier—both fall short of being conscious attempts to impose an ideology on science.

Jacobin ideology was only dominant for about a year, and was not synonymous with the French Revolution. Moreover, all sciences were not treated equally: important Jacobins were hostile to abstract, theoretical, and mathematical science, but were

favorable to natural history. After much debate about how to reform education and how to structure expertise in the Republic of Virtue, in 1793 the Academy was closed as an institution and its resources dispersed. But the issues here were not just science, and even not centrally science. Resentment of royal privilege and corporate prestige were mainly responsible for the closing of the Academy.[4] While the Academy was eliminated as a vestige of royalist corporatist elitism, the Jardin du Roi was kept intact.[5] Moreover, Academy scientists (for the most part) were subsequently employed by the government of Revolutionary France.

The case of Lavoisier, on the surface the purge of an unacceptable scientist who was hardly a strong advocate of Robespierre's government, upon inspection had even less to do with science than the closure of the Academy. Lavoisier was executed by the machinery of the Terror, but not for any reason connected to his positions in science. His association with the Tax Farm sealed his fate. Similarly, Condorcet died at the hands of the Revolution, not because of his mathematics, rather because of hostility to his rationalist Enlightenment views.[6]

The Terror prematurely ended the lives of several prominent scientists, and interrupted the careers of many others, forcing twenty academicians out of forty-eight into "exile" (most went to the provinces, and only about four actually emigrated during the Jacobin Republic). But other natural scientists in educational institutions, government branches, and other venues were actively recruited by the regime. In the case of the Jardin des Plantes an old royal corporate research institution was maintained, albeit with profound alterations (see below). During the period of revolutionary wars, members of the Academy of Sciences received government contracts to fulfill old academic projects like the metric system, gunpowder production, and military engineering. Most aided the Revolution with little grumbling, or even with little attention to other political events once the guillotine lost its central prominence.[7] In contrast, some scientists, like the chemist A. Fourcroy (1755–1809), not only did not resent the Jacobins, they were active members of the state administration.

During Napoleon Bonaparte's Empire, the state recruited scientists even more avidly than before, and many scientists were quite eager to serve Napoleon's "technocratic" regime and appreciated the return to stability he represented in their eyes. The state now poured an enormous amount of money into military institutions, and scientific and technical expertise were richly rewarded in new institutions like the École Polytechnique, which was founded under the Directory and blossomed under Napoleon.[8] Given the prominence of scientists in the advisory apparatus of the Old Regime, especially in the form of the Academy of Sciences, this development under the Directory and the Empire can be seen as a continuation of previously established practices. The persecution of "unacceptable" scientists seems more the aberration from French practice, and the recruitment of those old specialists who were acceptable continued as before, albeit in new institutions.

Since it is difficult to discern a clear ideology in the French case governing the selection of acceptable as opposed to unacceptable scientists, it is also difficult to find an ideological criterion for training new scientists. After the ideological excesses that led to the closure of the Academy had passed and some of the dust of the Terror had settled, France still faced a series of foreign wars, and sorely needed technical expertise and noticeably lacked qualified young practitioners.

The solution was the creation of a new system of Grandes Écoles, headed by the École Polytechnique, designed to inculcate military discipline into young minds drawn from all corners of France with the necessary technical and mathematical skills required by the emerging modern bureaucratic state and a conscripted army.[9] The plans for such a system of education originated in the Old Regime, whereby places in military schools would be attained through nationwide competition and nominations by local authorities. The subject matter of all schools, especially the Polytechnique, was heavily imbued with mathematics applied to concrete problems of military necessity.

The results were impressive: a new generation was trained, filling the officer ranks of the army, the upper levels of the bureaucracy, and other positions at the top of the modern nation-state. The ideology of discipline through mathematical scholarship and service to the State has been so well inculcated into French society that it is scarcely noticed how large an influence this, the most ideological of Napoleonic institutions, still plays in modern France.

The attack on Newtonianism and the retention of the Jardin du Roi as the ardin des Plantes after the dissolution of the Academy of Sciences are the two nost prominent candidates for ideology affecting science. Charles Gillispie and L. Pearce Williams have made the most forceful case for an anti-Newtonianism, arguing that an ideology of anti-elitist Romanticism, deriving from Rousseau and Diderot, motivated the Jacobins' hostility to an atomized, mathematized, Newtonian universe.[10] Many of the individuals in question expressed hostility to some aspects of the Newtonian worldview, and specific examples of ideology did seep into the way debates about the content of science were conducted.[11] Similarly, because of the French Revolution, an exact, scientific biology was displaced by a discursive descriptive natural history within the Jardin des Plantes.[12]

But these examples of ideology and science do not cut to the actual core of the politics of science during the French Revolution. It is unclear how much of a case can be made for the application of specific Rousseauvist anti-Newtonian ideals during the height of the Terror. Even the leading anti-Newtonian ideologue of the time, Marat, was motivated less by an ideological stance towards Newton than by revenge against the Académie des Sciences for its treatment of him during an earlier scientific dispute.[13] Perhaps more startling is the fact that these trends were not followed up once the Academy was closed and Lavoisier had been executed. Laplace wrote his great Newtonian masterpiece after the debates recorded by Williams and Gillispie, for example. But the fact that the attempt to reject or replace Newtonianism did not succeed does not mean that it should be dismissed.

The reformation of the Jardin du Roi into the Jardin des Plantes was more successful during this period. It is noteworthy that, despite the rhetoric of anti-corporatism that led to the demise of the Academy, the Jardin remained well financed and continued to perform the same sort of botanical research as it had under the Old Regime.[14] The secret of this persistence was the articulation by the botanists of a natural historical program they favored that also seemed to accord with the hostility to abstraction noted by Williams and Gillispie. As a result, this kind of empirical botany was encouraged. A particular type of ideologically-correct science was indeed adopted as a form of basic research, but basic research led the way here in defining what it meant to be ideologically correct.

ICS during the French Revolution was fleeting and superficial. At the same time that the Reign of Terror reached its height in Paris and de-Christianization was ravaging the French countryside, most scientists, who were products of the Old Regime, were put to work for the Revolution; others needed only to repackage their research in an ideologically congenial wrapper. The scientific community did have to make concessions to the new political order, and a few individual scientists (for example,

Condorcet and Lavoisier) suffered and died, but in general French science and scientists benefited from the French Revolution and Napoleon. The push for ICS did not produce a significant change in the content of research. However, ICS and the greater political and ideological currents of the French Revolution did create new scientific institutions and influence research programs and thereby led the way in terms of political control of scientists and scientists' accommodation.

The Soviet Union

The reconstruction of Russian science was well on its way to a socially more useful institution before 1917, but the chaos of the First World War and the transformation of Russian society by the Bolshevik Revolution also brought profound change to Russian science. Perhaps surprisingly in a nation where science had never penetrated much farther than a vanishingly small percentage of the literate intelligentsia, the Bolsheviks and their allied parties saw reform of higher education, and therefore indirectly, science as one of the first tasks on their agenda. At almost the first opportunity they approached the Imperial Academy of Sciences in Petrograd with both carrot and stick in hand.[15] On the one hand, the largely bourgeois membership of the Academy was reluctant to extend a badly needed hand to the new regime's quest for rapid economic improvement and technical expertise. On the other hand, compelling reasons kept the vast majority of Russian scientists from emigrating and induced them to cooperate, however reluctantly at first, with the regime.

It was by no means certain that the regime would last, and there seemed to be every hope of the regime moderating its rhetoric against bourgeois specialists (as indeed it did). Such hope gave those with a sincere patriotic bent an opportunity to actually put some of their practical suggestions to work. The thrust of the early period of science policy in the Soviet Union was more the establishment of new scientific institutions rather than the destruction of older ones. Moreover, the Bolsheviks offered unheard-of blandishments in the form of prestige, equipment, and funding in an effort to persuade the academicians to lend some assistance.[16] Finally, the Bolshevik regime seemed the lesser of several evils, as rabidly "anti-specialist" movements like *Proletkul't* hovered on the horizon.

During the period of War Communism (1918–21), *Proletkul't* was a vibrant cultural movement, one that marked the first effort to establish ideologically-correct-science in the Soviet context. *Proletkul't* was by no means directed exclusively, or even principally, towards science. It argued that a proletarian state required a proletarian culture, not the realism of capitalist art, the individualism of capitalist literature, or the technocracy of capitalist science.[17] Technocratic specialists and other remnants of tsarist capitalist culture would have to go, and a more democratic and proletarian science would be imposed in its stead.[18]

Thus for quite some time, bourgeois specialists did not know how they would be treated by the Bolsheviks, especially since there was now a "Communist Academy" alongside the traditional Academy of Sciences. But Lenin had little patience for such efforts to alienate much needed specialists, and his eventual suppression of *Proletkul't* and the closing of the Communist Academy served as a signal to bourgeois scientists that their kind would be tolerated as long as they were amenable to the new regime's demands.[19] In fact, except for an exile of some 200 dissident intellectuals, there was not much of a purge in science until the period 1928–31.

What it meant to be an "acceptable" scientist in the early Soviet Union fluctuated widely with the attitudes and needs of the fledgling regime. During the period of War Communism, bourgeois specialists were (officially) "unacceptable," but were nevertheless used. As the Soviet economy began to falter, the regime began to accommodate those in possession of needed technical skills. The heyday of the bourgeois specialist—the period of the New Economic Policy (NEP, 1921–27)—induced many scientists who were ideologically opposed to Bolshevism to make a temporary peace, while giving the Soviet regime time to lick its fiscal wounds and gear up for socialist industrialization.

The essence of the NEP attitude towards bourgeois scientists and engineers had already been expressed by Lenin in his opposition to the iconoclastic fury of *Proletkul't*. Technical expertise would always be necessary, and as long as those who had the knowledge would only share it if given sufficiently high salaries and ideological breathing room, then they should be afforded those luxuries.[20] This did not mean, however, that ideological constraints were put on hold. Ideologically "acceptable" scientists, like the young Lev D. Landau, for example, were actively encouraged by the regime and promoted over old bourgeois specialists who staffed the old universities.[21] Indeed, the communists made a fundamental distinction between research institutes where scientific ability was most important, even if not accompanied by appropriate political conviction, and the universities, where only politically reliable scientists would be used to train young scientists. It is interesting to note that National Socialists in Germany and to a certain degree the Communist Party in China made the same distinction.

For the time being, the Soviet state only mildly harassed those who chose to hold to their old views—provided their skills were truly indispensable to the industrialization of the new regime. The Shakhty trial of 1928 changed all this, however, when bourgeois specialists were accused of "wrecking" and industrial sabotage designed to cripple Soviet power.[22] The honeymoon had ended, and under the rising power of Joseph Stalin, bourgeois specialists were not tolerated during the years 1928–31. Thereafter the term "bourgeois specialist" was no longer used, and many former "specialists" quietly returned to the positions of prominence in science they had enjoyed before. The Soviet nuclear weapons project, which used both former specialists and younger scientists trained under the Soviet educational system, was typical in this regard.[23]

The generation of new cadres of ideologically suitable scientists and technicians constituted one of the most important aspects of early Soviet science policy. The splitting of research and education was the first stage in this development. Education was placed entirely in the hands of ideologically sanitized pedagogues within the People's Commissariat of Popular Enlightenment (Narkompros). Research institutes were left under various economic Commissariats, and permitted a more eclectic personnel. At the same time, the State transformed the research institutes, borrowing some aspects of Western organization for individual labs, but placing them all into a Soviet framework.[24]

While some bourgeois specialists were prominent in this framework, by the early 1930s and the conclusion of the first of Stalin's Five-Year Plans, most had either blended into the woodwork and adopted appropriate ideological colors or had been executed or exiled during the purges.[25] However, as late as the 1950s, non-Party scientists, former "bourgeois specialists," occupied the majority of high-level administrative positions in Soviet scientific research.

The Soviet Union grudgingly used its "bourgeois specialists" while simultaneously training new cadres of "red" scientists. But these new ideologically correct scientists had the same professional aspirations as their "bourgeois" mentors—concern for international scientific standards, the need for international contacts—and therefore sometimes clashed just as forcefully with the Stalinist regime's desire for ideologically fidelity.

The growth of cadres of communist researchers in the various fields of science was really quite extraordinary.[26] As more and more competent (and ideologically "clean") individuals were generated, they began to take over positions held by disgraced members of the older generation.[27] By the onset of the Second World War, Stalinists had essentially completed the ideological purification and installation of "red" specialists in almost all levels of the Soviet research empire.

Dialectical materialism, the official philosophy of science of the Soviet Union, complemented historical materialism—the Marxist theory of historical and economic development—to compose the complete orthodox set of beliefs about the social and natural world. The nature of dialectical materialism and the extent of its epistemological and ontological grasp had been a source of debate since the early interpretations of Engels' and Lenin's writings on the natural world during the 1920s.[28] But unlike the meaning of dialectical materialism, the unwritten "requirement" that scientists hold to some of its tenets (or at least not openly contradict them) was more or less constant through Soviet history—and reached some disastrous consequences during the Lysenko years. Yet the historian Loren Graham has pointed to another side of dialectical materialism, which he calls the "authentic phase."[29] Graham argues that dialectical materialism was sometimes used by scientists freely as a positive force for scientific reasoning.[30] Graham's persuasive argument thus makes the intriguing point that scientists often adopted ICS as basic research in the Soviet Union voluntarily (or semi-voluntarily) and occasionally used it to produce significant results.

Whereas dialectical materialism provides the most important example of ICS is basic research in the Soviet Union, the most famous instance of ICS as applied research is the well-known case of Lysenkoism.[31] Scion of a Ukrainian peasant family, Trofim Denisovich Lysenko began his work as an agronomist in the 1920s in an agricultural station near Baku. While there, he claimed to have discovered a biological process he dubbed "vernalization" (iarovizatsiia): the treatment of germinated seeds of various plants with abnormal conditions of eat, cold, and other forms of environmental exposure, in order to make plants develop in a more appropriate way—essentially a neo-Lamarckian biological program. Lysenko's attempts to present his results to the Soviet agronomic and genetics community were rebuffed as contrary to all known facts about genetics. The famed geneticist N.I. Vavilov initially supported Lysenko's research as potentially producing innovations in agronomic practice, but broke with Lysenko when he started to push his neo-Lamarckian views on plant breeding and genetics in the mid-1930s.

In the early 1930s, Lysenko teamed up with ideologist I.I. Prezent, who convinced Lysenko to link his neo-Lamarckian views of inheritance with Darwinism, and to couch both in a Marxist framework. This marriage of dialectical materialism and agronomic practice in opposition to genetic theory caught the attention of Stalin in the late 1930s, who praised Lysenko openly in various contexts as a means of supporting the regime's disastrous and bloody collectivization campaign in the countryside. Lysenko grew in power, becoming president of the Lenin All-Union Academy of Agricultural Sciences (VASKENIL) while its former president, Vavilov, was arrested on grounds of counter-revolutionary and anti-Soviet activity, and died of malnutrition shortly before being released from prison in the early 1940s. The actual banning of genetics did not happen until after the Second World War, when Lysenko's star actually seemed to be waning and the Cold War got underway. But in 1948, Lysenko read a speech (toned down but supported by the personal editing of Stalin)[32] condemning genetics as a "bourgeois" science and banning almost all research on it in the Soviet Union.

While Nikita Khrushchev liberalized much of the terror apparatus of Stalin's state after the latter's death in 1953 and especially after the Twentieth Party Congress in 1956, he liked Lysenko personally and continued to fund him lavishly and support a series of disastrous agricultural programs which his favorite proposed. After Khrushchev's fall in October 1964, however, Lysenko's days in power were numbered. Genetics was restored in 1965, but the recovery process was painful and the loss of Vavilov hard to forgive. The scars caused by Lysenkoism remain to this day.

Lysenkoism, which influenced Soviet science for decades, was obviously ideologically-correct-science. But it was eventually overthrown by forces within Soviet science and society, and the science it had discredited and dismantled, genetics, was reinstated and rebuilt. Lysenko and his followers also failed to extend their influence to other Soviet sciences. In particular, physics was able to rebuff Lysenko-inspired attacks on certain aspects of modern physics, both because of the relevance and irrelevance of physics to Stalin's foreign and domestic policies. At first physics was not very important to the Soviet leadership, for in contrast to Lysenko's theories, it promised neither to solve the country's problems, nor fit particularly well into Soviet ideology. When the Second World War began and the potential of nuclear weapons was clear, physics became far too important to purge or distort. Thus Lysenkoism was arguably exceptional, and reveals little about the other major purges of Soviet intellectuals that were attempted in the period after the Second World War.[33]

Germany

During the first year of National Socialist rule in Germany, a significant percentage of scientists (perhaps as much as 15 percent) were forced out on racial and political grounds.[34] This purge was not aimed particularly at scientists or science—the campaign against Albert Einstein is the exception that proves the rule—but rather was a consequence of the National Socialist "cleansing" of the entire civil service. This larger purge was itself apparently a largely unplanned, if not spontaneous reaction to the failure of the nationwide boycott of Jewish businesses in April 1933. The effect on science was tremendous, but this purge does not demonstrate any plan or intention on the part of the National Socialist leadership to create an Aryan science.

There is evidence for more direct interest by Hitler's government in the transformation of scientific institutions, but again a close look reveals different priorities. The universities were purged and transformed right away because they were educational institutions charged with the training of German youth.[35] Their transformation was profound, but also hidden. The structure of the university remained largely intact, at least on paper, but most autonomy was robbed by the introduction of the "leadership principle": a strict hierarchy, whereby one had to obey everyone above, but could order about everyone below. In principle, the faculty still met and voted, prepared lists of candidates for positions, and so on, but in practice the deans and rector—political appointees, of course—often had almost dictatorial power. Other scientific institutions, including the Kaiser Wilhelm Society and its research institutes, as well as the various Academies of Science, were transformed in a similar way, although significantly later in the Third Reich.[36]

The dates by which all Jewish members had been purged provides one of the most telling indications of how relatively unimportant these research institutes and academies were for the National Socialist leadership. Whereas most Jewish scientists in the universities had lost their positions in 1933, other Jewish scientists managed to remain at research institutions for many years. The Prussian Academy of Sciences, for example, under pressure from the Ministry of Education, finally asked the last Jewish member to resign in 1938, shortly before *Reichskrystallnacht,* the nationwide orgy of violence against German Jews. In 1933 the Education Ministry had been keen to publicize its treatment of Einstein. After he had already resigned from the Prussian Academy, the Ministry pressured its leadership to issue a press release, essentially saying good riddance. But in 1938 the Ministry wanted to keep the final purge of Jewish scientists quiet, so as not to publicize the fact that there were still Jews in the Prussian Academy.[37]

The purge of German science by the National Socialists makes clear how little interest Hitler and his followers had in scientific research. Their interest was aroused only when scientists demonstrated that modern science could serve National Socialism. It is striking and depressing to see how quickly the vacancies caused by the purge were filled by generally competent, racially and politically acceptable scientists eager to serve the new regime. For those who were already in place, it usually sufficed to demonstrate Aryan status and an apolitical attitude to keep their jobs, although they were pressured to yield greater political and ideological cooperation with the National Socialist movement.

Those moving up—i.e., who did not already have permanent positions—had to be both Aryan and willing participants in the political and ideological rituals introduced by the National Socialists into the universities. Such rituals included or encompassed attendance at "political" indoctrination camps, membership in National Socialist organizations, including, of course, the National Socialist German Workers Party (NSDAP), and other forms of participation in National Socialism. The authorities paid little attention to their research when it came to judging political acceptability, although this might well enter into whether or not they were hired.

Rearmament, especially beginning in 1936, offered great opportunities to scientists who had something to offer the regime; similarly, the new racial hygiene policies (sterilization, "euthanasia," restrictions on marriage) provided great opportunities for physicians, biologists, anthropologists, psychologists, and psychiatrists.[38] In general, most scientists did not make the transition to the racist, Aryan science, but did adapt themselves and their research in order to work under National Socialism.

Perhaps the greatest failure of National Socialist science policy concerned the training of the next generation of scientists. The politicization of education and emergence of National Socialist youth organizations like the Hitler Youth eroded both the quality and the quantity of scientific education. The creation of new institutions was not much more successful. The National Socialists did not really try very hard to create new scientific institutions. Even the *Ahnenerbe* ("Ancestral Heritage"), the scientific research arm of the SS, relied mainly on research contracts in order to encourage certain types of scientific work.[39] Most truly new research institutions created by the National Socialists had little to do with science, and were so ideological and politicized that they were really incapable of producing significant basic or applied research. Scientific work was done for the National Socialists, but it usually took place in institutions Hitler's movement had inherited. The rocket project is the exception that proves the rule: Army officers and engineers, not National Socialists, created and developed it during the Third Reich.

The politicization of the universities was compounded by the carnage of war, as very many students were offered up as cannon fodder during the Second World War. The result was a generation lost to German science: with few exceptions, only scientists who had entered the university system during the Weimar Republic survived the Third Reich. Despite the regime's attitudes towards women, by the middle of the war women made up a large percentage of university students because so many of their male counterparts were fighting and dying on the front.[40]

The regime wanted the help of scientists and physicians in order to provide a scientific basis for their racist, and eventually murderous race hygiene. But despite the active participation of scientists and physicians in this program, and the infamous experiments carried out at Auschwitz and elsewhere on unwilling concentration camp inmates,[41] researchers could not deliver scientific proof of the supremacy of the Aryan race. In retrospect, this was revealed by the infamous Nuremberg Laws in 1935. When the National Socialist State finally issued the binding legal definition of what "non-Aryan" meant, it had to fall back on to a religious, not racial definition of who was a Jew.

However, there were also scientists who claimed to be practicing Aryan science when carrying out their basic research. These were not race hygienists—although a biologist claimed to be fighting for an "Aryan Biology" (*Deutsche Biologie*)—but rather physicists, mathematicians, psychologists, chemists, and engineers. These Aryan movements in German science and technology[42] eventually failed in their efforts to seize control of their disciplines precisely because they were barren in the National Socialist sense. The rulers of Germany wanted science useful to them, and it was the leaders of the established scientific communities, not the Aryan scientists, who were able to gain and retain the backing of influential patrons in the National Socialist state.

Thus the calls by the Nobel laureates and "Aryan Physicists" Philipp Lenard and Johannes Stark to eliminate the influence of Jews in physics, and the "Aryan Mathematician" Ludwig Bieberbach's assertion that Aryans and Jews made different types of mathematics eventually fell on deaf ears, while the applied mathematician Ludwig Prandtl and the theoretical physicist Werner Heisenberg offered their expertise in designing wind tunnels and nuclear weapons, respectively. Established scientists like Heisenberg and Prandtl were thereby able to sideline or neutralize their "Aryan" colleagues by convincing political leaders that their basic and applied research might facilitate both military conquest by creating new and improved weapons[43] as well as the racial engineering of Europe by providing new methods for distinguishing Aryans and non-Aryans.

Ideologically-correct-science had a profound effect on German science, far beyond the well-known case of the ideological attacks by Aryan Physics on modern physics. But the end effect was almost always defeat for the ideologues, is the established scientific communities sought refuge and support from leading National Socialists. The main effect of "Aryan Science" was to drive most other scientists further and faster down into the arms of National Socialism, making themselves more useful and relevant for the often murderous policies of the regime.

The very anti-intellectual climate within the ranks of leading National socialists worked against the advocates of Aryan Science. A National Socialist Ideologue disdainful of science also had little interest in its Aryan variant, while the technocrats scattered throughout the National Socialist hierarchy naturally threw their support behind established scientists who could deliver the goods.

Scientists helped build rockets[44] and jet planes,[45] researched new biological and chemical weapons (which fortunately were not used), and dangled the prospect of "Wonder Weapons" like the atom bomb[46] before leading National Socialists. Researchers ranging from physicians, biologists, psychiatrists, psychologists, anthropologists, economists and geographers[47] helped implement the murderous race hygiene policies of "euthanasia" and "Germanization," and finally murder and genocide by helping to select the victims and create new and more effective ways to torture and kill them.

Japan

During the Second World War in Japan there was no need for a racial or ethnic cleansing in the sciences because, with the exception of a few notable Koreans, all scientists were Japanese and shared essentially the same racial, ethnic, and cultural identity. Moreover, because of the relative paucity of scientists with advanced scientific training in Japan, even Koreans—whose homeland Japan had occupied before the war and who were considered second-class citizens in Japan—were allowed to retain their positions at the university.[48]

There were also few incidents in which scientists were jailed for expressing anti-imperialist views. In one case, several members of an academic research group, including the physicist Taketani Mitsuo, were arrested for advocating resistance to Japanese imperialism through their serial publication, *Sekai bunka* (World Culture). Taketani was detained—allegedly for his research activities on natural dialectics—accused of helping to promote the Communist Party in Japan, and forced to state that he had acted under instructions from the Comintern. The judge who reviewed his case, however, suspended prosecution and released him to the custody of his colleague and close friend, Yukawa Hideki.[49] On the whole, such instances were rare.

Under military influence, the government enacted numerous laws in the 1930s to acquire greater control over the people and the economy. As part of the militarization of the nation, institutions of scientific research, such as the imperial universities and the prestigious Institute for Physical and Chemical Research (also known by its Japanese acronym "Riken"), were also brought under the aegis of the military. By influencing budget allotments for basic and applied research, as well as the production orders for the resulting manufactured items, the military began to have a significant impact on scientific research. As a result, there was no need to transform such institutions; only greater administrative and economic control proved necessary.[50]

Nevertheless, the recruitment of first-rate scientists became a critical problem for the military. Nearly all of the nation's most famous and competent scientists had spent years abroad studying in Western nations, and they were thus considered suspect by the military leadership, which was by and large xenophobic in its worldview. Moreover, most scientists at the university had little interest in suspending their own research for military projects.[51] Out of necessity, however, the military consulted leading scientists at the universities when those in uniform—who were usually little more than higher-school educated technicians—proved incapable of advanced level research. Both the army and navy, for example, turned to physicist Nishina Yoshio, Director of the Physics Department at the Riken, to complete a feasibility study of the possible exploitation of nuclear energy for military purposes.[52]

Like Nazi Germany, Japan proved terribly shortsighted in preparing the nation's scientific infrastructure for a prolonged and total war. One particularly stable area of failure was in the training of new scientists and the creation of new institutions of science and technology. The military continued to draft students from university science programs and departments throughout the war. This was halted only through the concerted efforts of senior scientists like Nishina, who insisted upon military deferments for designated students of exceptional ability in exchange for agreeing to conduct research for the military. Only by such means was the older generation able to preserve the next generation of scientists.[53]

As for new institutions of science, by the early 1940s, the resources for their construction and maintenance were dwindling and the move was toward consolidation and rationalization, not expansion or the creation of new scientific institutions. The capstone of the trend came in early 1942 with the establishment of the Board of Technology. Roughly analogous to the Office of Scientific Research and Development in the United States, the purpose of the Board of Technology was to coordinate scientific research and development of new technologies between civilian and military institutions, as well as between the army and navy.

The organization looked impressive on paper, but in reality, the Board of Technology proved a dismal failure. It never acquired sufficient authority or capability to supersede the numerous administrative boundaries that such a task entailed. It could not overcome the substantial compartmentalization of civilian and military research. Neither could it redirect the complex network of financial arrangements, production contracts, and social ties that each military service had to its preferred *zaibatsu* (industrial combines) and university cliques. The board had little success in overcoming the bitter enmity that existed between the army and navy to convince them to collaborate on key projects until the war was already all but lost.[54]

There was no single ideology in wartime Japan comparable to Aryan Physics a Nazi Germany, nor was there any ideological movement in the sciences that gained such comprehensive state support and promotion as Lysenkoism did in the Soviet Union. Beyond the ubiquitous rhetoric of national militarism that emphasized sacrifice and service to the Emperor and nation, there was no prevailing ideology to impact science as there was in Germany or the Soviet Union. Yet there was a call for a distinctly Japanese form of technological development based on the nation's situational imperatives, that is, the rise in demand for military production in the face of rapidly diminishing raw materials.

Japan's progress in science and technology had been dependent upon the cooperation of Western nations and foreign teachers since the late nineteenth century, but as the military leaders of Japan drove the nation inexorably toward world war in the 1930s, increased hostility toward the West compelled bureaucrats and intellectuals alike to question the West as a model for Japan's technological development. Having no indigenous model as a substitute, a distinction had to be drawn between American and German paths of development. The rationalized "German path," it was argued, was better suited to Japan, as it aimed to limit the use of raw materials and promoted the use of substitutes. When its German ally proved niggardly in technology transfers, however, Japan was forced again to look inward.

Imitation of the West was to be rejected in favor of a uniquely Japanese path of technological development in accord with the nation's paucity of natural resources.[55] As one ideologue stated:

> The resources of the Greater East Asia Co-Prosperity Sphere are awaiting the creation of the new technologies that

will make most effective use of them. It is only then that these resources will acquire value. The existence of scientific research, which may give birth to this new technological creativity, will provide a firm basis for the cultivation of the Co-Prosperity Sphere, and for this reason the promotion of such research is currently an urgent necessity.[56]

Despite the flurry of mobilization and rationalization laws and measures that were enacted to realize this vision, no distinctively "Japanese" science emerged, nor was Japan able to free itself of its pattern of technological borrowing and dependence upon the West. Toward the end of the war, when necessity and desperation drove the nation's leadership to extremes, the government and military called upon scientists and engineers to draw inspiration from Japan's traditional past and the unique characteristics of the Japanese people, all in the effort to create an ideological rallying point for Japanese scientists in the development of some new weapons technology that could turn the tide of war in their favor.[57]

By 1943 the National Socialist state was also calling upon its scientists, engineers, and even inventors to create "wonder weapons" which would use qualitative superiority to overcome the quantitative superiority of its opponents. The Japanese military's answer, however, was no miracle weapon of science, such as a rocket or an atomic bomb. Rather, it responded with crude suicide craft, such as the *Ohka* piloted missile and the *Kaiten* midget submarine. No "Japanese" style science emerged from the war, and the ultimate "Japanese path" of technological development resulted in death for many of the nation's youth.

In the case of Japan, there was no readily identifiable ideologically-correct-science. Rather there was only a vague policy objective to guide technological development that was in accord with the nation's situational imperatives. The lack of natural resources, coupled with the nation's Spartan industrial infrastructure and its limited scientific and technological capacity, predestined Japan's fate in a total war against the United States. Japan proved incapable of producing such wonder weapons as the atomic bomb, long-range guided missiles, and advanced radar, but its army did, for a time, strongly support the development of biological weapons and made significant advances in this field.[58]

Biological weapons were easy to mass-produce with few resources, and thus fit within the military's

vision of a new weapon derived from a "Japanese technological path." Ultimately, however, the fear of a response in kind from the United States and the ever-present possibility of a boomerang effect appear to have deterred Japan's use of biological weapons on a wider scale beyond the war in China. As a result, biological weapons never became a significant factor in determining the outcome of the Pacific War, and the military never acquired a uniquely Japanese wonder weapon of his own.

United States

If communism served as a powerful ideology of science in the East, especially in the Soviet Union and China, did anti-communism play the same role in the West during the Cold War? Since American McCarthyism in the late 1940s and 1950s represented the peak of this political and cultural phenomenon, we might expect to see signs of the search for an ideologically correct science.[59] However, there were very few unambiguous examples of efforts to influence scientific content that were motivated or constrained by anti-communism as a political ideology. Rather, what does emerge clearly is the pervasive influence of anti-communism on the *political* roles of scientists whose professional identity assumed significant but not overwhelming importance.

Scholars now generally agree that the two characteristics of McCarthyism, domestic anti-communism and the denial of due process to those accused of communism, existed both before and after the period when Senator Joseph McCarthy made the cause his personal crusade in 1950—4.[60] The Cold War ideology of anti-communism not only saw a direct threat to American security in potential Soviet expansion on the international front, but also from perceived communist subversion at the domestic front. The national security state organized national life around national defense and fed on the Cold War ideology of anti-communism. It dominated science and technology policy, and thus indirectly but powerfully shaped American scientists' political and scientific activities.

On the one hand, a large number of American scientists engaged directly in the making and testing of nuclear weapons in the national laboratories of the Atomic Energy Commission (AEC). Those in academic and industrial settings also came to depend on the defense establishment for funding. As Paul Forman has argued, this dependence tended to make scientists choose, consciously or unconsciously, research directions that would benefit their patrons.[61]

On the other hand, scientists were persecuted during the McCarthy era because their past association, political opinions, or policy advice deviated from the political orthodoxy prevailing at the time.

The security clearance case of J. Robert Oppenheimer was perhaps the best-known example of McCarthyist attacks on scientists. As the famous director of the Los Alamos laboratory that created the atomic bomb during Second World War, Oppenheimer was nevertheless stripped of his security clearance in 1954 by the Atomic Energy Commission (AEC). The AEC decision cited past association with radical causes, opposition to the hydrogen bomb, and "defects in character."[62] Hundreds of other, less-well known scientists suffered similar or worse treatment both before and after the Oppenheimer case.[63] The U.S. State Department denied passports to a number of American scientists with liberal reputations so they could not travel abroad. It also refused to issue visas to some foreign scientists who wanted to visit the U.S.A.[64]

The Oppenheimer case evoked a most vehement protest from the scientific community, which generally blamed the injustice on a paranoid security system.[65] It undoubtedly marked a profound deterioration in the relationship between many American scientists and the national security state. The far-reaching repercussions did not escape the top government officials. Eisenhower, while agreeing with the AEC's decision, nevertheless worried about the case's effect on scientists in various defense projects. Aware of the potentially explosive impact on scientists and dangerous exploitation by McCarthy, Eisenhower told his aides that "we've got to handle this [Oppenheimer case] so that all our scientists are not made to be Reds."[66] Eisenhower wrote to New York writer Robert Sherwood shortly before the AEC decision, stating that because he was "so acutely conscious of the great contributions the scientists of our country have made to our security and welfare," he shared the hope that Oppenheimer could be cleared.[67]

It was to the President's and the AEC's relief that the "mass exodus" from weapons laboratories, as predicted by various scientific groups, failed to materialize in the wake of the Oppenheimer case.[68] While Eisenhower may have feared losing the services of scientists, there was no lasting damage to the Cold War partnership between science and the State. Scientists had warned since the late 1940s that unfair security procedures would lead scientists to desert government positions, but the threat was more rhetoric than reality.[69]

The fact that few scientists left their government research positions in the aftermath of the Oppenheimer case indicates that scientists, especially younger scientists, learned to live with the new Cold War political economy of science, what Eisenhower would call the "military-industrial complex." As physicist Herbert York, first director of the AEC's Livermore Laboratory on nuclear weapons in the 1950s, later reflected, young scientists had the practical needs of finding jobs that matched their training and supported their families.[70] Others simply recognized the need to combat two "Joes," both Joe McCarthy and Joe Stalin, especially after the outbreak of the Korean War.[71] In any case, since the newly-founded National Science Foundation only slowly gained substantial budgets, the funding structure of American science was skewed toward the military, which left few alternative sources of financial support to scientists, including the training of new ones.[72] Consciously or unconsciously, scientists were integrated into the national security state.

In the United States during the McCarthy era, there was a general acceptance, either tacit or explicit, by both individuals and institutions, of loyalty oaths and security clearances. Most universities, for example, refused to hire known communists. In 1950 the regents of the University of California adopted a requirement that all university employees had to sign a loyalty oath stating that they were not members of the Communist Party. As a result, dozens of faculty members, many of them scientists, left for other institutions in protest while some others, including tenured professors, were fired from the university.[73]

The same year, scientific organizations such as the Federation of American Scientists and the National Academy of Sciences fought successfully to remove amendments in the National Science Foundation Bill which would have required applicants to the foundation for unclassified research to undergo security clearance and an FBI background check. But, with dismay, they felt that they had to accept the requirement of loyalty oath in the bill as a compromise to get it passed.[74] There were a few nuclear scientists who deliberately switched to fields where they did not need security clearance. Leo Szilard, for example, turned to molecular biology.[75] Others, like Philip Morrison, continued in their fields but avoided work that would have required them to apply for a security clearance.[76]

Without belittling the pains and the injustice that the victims of Cold War anti-communism suffered, a distinction should be made between Stalinist/ Maoist communism and American McCarthyism. Few American scientists were persecuted through State-sponsored violence for their particular beliefs in science, in contrast to what happened in Stalinist Russia and Maoist China, although they were repressed in other ways. While Vavilov starved to death in Stalin's prison for resisting Lysenkoist theory and while dozens of prominent Chinese scientists were killed during the Cultural Revolution for being bourgeois "reactionary academic authorities" (see below), the worst that happened to American scientist-victims of anti-communism was, with few exceptions, that they lost their jobs, or as in Oppenheimer's case, their security clearances.

The consequences of McCarthyism went beyond the harm done to individuals, however. In fighting dictatorial communism, American anti-communist crusaders adopted the same anti-democratic tactics employed by the enemy. The American left, including that in science, was largely silenced and social reforms aborted. Fearing the charge of "being soft on communism," Cold War liberals energetically led the U.S.A. into a costly and misguided war in Vietnam.[77] Few of those American scientists who served as major advisors to the government in the 1950s and 1960s, for example, represented the liberal and left-wing positions briefly influential in the immediate post-Second World War years.[78] The significance of McCarthyism "may well have been in what did not happen rather than in what did."[79] In science, for example, we will never know what scientific research could have been pursued during the Cold War had there not been pressure to work in the national security state system. Likewise, we can only speculate on whether scientific research more closely related to civilian technology could have been advanced further and earlier than it did.

Scientists, especially nuclear scientists associated with government labs, were certainly subject to special scrutiny from the government because of their perceived access to "atomic secrets." Yet, the ideological impact of anti-communism on American science was only general and indirect. In contrast to Marxism, which attempted to function in Stalinist Russia and Maoist China as an all-encompassing ideology with specific doctrines governing science and philosophy, American anti-communism was primarily a political ideology. It guided American foreign policy and influenced domestic politics, but never set down a number of doctrines to be followed in science and philosophy.

The excesses of the McCarthy era in the United States had a profound effect on scientists and their relationship with the state, but it did not lead directly to a new type of an ideologically-correct "anti-communist" science like Aryan Physics or Lysenkoist biology. In the United States, ICS manifested itself as the connection between research and the goals of the State, in particular the integration of science into the national security state, rather than the establishment of ideological tests for the content of science. Not even the "Oppenheimer affair" could halt or even slow the flow of scientists into military-related research.

China

As with the Soviets, science occupied a special position in the ideology of the Chinese Communist Party. The founders of the party in the 1920s turned to Marxism as a "scientific" explanation of history and communism as a natural course of social changes. Mao Zedong later developed an essentially instrumentalist ideology of science that might be called "revolutionary utilitarianism." "Revolutionary" referred to the goal of the Communist Party's science policy to ensure the political loyalty of the scientists, while "utilitarian" spoke to the pressure on scientists to produce immediate, practical results.

Even before their take-over of mainland China in 1949, Mao and his followers had launched a notorious "rectification" campaign in their stronghold in Yanan during the early 1940s. It resulted in a complete re-structuring of the scientific establishment under communist control. A number of scientists and science administrators had insisted on the priority of basic science in education and research. Now they were removed from their positions and punished in favor of those who advocated the re-orientation toward meeting immediate needs in production and military technology. Some of those suspected of harboring bourgeois thoughts—that is, thought not sufficiently revolutionary—were "sent down" to the countryside to learn from the peasants and thereby set an ominous precedent. As the historian James Reardon-Anderson points out, the narrow enforcement of revolutionary utilitarianism marginalized fundamental science "with long-term repercussions on the modernization of China."[80]

Ideological purification of the scientists, often Western-trained, started almost immediately in the People's Republic of China (PRC), and intensified during the "thought reform" campaign at the height of the Korean War in 1952–3. In a scene paralleling that in the U.S.A., war hysteria turned into a hunt for internal enemies, resulting in the suicides of a number of scientists in Shanghai.[81] Many scientists were accused of following bourgeois scientific theories, such as Mendelian genetics, cybernetics, resonance chemical theory, and Gestalt psychology, and forced to renounce them. The Anti-Rightist Campaign of 1957, however, outdid all these previous purges. Attacking those scientists who had criticized the party's mishandling and distrust of scientists as part of a conspiracy to overthrow the new government, Mao ordered the purge of hundreds of thousands of intellectuals, including scientists.[82] Many of the brightest scientists were thus taken away from science and education and placed in forced labor for many years. Despite periods of relaxation, the pressure on scientists for ideological purification never completely relented over the next two decades.[83]

Mao's distrust of scientists reached a crescendo during the Cultural Revolution of 1966–76. He and his supporters unleashed a harsh reign of terror by the radical Red Guards against anyone, including scientists, who could be accused of deviating from Mao's correct political line. Along with other intellectuals, scientists were again purged for their bourgeois ideology and their elitism; they had to be cleansed and reformed. Red Guards and other rebels took over scientific and educational institutions and stopped virtually all research. Scientists, especially those formerly in administrative positions, were criticized and persecuted, and sometimes beaten, tortured, and killed. By 1969, many scientists who survived the ordeal were sent to the countryside or factories to perform physical labor and help make a "people's science." Only after Mao's death in 1976 was it possible for a full-scale restoration of utilitarian science policy under the leadership of Deng Xiaoping.[84]

Scientists were usually purged in the Mao era not so much for the ideological content of their scientific theories as for their political opinions and even personal background, such as training in the West and working under the Nationalists before 1949. Yet, in a few cases, notably Lysenkoism, persecution did fall on those with the "wrong" beliefs. There were numerous cases where the pursuit of basic research by itself could bring on the indictment that one was ignoring the practical duties of a scientist and thus deviating from the correct Maoist model of integrating theory with practice. Perhaps the most striking feature of the treatment of Chinese scientists under Mao was the wide swings between liberalization and harsh tightening, which reflected the divisions within the Communist party leadership over the future course for China and the complexities of modernization.

In 1948–9, communist leaders encouraged scientists who had worked under the rival Nationalists to stay where they were, instead of following the fleeing Nationalist forces to Taiwan. After the establishment of the PRC in October 1949, many of these carry-over scientists attained important administrative positions in the reconstructed scientific institutions of the new regime, including the Chinese Academy of Sciences. Characteristically, however, the most important policies on personnel and research directions were determined by party officials and the few scientists who were also party members.[85] In the early 1950s, the party and government pursued a policy of encouraging scientific research and education, in part to persuade those who had worked under the Nationalists to stay with the new regime, and in part to attract those Chinese students and scientists training or working in the West to return to their homeland. Despite the various political purges in this period, by and large, the latter succeeded as thousands of them overcame obstacles in the West to return to China.[86]

Distrusting these carry-over and returnee scientists, however, the party launched efforts to train its own "red" experts almost immediately following the establishment of the PRC. Typically, the new recruits undertook narrowly focused undergraduate studies in one of the Chinese universities, which were radically restructured in the early 1950s according to the Soviet model to emphasize specialized technical fields, such as metallurgy or geology. The best of these students—in terms of both technical competence and political loyalty—were sent to the Soviet Union for graduate study. Upon their return, they were expected to become leaders in the Chinese scientific enterprise. For example, Zhou Guangzhou, a physics student, followed this path. He would later become a leader in the Chinese nuclear weapons project and president of the Chinese Academy of Sciences in the 1980s.[87]

During the early stage of the Cultural Revolution, there was little effort to train new scientists, except for what existed in the nuclear and military space projects. Most of the universities were shut down from 1966 to about 1971, with no students admitted or graduated. In 1971, universities were re-opened and operated under a radical new direction: freshmen were to come not from high school graduates based on national entrance examinations, as before, but from peasants, workers, and soldiers with practical experience but with junior high school preparation, selected on political criteria. The standards in this new educational regime proved so low that years later, following the end of the Cultural Revolution, the so-called peasant-worker-soldier students had to be retrained after graduation to reach university level.[88]

The experiment in recruitment and enlistment in science turned out to be a complete failure: it may have produced a "red," but not by any means "expert," generation. When the pragmatist party leader Deng Xiaoping, who was purged by Mao during the Cultural Revolution, returned to power following Mao's death in 1976, he brought the returnee/carryover generation of scientists back into power. The older scientists often bypassed the peasant-worker-soldiers of the Maoist era and began to train a new generation of scientists who came through a restored educational system. Many of the latter also began to pursue studies abroad, especially in the U.S.A. and Western Europe.

Deng's advocacy of utilitarianism continued to dominate Chinese science policy in the 1980s and 1990s as market-driven economic reform brought another wave of structural changes to Chinese science. A number of scientists did run afoul of the regime in this period, but, again, because of their political beliefs and activities, not their scientific theories. Fang Lizhi, the prominent astrophysicist and political dissident, lost his position as vice president of the University of Science and Technology of China, Hefei, and was expelled from the party when he was blamed for student unrest. However, in a move that echoed that of the Soviet Communist Party, Fang was allowed to work at an observatory in isolation from students. In fact, the regime intentionally publicized his research to indicate that it continued to value science although it discouraged political dissent.[89]

Despite the claims by the Red Guards and other Maoists at the time, the Cultural Revolution produced few, if any, ideologically correct scientific theories. Much of the energy of the radical Maoist theorists was focused on attacking what they viewed as bourgeois scientific beliefs within and outside of China, rather than constructing new ones. These included Albert Einstein's relativity theory, which the Maoists denounced as politically capitalistic and philosophically idealistic.[90] Perhaps the only plausible case of ideologically-correct-science in China took place on the eve of the Cultural Revolution.

On August 18, 1964, Mao Zedong invited several Chinese philosophers of science to his residence in Beijing's Forbidden City to chat about the philosophical implications of new theories of elementary particles. "The world is infinite," he asserted, "time and space are infinite. Space is infinite at both the macro and micro levels. Matter is infinitely divisible." Mao's comments, which were prompted by a recent article on elementary particles written by the Japanese Marxist and physicist Shoichi Sakata, led to widespread, officially sponsored discussions on philosophy of science. A number of physicists participated in the discussion, at the end of which they concluded that, based on Mao's version of dialectical materialism, elementary particles were not "elementary," but could be further divided into constituent parts—stratons—to signify the infinite stratification of matter.[91] Starting in 1965, these physicists, many of whom had participated in the making of the Chinese hydrogen bomb, began to construct a theoretical model of how stratons made up hadrons (protons and neutrons).[92]

At the time, there were already models of hadrons in the West, which assumed that they were composed of quarks, but it was not clear whether quarks were merely mathematical devices or real particles. The Beijing group claimed that its model differed from the quark model in that it was relativistic and assumed that there were real sub-particles that were made up of hadrons. However, the model was essentially compatible with, and later subsumed under the quark model, which eventually did assume that quarks represented not merely mathematical constructs but real physical entities in what is now called the Standard Model.[93]

There seems little doubt that the initiation and the philosophical interpretation of the straton theory had much to do with Mao's pronouncements on the infinite divisibility of matter in the Chinese context. Maoism provided both the political and ideological justification of a fairly esoteric branch of science that could easily have been branded a bourgeois exercise in an ivory tower. One might argue that these scientists would probably have received protection due to their contributions to the Chinese bomb project, as in the case of Soviet nuclear physicists described by David Holloway. But the fact that some of these scientists were criticized and not allowed to present the theory to an international gathering of scientists in Beijing after the start of the Cultural Revolution, despite the apparent ideological correctness of their work, seems to indicate that the "nuclear umbrella" worked less in Maoist China than in Stalinist Russia.[94] The straton model was thus more a case of protective coloration than a genuine scientific creation of dialectical materialism. Yet, perhaps more than any other scientific endeavors in China in the Maoist years, the straton model represented an attempt at ideologically-correct-science.

Not surprisingly, in comparison with the often critical reception of theoretical research by Maoist "revolutionary utilitarianism," applied research fared much better. In addition to the nuclear weapons project, which was both politically correct and utilitarian to the extreme, a number of other fields also received ample moral and material support in the Mao era. The pivotal Twelve-Year Plan in Science and Technology, formulated by hundreds of scientists under the leadership of Zhou Enlai in 1956, for example, identified about fifty-six applied research areas for heavy investment in resources and personnel. Only as an afterthought was "major theoretical problems in natural sciences" tagged on as the last item in the plan. Even in these basic research fields, applications were the major motivation.[95]

The Communist Party provided lavish rewards for achievements in applied research. Since the early 1990s, the faith in science and technology as the basis of China's modernization has flourished in China under Party leader Jiang Zemin, who had trained as an engineer. By applying modern science and technology to the economy and management, Jiang and his supporters hoped for both a robust economy and a stable social order under Communist rule.

Perhaps no modern state has purged and persecuted its scientists as ruthlessly and repeatedly as the People's Republic of China, but these waves of purges, as well as the intervals of relative tranquility which followed them, had little to do with ideologically-correct-science. Mao and his successors have wanted first and foremost obedience from their scientists. Ideologically correct conduct is part of this, but in this regard scientists have been no different than other Chinese.

Conclusion

If one assumes that the state (or forces or individuals within it) was imposing ideology on science and scientists, then this implies that science and ideology are separable. In fact, as the above examples demonstrate, even in the case of ICS science is not being determined by ideology, but also is not free of it. No political regime has ever tried consistently and comprehensively to impose ICS on its scientists. There have been individuals or portions of a regime that have tried to impose some ideologies on some aspects of science and on some scientists. However, such individual cases are often "overdetermined": there are alternative explanations other than ICS, both for the specific ideological attacks and the efficacy of the victims' resistance. On the other hand, ICS did happen. Just because the entire state was not behind it for the entire period does not contradict that. Although ICS in the ideal sense always failed, it also always had a significant effect, whether an intended one, or not.

No single ideology, including liberal democracy, has historically proven more effective than another in driving science or leading to intended results. Communist regimes appear to have been more likely to try and impose ideological standards on their science, perhaps because Marxism is so comprehensive a political philosophy. In other regimes, ideological attacks on science remained more crude and overt, like National Socialist calls to eliminate "Jewish Science" or anti-communists in the United States during the 1950s denouncing "internationalism" in science.

Although communist regimes were sometimes the most ruthless oppressors of their own scientists, they were also sometimes more flexible and pragmatic.[96] Thus in the Soviet Union and the Peoples' Republic of China, scientific representatives of the ideological enemy, the "bourgeois specialists," were kept on for a while, sometimes even pampered, in order to bridge the gap of time between the beginning of the revolution and the point at which the new regime had trained its own, ideologically acceptable (or more ideologically acceptable) scientists. In some cases, these holdovers from the previous regime held on long enough to shed their label and remain productive and sometimes integral parts of their national scientific efforts.

This would have been inconceivable in Hitler's Germany. The few exceptions of Jewish or part-Jewish scientists who managed to survive the Third Reich working for the regime are the exceptions that prove the rule. The overwhelming majority of "non-Aryan" scientists were purged because of their race and irregardless of their expertise or the scientific needs of the regime. Similarly, scientists suspected or accused of communist sympathies during the McCarthy period were damaged goods and not to be used by the government. It should also be noted that, whatever the actual loss of scientific manpower to Hitler's Germany and Eisenhower's America, in both cases there were sufficient numbers of ambitious, competent, and politically and ideologically acceptable scientists to fill the vacancies.

It is probably no coincidence that all of these examples also deal with military conflict or preparations for it. The French Revolution and National Socialism unleashed war, the Soviet Union and the People's Republic of China were born of war, Japan was fighting the Second World War, and in the 1950s America and the Soviet Union were battling in the Cold War. The militarization of society and of science bring the two closer together: the former's chances for success may hinge on the military potential of its science and technology; while the latter's service to its nation at war may entangle and immerse it more deeply in the regime's political and ideological goals. One pattern thus emerges quite often in the examples studied by this essay: the military potential of science and scientists outweigh and overrule attempts to purify science ideologically.

It was never easy to introduce ideologically driven and compliant science and technology. Purges of scientists or attacks on scientific theories on ideological grounds were often self-defeating and short-lived. Scientists sometimes suffered, but not because they were scientists, rather because they were part of a greater ideological or racial group perceived by the regime to be a potential threat. When challenged, scientists did not defend themselves by claiming the ideological neutrality of their work and demanding intellectual freedom. Instead they strove to blunt attacks by either winning over their critics or enlisting other patrons. Of course, this meant that they had to work closely, or at least more closely with at least some forces in the state and demonstrate their usefulness to them.

The end result was usually partial success for the established (and now sometimes embattled) scientific community, for their critics were eventually silenced and they did safeguard some of their professional prerogatives, but also partially entailed an accommodation to, and collaboration with ideological aspects of the regime, sometimes directly related to their scientific work, other times in a more general form.

This is not what the sociologist Robert Merton or physicist Samuel Goudsmit predicted in the aftermath of the Second World War and the beginning of the Cold War.[97] Science is not especially suited to democracy, as Goudsmit claimed, and is able to compromise some of Merton's norms for scientific work. Indeed it is striking, that, despite the great differences between these often very different examples, most of the scientists responded in a similar way when their apolitical science was threatened. ICS tells us that (1) no matter how ruthless, totalitarian, racist, or intolerant a regime might be, when it needs its scientists, it will do what it has to in order to harness them; and (2) whether they support the regime whole-heartedly or not, most scientists, or perhaps better put, scientific communities, will do what they have to in order to be able to do science. Thus science is independent of particular political and ideological regimes: just not in the way most people believe it is.

Notes

1. Society of Fellows, Harvard University and History Department, Princeton University; Department of History, Bowling Green State University; Union College; California State Polytechnic University, Pomona. We would like to thank the colleagues who read and commented on a preliminary version of our manuscript: Mitchell Ash, Charles C. Gillispie, Loren Graham, Alexei Kojevnikov, and Jessica Wang.

2. Jessica Wang, *American Science in an Age of Anxiety: Scientists, Anti-Communism, and the Cold War* (Chapel Hill, NC: University of North Carolina Press, 1998).

3. Doug Weiner, *Models of Nature* (Bloomington, IN: Indiana University Press, 1988).

4. Harold T. Parker, "French Administrators and French Scientists during the Old Regime and the Early Years of the Revolution," in Richard Herr and Harold T. Parker (eds), *Ideas in History* (Durham, NC: Duke University Press, 1965), 85–109.

5. Roger Hahn, *The Anatomy of a Scientific Institution: The Paris Academy of Sciences, 1666–1803* (Berkeley, CA: University of California Press, 1971); Charles C. Gillispie, *Science and Polity in France at the End of the Old Regime* (Princeton, NJ: Princeton University Press, 1980), 81–99.

6. Keith Michael Baker, *Condorcet* (Chicago: University of Chicago Press, 1975), 350–2.

7. Henry Moss, "Scientists and Sans-culottes: The Spread of Scientific Literacy in the Revolutionary Year II," *Fundamenta Scientiae*, 4 (1983): 101–15; Gillispie, *Science and Polity in France at the End of the Old Regime*, 143–84.

8. Ken Alder, *Engineering the Revolution: Arms and Enlightenment in France, 1763–1815*

(Princeton, NJ: Princeton University Press, 1997), ch. 8; Terry Shinn, *L'École Polytechnique, 1794–1914* (Paris: Presses de la fondation nationale des sciences politiques, 1980); Ambroise Fourcy, *Histoire de l'Ecole Polytechnique* (Paris: Belin, 1987); Nicole et Jean Dhombres, *Naissance d'un nouveau pouvoir: Sciences et Savants en France, 1793–1824* (Paris: Payot, 1989); Janis Langins, "The *Ecole Polytechnique* and the French Revolution: Merit, Militarization, and Mathematics," *LLULL*, 13 (1990): 91–105; and Janis Langins, *La République avait besoin de savants* (Paris: Belin, 1987).

9. For a short account, see Langins, "The *Ecole Polytechnique*". More detailed treatments are provided by the Alder and Langins works cited above.

10. See Charles C. Gillispie, "The *Encyclopédie* and the Jacobin Philosophy of Science: A Study in Ideas and Consequences," in Marshall Clagett (ed.), *Critical Problems in the History of Science* (Madison, WI: University of Wisconsin Press, 1959), 255–89; and L. Pearce Williams, "The Politics of Science in the French Revolution," in Clagett, 291–320.

11. Jessica Riskin, "Rival Idioms for a Revolutionized Science and a Republican Citizenry," *Isis*, 89 (1998): 203–32.

12. Charles C. Gillispie, "De l'Histoire naturelle à la biologie: Relations entre les programmes des recherche de Cuvier, Lamarck, et Geoffrey Saint-Hilaire," in *Collecter, observer, classer* (forthcoming).

13. J.W. Dauben, "Marat: His Science and the French Revolution," *Archives Internationales de l'Histoire des Sciences*, 22 (1969): 235–61; Gillispie, *Science and Polity in France at the End of the Old Regime*, 290–330.

14. See Hahn, *passim*, for references to primary literature on the Jardin.

15. Loren R. Graham, *The Soviet Academy of Sciences and the Communist Party, 1927–1932* (Princeton, NJ: Princeton University Press, 1967); and Alexander Vucinich, *Empire of Knowledge: The Academy of Sciences of the USSR (1917–1970)* (Berkeley, CA: University of California Press, 1984).

16. Graham, *The Soviet Academy of Sciences*, 24–79; and Kendall E. Bailes, *Technology and Society under Lenin and Stalin: Origins of the Soviet Technical Intelligentsia, 1917–1941* (Princeton, NJ: Princeton University Press, 1978).

17. Lynn Mally, *Culture of the Future: The Proletkult Movement in Revolutionary Russia* (Berkeley, CA: University of California Press, 1990). Katerina Clark, "The Changing Image of Science and Technology in Soviet Literature," in Loren R. Graham (ed.), *Science and the Soviet Social Order* (Cambridge, MA: Harvard University Press, 1990), 266–67.

18. Mally, chapter 6.

19. On specialists during this period, see Bailes.

20. Jeremy R. Azrael, *Managerial Power and Soviet Politics* (Cambridge, MA: Harvard University Press, 1966). On the economics and politics of the NEP period, see Stephen F. Cohen, *Bukharin and the Bolshevik Revolution: A Political Biography, 1888–1938* (Oxford: Oxford University Press, 1980), and Robert V. Daniels, *Conscience of the Revolution: Communist Opposition in Soviet Russia* (Boulder, CO: Westview Press, 1988).

21. On the tense compromise between ideological promotion of young Soviet acolytes like Landau and the surviving tsarist professorate, see Karl Hall, "Purely Practical Revolutionaries," (Ph.D., Harvard University, 1999).

22. For a detailed account of the Shakhty Trial and the subsequent Industrial Party Affair, see Bailes, chapter 3.

23. David Holloway, *Stalin and the Bomb* (New Haven: Yale University Press, 1994).

24. On the education system generally in this period, see Sheila Fitzpatrick, *The Commissariat of Enlightenment: Soviet Organization of Education and the Arts under Lunacharsky, October 1917–1921* (Cambridge: Cambridge University Press, 1970). On the reform of institutes, see Loren R. Graham, "The Formation of Soviet Research Institutes: A Combination of Revolutionary Innovation and International Borrowing," *Social Studies of Science, 5* (1975): 303–29. On Leningrad Physico-Technical Institute in this period, see Paul Josephson, *Physics and Politics in Revolutionary Russia* (Berkeley, CA: University of California Press, 1991).

25. See Graham, *The Soviet Academy of Sciences*, 24–79; Vucinich, 72–129; and Azrael, chapters 3–4.

26. Bailes.

27. Sheila Fitzpatrick, *The Cultural Front: Power and Culture in Revolutionary Russia* (Ithaca, NY: Cornell University Press, 1992).

28. David Joravsky, *Soviet Marxism and Natural Science, 1917–1932* (London: Routledge and Kegan Paul, 1961).

29. Loren R. Graham, *Science in Russia and the Soviet Union: A Short History* (Cambridge: Cambridge University Press, 1993), ch. 5.

30. Loren R. Graham, *Science, Philosophy, and Human Behavior in the Soviet Union* (New York: Columbia University Press, 1987), chaps 1–2.

31. David Joravsky, *The Lysenko Affair* (Cambridge, MA: Harvard University Press, 1970); Krementsov; Valery N. Soyfer, *Lysenko and the Tragedy of Soviet Science* (New Brunswick, NJ: Rutgers University Press, 1994); Zhores A. Medvedev, *The Rise and Fall of T.D. Lysenko,* (New York: Columbia University Press, 1969); David Joravsky, "Soviet Marxism and Biology Before Lysenko," *Journal of the History of Ideas,* 20 (1959): 85–104; Dominique Lecourt, *Proletarian Science? The Case of Lysenko* (Norfolk, VA: NLB, 1977); Nikolai L. Krementsov, *Stalinist Science* (Princeton, NJ: Princeton University Press, 1997).

32. Kirill O. Rossianov, "Stalin as Lysenko's Editor: Reshaping Political Discourse in Soviet Science," *Configurations,* 3 (1993): 439–56; and Kirill O. Rossianov, "Editing Nature: Joseph Stalin and the 'New' Soviet Biology," *Isis,* 84 (1993): 728–45.

33. Alexei Kojevnikov, "Rituals of Stalinist Culture at Work: Science and the Games of Intraparty Democracy circa 1948," *Russian Review,* 57, 1 (1998): 25–52.

34. Mitchell Ash and Alfons Söllner, "Forced Migration and Scientific Change after 1933," in Mitchell Ash and Alfons Söllner (eds), *Forced Migration and Scientific Change: Emigré German-Speaking Scientists and Scholars after 1933* (Cambridge: Cambridge University Press, 1996), 1–19.

35. Alan Beyerchen, *Scientists under Hitler: Politics and the Physics Community in the Third Reich* (New Haven, CT: Yale University Press, 1977).

36. Rudolf Vierhaus and Bernhard vom Brocke (eds), *Forschung im Spannungsfeld von Politik und Gesellschaft–Geschichte und Struktur der Kaiser-Wilhelm/Max-Planck-Gesellschaft* (Stuttgart: DVA, 1990); Kristie Macrakis, *Surviving the Swastika: Scientific Research in Nazi Germany* (Cambridge, MA: Harvard University Press, 1993).

37. Mark Walker, *Nazi Science* (New York: Plenum, 1995).

38. Robert Proctor, *Racial Hygiene: Medicine under the Nazis* (Cambridge, MA: Harvard University Press, 1988); Paul Weindling, *Health, Race, and German Politics* between *National Unification and Nazism, 1870–1945* (Cambridge: Cambridge University Press, 1989); Uwe Hoßfeld, "Staatsbiologie, Rassenkunde und Moderne Sythese in Deutschland während der NS-Zeit," in Rainer Brömer, Uwe Hoßfeld, and Nicolaas Rupke (eds), *Evolutionsbiologie von Darwin bis heute* (Berlin: VWB, 2000), 249–305; Uwe Hoßfeld and Thomas Junker, "Synthetische Theorie und 'Deutsche Biologie': Einführender Essay," in Brömer, Hoßfeld, and Rupke, 231–48; Thomas Junker, "Synthetische Theorie, Eugenik und NS-Biologie," in Brömer, Hoßfeld, and Rupke, 307–60.

39. Walker, *Nazi Science.*

40. Jacques Pauwels, *Women, Nazis, and Universities: Female University Students in the Third Reich* (New Haven, CT: Greenwood Press, 184).

41. Alexander Mitscherlich and Fred Mielke, *Doctors of Infamy: The Story of the Nazi Medical Crimes* (New York: Henry Schuman, 1949); Robert J. Lifton, *The Nazi Doctors: Medical Killing and the Psychology of Genocide* (New York: Basic Books, 1986); Benno Müller-Hill, *Murderous Science: Elimination by Scientific Selection of Jews, Gypsies, and Others, Germany 1933–1945* (Oxford: Oxford University Press, 1985).

42. For biology, Ute Deichmann, *Biologists under Hitler* (Cambridge, MA: Harvard University Press, 1996); for physics, Beyerchen, David Cassidy, *Uncertainty: The Life and Science of Werner Heisenberg* (New York: Freeman, 1991), Mark Walker, *German National Socialism and the Quest for Nuclear Power, 1939–1949* (Cambridge: Cambridge University Press, 1989), and Walker, *Nazi Science;* for mathematics, Herbert Mehrtens, "Ludwig Bieberbach and 'Deutsche Mathematik,'" in E. Phillips (ed.), *Studies in the History of Mathematics* (Washington, DC: American Mathematical Association, 1987), 195–247; for psychology see Ulfried Geuter, *The Professionalization of Psychology in Nazi Germany* (Cambridge: Cambridge University Press, 1992), Mitchell Ash, "From 'Positive Eugenics' to Behavioral Genetics: Psychological Twin Research under Nazism and Since," *Historia Pedagogica—International Journal of the History of Education, Supplementary Series,* 3 (1998): 335–58, and Mitchell Ash, "Constructing Continuities: Kurt

Gottschaldt und Psychological Research in Nazi and Socialist Germany," in Kristie Macrakis and Dieter Hoffmann (eds), *Science and Socialism in the G.D.R. in Comparative Perspective* (Cambridge, MA.: Harvard University Press, 1999), 286–301, 360–65; for chemistry Ute Deichmann, *Flüchetn, Mitmachen, Vergessen* (Weinheim: Wiley- VCH, 2001); for technology, Karl' Heinz Ludwig, *Technik und Ingenieure im Dritten Reich* (Düsseldorf: Droste Verlag, 1974) and Helmut Maier, "National sozialistische Technikideologie und die Politisierung des 'Technikerstandes': Fritz Todt und die Zeitschrift 'Deutsche Technik,'" in Burkhard Dietz, Michael Fessner, and Helmut Maier (eds), *Technische Intelligenz und "Kulturfaktor Technik"* (Münster: Waxmann, 1996), 253–68.

43. Cassidy; Walker, *Nazi Science.*

44. Michael Neufeld, *The Rocket and the Reich: Peenemünde and the Coming of the Ballistic Missile Era* (New York: Free Press, 1995).

45. Helmuth Trischler, *Luft- und Raumfahrtforschung in Deutschland 1900–1970* (Frankfurt am Main: Campus Verlag, 1992); Ulrich Albrecht, "Military Technology and National Socialist Ideology," in Renneberg and Walker (eds), 88–125, 358–63.

46. Walker, *German National Socialism and the Quest for Nuclear Power.*

47. For physicians, Lifton, Weindling, Proctor; for biologists, Deichmann, Junker; for psychologists, Geuter, Mitchell Ash, "Denazifying Scientists and Science," in Matthias Judt and Burghard Ciesla (eds), *Technology Transfer out of Germany* (Amsterdam: Harwood, 1996), 61–80; for anthropologists, Hossfeld; for geographers, Mechtild Rössler, "'Area Research' and 'Spatial Planning' from the Weimar Republic to the German Federal Republic: Creating a Society with a Spatial Order under National Socialism," in Renneberg and Walker (eds), 126–38, 363–6.

48. Two examples are Dr. Pak Ch'ul Jai, an X-ray physicist, and Dr. Lee Tai Kyu, a professor of chemistry, both of whom held positions at Kyoto Imperial University during the war. See Walter E. Grunden, "Hungnam and the Japanese Atomic Bomb: Recent Historiography of a Postwar Myth," *Intelligence and National Security,* 13 (Summer 1998): 32–60.

49. Yukawa was awarded the Nobel Prize in physics in 1949. Taketani Mitsuo, "Methodological Approaches in the Development of the Meson Theory of Yukawa in Japan," in Nakayama Shigeru, David L. Swain, and Yagi Eri (eds), *Science and Society in Modern Japan: Selected Historical Sources* (Tokyo: University of Tokyo Press, 1973), 24–38.

50. Hiroshige Tetu, "The Role of the Government in the Development of Science," *Journal of World History,* 9 (1965): 320–39. Itakura Kiyonobu and Yagi Eri, "The Japanese Research System and the Establishment of the Institute of Physical and Chemical Research," in Nakayama, Swain, and Yagi (eds), *Science and Society in Modern Japan,* 158–201. Kamatani Chikayoshi, "The History of Research Organization in Japan, *Japanese Studies in the History of Science,* 2 (1963): 1–79.

51. S. Watanabe, "How Japan Has Lost a Scientific War" (September 1945), File: "Historical Information," RG 38, Box 111, U.S. National Archives, Washington, DC.

52. Yomiuri Shimbunsha (ed.), "Nihon no genbaku" (Japan's Atomic Bomb), in *Shôwa shi no Tennô* (The Emperor in Shôwa History), vol. 4 (Tokyo: Yomiuri Shimbunsha, 1968), 78–229.

53. John W. Dower, "'NI' and 'F': Japan's Wartime Atomic Bomb Research," *Japan in War and Peace: Selected Essays* (New York: New Press, 1993), 55–100. Morris Fraser Low, "Japan's Secret War? 'Instant' Scientific Manpower and Japan's World War II Atomic Bomb Project," *Annals of Science,* 47 (1990): 347–60.

54. Walter E. Grunden, "Science under the Rising Sun: Weapons Development and the Organization of Scientific Research in World War II Japan" (Ph.D., University of California at Santa Barbara, 1998).

55. Tessa Morris-Suzuki, *The Technological Transformation of Japan: From the Seventeenth to the Twenty-First Century* (Cambridge: Cambridge University Press, 1994), 144–5.

56. Morris-Suzuki, 145.

57. Excerpts of Lieutenant-General Tada Reikichi, President of the Board of Technology, from Captain George B. Davis to Major Francis J. Smith, "Transmittal of Items from the 'Daily Digest of World Broadcasts'" (27 July 1945), "Japan, Misc" RG 77, Entry 22, Box 173, Folder #44.70, U.S. National Archives, College Park, MD; See also, Major H.K. Calvert to Major F.J. Smith, "Japanese Militarists Want Miracle Weapon" (29 May 1945) "Japan, Misc" RG 77, Entry 22, Box 173, Folder #44.70, U.S. National Archives, College Park, MD.

58. Peter Williams and David Wallace, *Unit 731: Japan's Secret Biological Warfare in World War II* (New York: The Free Press, 1989).

59. For a concise treatment of McCarthyism in general, see Ellen Schrecker, *No Ivory Tower: McCarthyism and the Universities* (New York: Oxford University Press, 1986) and Ellen Schrecker, *The Age of McCarthyism: A Brief History with Documents* (Boston: Bedford Books, 1994).

60. Lawrence Badash, *Scientists and the Development of Nuclear Weapons: From Fission to the Limited Test Ban Treaty, 1939–1963* (Atlantic Highlands, NJ: Humanities Press, 1995), 102.

61. Paul Forman, "Behind Quantum Electronics: National Security as Basis for Physical Research in the United States, 1940–1960," *Historical Studies in the Physical and Biological Sciences,* 18 (1987): 149–229.

62. Richard G. Hewlett and Jack M. Holl, *Atoms for Peace and War, 1953–1961: Eisenhower and the Atomic Energy Commission* (Berkeley, CA: University of California Press, 1989), ch. 4, "The Oppenheimer Case," 73–112.

63. Wang, *Anti-Communism,* 289–95.

64. Wang, *Anti-Communism,* 274–9.

65. "Scientists Affirm Faith in Oppenheimer," *Bulletin of the Atomic Scientists,* 10 (May 1954): 188–91. "Scientists Express Confidence in Oppenheimer," *Bulletin of the Atomic Scientists,* 10 (September 1954): 283–6.

66. Steven E. Ambrose, *Eisenhower: The President* (New York: Simon and Schuster, 1984), 166.

67. Eisenhower to Robert E. Sherwood, April 21, 1954, in *The Diaries of Dwight D. Eisenhower,* reel 4, 193. See also Dwight D. Eisenhower, *Mandate for Change, 1953–1956* (Garden City, NY: Doubleday and Company, Inc., 1963), 314.

68. Hewlett and Holl, 111–12.

69. Wang, *Anti-Communism.*

70. Interview with Herbert York by Zuoyue Wang, La Jolla, CA, July 18, 1992.

71. Daniel Kevles, "Cold War and Hot Physics: Science, Security, and the American State, 1945–56," *Historical Studies in the Physical and Biological Sciences,* 20 (1990): 239–64.

72. Wang, *Anti-Communism,* 261–2, 281.

73. Schrecker, *No Ivory Tower,* and Schrecker, *The Age of McCarthyism;* interview with Wolfgang Panofsky by Zuoyue Wang, Stanford, CA, March 5, 1992.

74. Schrecker, *No Ivory Tower;* Wang, *Anti-Communism,* 254–261.

75. William Lanouette, *Genius in the Shadows: A Biography of Leo Szilard, the Man Behind the Bomb* (Chicago: University of Chicago Press, 1994).

76. Wang, *Anti-Communism,* 277.

77. Wang, *Anti-Communism,* 289–295, Schrecker, *The Age of McCarthyism,* 92–94.

78. Zuoyue Wang, *In the Shadow of Sputnik: American Scientists and Cold War Science Policy* (forthcoming). Wang, *Anti-Communism.*

79. Schrecker, *No Ivory Tower;* Schrecker, *The Age of McCarthyism,* 92.

80. James Reardon-Anderson, *The Study of Change: Chemistry in China, 1840–1949* (Cambridge: Cambridge University Press, 1991), 339–64, on 359.

81. Yao Slurping *et al., Zhongguo kexueyuan* (Chinese Academy of Sciences), 3 volumes (Beijing: Contemporary China Press, 1994), vol. 1, 26–7.

82. Jonathan Spence, *The Search for Modern China* (New York: W.W. Norton 1990) 572.

83. Yao *et al.,* 87–9.

84. See, for example, Peter Neushul and Zuoyue Wang, "Between the Devil and the Deep Sea: C.K. Tseng, Ocean Farming, and the Politics of Science in Modern China," *Isis,* 91, 1 (2000): 59–88.

85. See, for example, several articles in the Chinese official journal *Bai nian chao* (Hundred Year Tide), June 1999, especially Li Zhenzhen, "Interview with Yu Guangyuan and Li Peishan," 23–30, and "Interview with Gong Yuzhi," 31–27.

86. See Jin Chongji, *Zhou Enlai zhuan* (A Biography of Zhou Enlai), 2 volumes (Beijing: Central Documentation Press, 1998), vol. 1, 234.

87. Dai Minghua, *et al,* "Zhou Guangzhao" in Lu Jiaxi (ed. in chief), *Zhongguo xiandai kexuejia zhuanji* (Biographies of Modern Chinese Scientists), 6 volumes (Beijing: Science Press, 1991–4), vol. 6, 187–96.

88. Wu Heng, *Keji zhanxian wushinian* (Fifty Years on the Scientific and Technological Front) (Beijing: Science and Technology Documentation Press, 1992), 348–9.

89. See H. Lyman Miller, *Science and Dissent in Post-Mao China: The Politics of Knowledge* (Seattle: University of Washington Press, 1996).

90. Xu Liangying and Qiu Jingcheng, "Guanyu wuoguo 'wenhua dageming' shiqi pipan ai-in-si-tan

he xiangduilun de chubu kaocha" (A Preliminary Study of the Movement to Denounce Einstein and Relativity During the Period of the Great Cultural Revolution in Our Country), in Xu (ed. in chief), *Ai-in-si-tan yanjiu* (Einstein Studies), 1 (1989): 212–50.

91. Gong Yuzhi, "Mao Zedong yu zhiran kexue" (Mao Zedong and the Natural Sciences), in Gong, *Zhiran bianzhengfa zai zhongguo* (Dialectics of Nature in China) (Beijing: Peking University Press, 1996), 87–112.

92. He Zuoxiu, "A History of the Establishment of the Straton Model," *Yuanshi zhiliao yu yanjiu* (Documentation and Research in the History of the Chinese Academy of Sciences), 1994, issue no. 6, 16–30.

93. He.

94. Holloway.

95. See "1956–1967 nian kexue jishu fazhan guihua gangyao" (Outline of a Long-Term Plan for the Development of Science and Technology, 1956–1967), in Chinese Communist Party Central Documentation Institute (ed.), *Jianguo yilai zhongyao wenxián xuanbian* (Selected Key Documents Since the Founding of the Country) (Beijing: Central Documentation Press, 1992), vol. 9, 436–535.

96. Also see for the example of the German Democratic Republic the following articles: Mitchell Ash, "Wissenschaft, Politik und Modernität in der DDR: Ansätze zu einer Neubetrachtung," in K. Weisemann, P. Kroener, and R. Toellner (eds), *Wissenschaft und Politik: Genetik und Humangenetik in der DDR (1949–1989)* (Münster: LIT-Verlag, 1997), 1–26; Mitchell Ash, "1933, 1945 und 1990—drei Bruchstellen in der Geschichte der deutschen Universität," in A. Söllner (ed.) *Ostblicke—Perspektiven der Hochschulen in den neuen Bundesländern* (Opladen: Westdeutscher Verlag, 1998), 212–37; Mitchell Ash, "Scientific Changes in Germany 1933, 1945 and 1990: Towards a Comparison," *Minerva*, 37 (1999): 329–54.

97. See the introduction to this volume, as well as Jessica Wang, "Merton's Shadow: Perspectives on Science and Democracy since 1940," *Historical Studies in the Physical and Biological Sciences,* 30, No. 1 (1999): 279–306, and David Hollinger, "The Defense of Democracy and Robert K. Merton's Formulation of the Scientific Ethos," in David Hollinger, *Science, Jews, and Secular Culture: Studies in Mid-Twentieth Century American Intellectual History* (Princeton, NJ: Princeton University Press, 1996), ch. 5.

Introduction: How Users and Non-Users Matter

Nelly Oudshoorn and Trevor Pinch

New uses are always being found for familiar technologies. Sometimes these changes in use are dramatic and unexpected. Before September 11, 2001, no one foresaw that an airliner could be turned by a small number of its occupants into a giant Molotov cocktail. After the Gulf War of 1991, it was discovered that an effective way to put out oil-rig fires was to strap down captured Mig jet fighters and blow out the fires using their exhaust. Such examples remind us that we can never take the use of a technology for granted.

Susan Douglas (1987) has pointed out how amateur operators discovered new uses to which the emerging technology of radio could be put, and how commercial operators soon followed the amateurs' lead. Claud Fischer (1992) and Michele Martin (1991) have drawn attention to the use of the telephone by rural women to overcome their isolation—a use not foreseen by telephone companies, which conceived of the telephone mainly as a business instrument.

Our concern in this book is with the role of users in the development of technology in general. We are interested in how users consume, modify, domesticate, design, reconfigure, and resist technologies. In short, our interest is in whatever users do with technology.

There is no one correct use for a technology. "What is an alarm clock for?" we might ask. "To wake us up in the morning," we might answer. But just begin to list all the uses to which an alarm clock can be put and you see the problem. An alarm clock can be worn as a political statement by a rapper; it can be used to make a sound on a Pink Floyd recording; it can be used to evoke laughter, as Mr. Bean does in one of his comic sketches as he tries to drown his alarm clock in his bedside water pitcher; it can be used to trigger a bomb; and, yes, it can be used to wake us up. No doubt there are many more uses. Of course, there may be one dominant use of a technology, or a prescribed use, or a use that confirms the manufacturer's warranty, but there is no one essential use that can be deduced from the artifact itself. This is an axiomatic assumption for the scholars whose work we collect here. All the contributors follow the research path of studying technologies in their "context of use"—the society and the web of other artifacts within which technologies are always embedded. In short, we look at how technologies are actually used in practice.

In addition to studying what users do with technology, we are interested in what technologies do to users. Users of technologies do not arrive de novo. Think of the camera. When George Eastman developed his revolutionary new technology of roll film and a cheap camera, he had one outstanding problem: There were as yet no users for it. Photography was seen as a high-end activity practiced by a small group of skilled professionals. Eastman had to define explicitly who the new users might be, and he had to figure out how to recruit them to his new technology. He had to redefine photography and the camera. After he did, photography became something that anyone could participate in, and cameras became usable by all (Jenkins 1975). Working out who the new users are and how they will actually interact with a new technology is a problem familiar to many innovators of new technologies. Some fields, including information technology, are particularly cognizant of the problem of users. It has long been recognized that the most sophisticated and complex computer hardware and software will come to naught if users don't known how to use them. Studies of human-computer interaction, of work practices, and of user interfaces are often carried out by the computer industry, and they have become important not only for that industry but also for developing new ideas of how the user-technology nexus should be conceptualized (Suchman 1994; Woolgar 1991).

One important research question addressed in this book is how users are defined and by whom. For instance, are users to be conceived of as isolated autonomous consumers, or as self-conscious groups? How do designers think of users? Who speaks for them, and how? Are users an important new political group, or a new form of social movement? In short, what general lessons are to be drawn from a renewed focus on users in today's technologically mediated societies?

Different Approaches to Users

Users and technology are too often viewed as separate objects of research. This book looks for connections between the two spheres. Users and technology are seen as two sides of the same problem—as co-constructed. The aim is to present studies of the co-construction of users and technologies that go beyond technological determinist views of technology and essentialist views of users' identities.

In this introduction we discuss several influential approaches to user-technology relations, focusing in particular on the conceptual vocabulary developed within the different approaches and on the similarities and differences between them.

The SCOT Approach: Users as Agents of Technological Change

In the 1980s and the 1990s, the old view of users as passive consumers of technology was largely replaced in some areas of technology studies, and along with it the linear model of technological innovation and diffusion. One of the first approaches to draw attention to users was the social construction of technology (SCOT) approach.

Pinch and Bijker (1984), in defining the SCOT approach, conceived of users as a social group that played a part in the construction of a technology. Different social groups, they noted, could construct radically different meanings of a technology. This came to be known as a technology's interpretive flexibility. In a well-known study of the development of the bicycle, it was argued that elderly men and women gave a new meaning to the high-wheeled bicycle as the "unsafe" bicycle, and that this helped pave the way for the development of the safety bicycle. The SCOT approach specifies a number of closure mechanisms—social processes whereby interpretative flexibility is curtailed. Eventually, a technology stabilizes, interpretative flexibility vanishes, and a

predominant meaning and a predominant use emerge (Bijker and Pinch 1987; Bijker 1995). The connection between designers and users was made more explicit with the notion of a technological frame (Bijker 1995). Users and designers could be said to share a technological frame associated with a particular technology.

Many of the classic SCOT studies were of the early stages of technologies. For example, there were studies of how the bicycle, fluorescent lighting, and Bakelite moved from interpretative flexibility to stability. Early on, social groups were seen as the shaping agents. Not until later, with notions such as that of sociotechnical ensembles, did SCOT fully embrace the idea of the co-construction or mutual shaping of social groups and technologies (Bijker 1995b). The SCOT approach was rightly criticized for its rather cavalier attitude toward users—it closed down the problem of users too early, and it did not show how users could actively modify stable technologies (Mackay and Gillespie 1992). Kline and Pinch (1996) remedied this with their study of how a stable technology, the Model T automobile, could be appropriated and redesigned by groups such as farmers who used cars as stationary power sources. Kline and Pinch referred so such users as "agents of technological change." Also attempting to correct SCOT's neglect of gender, Kline and Pinch argued that users should be studied as a crucial location where often-contradictory genders identities and power relationships were woven around technologies. Bijker (1995) argued for a semiotic conception of power whereby power is embedded and mediated by artifacts as well as by frames and social groups. However, this semiotic notion of power (like most semiotic approaches within technology studies) seems inevitably to leave out invisible actors and social groups, which in the SCOT approach might be termed "non-relevant social groups."

Feminist Approaches: Diversity and Power

Feminist scholars have played a leading role in drawing attention to users. Their interest in users reflects concerns about the potential problematic consequences of technologies for women and about the absence of women in historical accounts of technology. Since the mid 1980s, feminist historians have pointed to the neglect of women's role in the development of technology. Because women were historically underrepresented as innovators of technology, and because historians of technology often focused exclusively on the design and production of technologies, the history of technology came to be dominated by stories about men and their machines. Moreover, these stories represented a discourse in which gender was invisible. Historian did not consider it relevant in settings where women were absent, thus reinforcing the view that men had no gender. Feminist historians suggested that focusing on users and use rather than on engineers and design would enable historians to go beyond histories of men inventing and mastering technology (Wajcman 1991; Lerman et al. 1997). In response to this criticism, users were gradually included in the research agenda of historians of technology. This "turn to the users" can be traced back to Ruth Schwartz Cowan's exemplary research on user-technology relations. In the late 1970s, Cowan brought the fields of history of technology and women's history together, emphasizing that women as users of technology perceive technological change in significantly different ways from men (Pursell 2001). Cowan's notion of "the consumption junction," defined as "the place and time at which the consumer makes choices between competing technologies" (Cowan 1987: 263), was a landmark. Cowan argued that focusing on the consumer and on the network relations in which the consumer is embedded enables historians and sociologists of technology to improve their understanding of the unintended consequences of technologies in the hands of users. Focusing on users would enrich the history of technology with a better understanding of the successes and failures of technologies (ibid.: 279). In contrast to actor-network theory (which we will discuss below), Cowan urged historians and sociologists of technology to choose the user, rather than the artifact or the technologist, as a point of departure in network analyses of technology, and to look at networks from the consumer's point of view (ibid.: 262). The scholarship that Cowan inspired rejects the idea that science and technology begin or end with the actions of scientists and engineers. Scholars in the field of Science and Technology Studies (STS) were urged to follow technologies all the way to the users (Rapp 1998: 48). An exemplary study is Cynthia Cockburn and Susan Ormrod's 1993 book on the microwave oven in the United Kingdom, which analyzes the design, the production, and the marketing as well as the use of a new technology.

Gender studies, like technology studies in general, reflect a shift in the conceptualization of users from passive recipients to active participants. In the early feminist literature, women's relation to technology had been conceptualized predominantly in terms of victims of technology. The scholarship of the last two decades, however, has emphasized women's active role in the appropriation of technology. This shift in emphasis was explicitly articulated in the first feminist collection of historical research on technology, *Dynamos and Virgins Revisited* (Trescott 1979), which included a section on "women as active participants in technological change" (Lerman et al. 1997: 11). The authors of the essays in that section argued that feminists should go beyond representations of women as essentially passive with respect to technology. Having accepted that challenge, feminist historians, anthropologists, and sociologists have published numerous accounts of how women shape and negotiate meanings and practices in technology, including studies of the relationship between reproductive technologies and women's health and autonomy, of the gendered medicalization of bodies, of women's relations to computers and the impact of computer technologies on women's work, of the consequences of household technologies for women's lives, and of the exclusion of women from technologies. Granting agency to users, particularly women, can thus be considered central to the feminist approach to user-technology relations.

Another important concept in feminist studies of technology is diversity. As Cowan (1987) suggested, users come in many different shapes and sizes. Medical technologies, for example, have a wide variety of users, including patients, health professionals, hospital administrators, nurses, and patients' families. "Who is the user?" is far from a trivial question. The very act of identifying specific individuals or groups as users may facilitate or constrain the actual roles of specific groups of users in shaping the development and use of technologies. Different groups involved in the design of technologies may have different views of who the user might or should be, and these different groups may mobilize different resources to inscribe their views in the design of technical objects (Oudshoorn et al., forthcoming). And these different type of users don't necessarily imply homogeneous categories. Gender, age, socio-economic, and ethnic differences among users may all be relevant. Because of this heterogeneity, not all users will have the same position in relation to a specific technology. For some users, the room for maneuvering will be great; for others, it will be very slight. Feminist sociologists thus emphasize the diversity of users and encourage scholars to pay attention to differences in power relations among the actors involved in the development of technology.

To capture the diversity of users and the power relations between users and other actors in technological development, feminist sociologists have differentiated "end users," "lay end users," and "implicated actors." End users are "those individuals and groups who are affected downstream by products of technological innovation" (Casper and Clarke 1998). The term "lay end users" was introduced to highlight some end users' relative exclusion from expert discourse (Saetnan et al. 2000: 16). Implicated actors are "those silent or not present but affected by the action" (Clarke 1998: 267). And there are two categories of implicated actors: "those not physically present but who are discursively constructed and targeted by others" and "those who are physically present but who are generally silenced/ignored/made invisible by those in power" (Clarke, forthcoming). All three terms reflect the long-standing feminist concern with the potential problematic consequences of technologies for women and include an explicit political agenda: the aim of feminist studies is to increase women's autonomy and their influence on technological development. A detailed understanding of how women as "end users" or "implicated actors" matter in technological development may provide information that will be useful in the empowerment of women or of spokespersons for them, such as social movements and consumer groups.

The concept of the implicated actor also reflects a critical departure from actor-network approaches in technology studies. Feminists have criticized the sociology of technology, particularly actor-network theory, for the almost exclusive attention it gives to experts and producers and for the preference it gives to design and innovation in understanding socio-technical change. This "executive approach" pays less attention to non-standard positions, including women's voices (Star 1991; Clarke and Montini 1993: 45; Clarke 1998: 267). Moreover, the "executive approach" implicitly assumes a specific type of power relations between users and designers in which designers are represented as powerful and users as disempowered relative to the experts. Feminist sociologists suggest that the distribution of power among the multiple actors involved in socio-technical networks should be approached as an empirical question (Lie and Sørensen 1996: 4, 5; Clarke 1998: 267; Oudshoorn et al., forthcoming). Thus, the notion of the implicated actor was introduced to avoid silencing invisible actors and

actants and to include power relations explicitly in the analysis of user-expert relations.

Another important word in the feminist vocabulary is "cyborg." Donna Haraway was the first to use this word to describe how by the late twentieth century humans had become so thoroughly and radically merged and fused with technologies that the boundaries between the human and the technological are no longer impermeable. The cyborg implies a very specific configuration of user-technology relations in which the user emerges as a hybrid of machine and organisms in fiction and as lived experience. Most important, Haraway introduced the cyborg figure as a politicized entity. Cyborg analyses aim to go beyond the deconstruction of technological discourses. On page 149 of her "Cyborg Manifesto" (1985), Haraway invites us to "question that which is taken as 'natural' and 'normal' in hierarchic social relations." Haraway writes of cyborgs not to celebrate the fusion of humans and technology, but to subvert and displace meanings in order to create alternative views, languages, and practices of technosciences and hybrid subjects. In the 1990s, the concept of the cyborg resulted in an extensive body of literature that described the constitution and transformation of physical bodies and identities through technological practices.

Semiotic Approaches to Users: Configuration and Script

An important new approach to user-technology relations was introduced by STS scholars who extended semiotics, the study of how meanings are built, from signs to things. The concept of "configuring the user" is central to this approach. Exploring the metaphor of machine as text, Steve Woolgar (1991: 60) introduced the notion of the user as reader to emphasize the interpretive flexibility of technological objects and the processes that delimit this flexibility. Although the interpretative flexibility of technologies and questions concerning the closure or stabilization of technology had already been addressed in the SCOT approach, Woolgar focused on the design processes that delimit the flexibility of machines rather than on the negotiations between relevant social groups. He suggested that how users "read" machines is constrained because the design and the production of machines entails a process of configuring the user. For Woolgar, "configuring" is the process of "defining the identity of putative users, and setting constraints upon their likely future actions" (ibid.: 59). He describes the testing of a new range of microcomputers as "a struggle to configure (that is to define, enable, and constrain) the user," a struggle that results in "a machine that encourages only specific forms of access and use" (ibid.: 69, 89). In this approach, the testing phase of a technology is portrayed as an important location in which to study the co-construction of technologies and users. In contrast to the approaches discussed thus far, this semiotic approach draws attention to users as represented by designers rather than to users as individuals or groups involved or implicated in technological innovation.

In recent debates, the notion of the configuration of users by designers has been extended to better capture the complexities of designer-user relations. Several authors criticized Woolgar for describing-configuration as a one-way process in which the power to shape technological development is attributed only to experts in design organizations. For example, Mackay et al. (2000: 752) suggested that "designers configure users, but designers in turn, are configured by both users and their own organizations," and that this is increasingly the case in situations where designer-user relations are formalized by contractual arrangements (ibid.: 744). The capacity of designers to configure users can be further constrained by powerful groups within organizations that direct design projects. In large organizations, designers usually have to follow specific organizational methods or procedures that constrain design practices (ibid.: 741, 742, 744; Oudshoorn et al. 2003). In many companies in the information and communication technologies sector, for example, designers are allowed to test prototypes of new products only among people who work in the organization. In this highly competitive sector, companies are reluctant to test new products among wider groups of users for fear that other firms will become aware of their plans at an early phase of product development (European Commission 1998: 22; Oudshoorn et al. 2003).

Another criticism and extension of the configuration approach was introduced by scholars who questioned who was doing the configuration work. In Woolgar's studies, configuration work was restricted to the activities of actors within the company who produced the computers. Several authors broadened this rather narrow view of configuration to include other actors and to draw attention to the configuration work carried out by journalists, public-sector agencies, policy makers, and social movements

acting as spokespersons for users (van Kammen 2000a; van Kammen, this volume; Epstein, this volume; Parthasarathy, this volume; Oudshoorn 1999; Rommes 2002). Other scholars attempted to broaden the scope of the analysis by including the agency of users. Whereas Woolgar explored the metaphor of machine and text to highlight "encoding," thus focusing attention on the work performed by the producers of texts and machines, a more symmetrical use of the metaphor requires that we also focus on the processes of "decoding," the work done by readers and users to interpret texts and machines (Mackay et al. 2000: 739, 750, 752). A similar criticism of the asymmetry of Woolgar's work was voiced by scholars who had adopted domestication approaches to technology.

A second central notion in the semiotic approaches to user-technology relations is the concept of script. Madeleine Akrich and Bruno Latour, in theorizing relationships between users and technology, use this term to describe the obduracy of objects. The concept of script tries to capture how technological objects enable or constrain human relations as well as relationships between people and things. Comparing technologies to film, Akrich (1992: 208) suggested that "like a film script, technical objects define a framework of action together with the actors and the space in which they are supposed to act." To explain how scripts of technological objects emerge, she drew attention to the design of technologies. Akrich suggested that in the design phase technologists anticipate the interests, skills, motives, and behavior of future users. Subsequently, these representations of users become materialized into the design of the new product. As a result, technologies contain a script (or scenario): they attribute and delegate specific competencies, actions, and responsibilities to users and technological artifacts. Technological objects may thus create new "geographies of responsibilities" or transform or reinforce existing ones (ibid.: 207, 208). Rooted in actor-network theory, Akrich and Latour's work challenges social constructivist approaches in which only people are given the status of actors. The script approach aims to describe how technical objects "participate in building heterogeneous networks that bring together actants of all types and sizes, whether humans or nonhumans" (ibid.: 206).

In the 1990s, feminist scholars extended the script approach to include the gender aspects of technological innovation. Adopting the view that technological innovation requires a renegotiation of gender relations and an articulation and performance of gender identities, Dutch and Norwegian feminists introduced the concept of genderscript to capture all the work involved in the inscription and de-inscription of representations of masculinities and femininities in technological artifacts (Berg and Lie 1993; Hubak 1996; van Oost 1995; van Oost, this volume; Oudshoorn 1996; Oudshoorn et al. 2003; Oudshoorn et al., forthcoming; Rommes et al. 1999; Spilkner and Sørensen 2000). This scholarship emphasizes the importance of studying the inscription of gender into artifacts to improve our understanding of how technologies invite or inhibit specific performances of gender identities and relations. Technologies are represented as objects of identity projects—objects that may stabilize or de-stabilize hegemonic representations of gender (Oudshoorn, this volume; Saetnan et al. 2000). Equally important, the genderscript approach drastically redefines the exclusion of specific groups of people from technological domains and activities. Whereas policy makers and researchers have defined the problem largely in terms of deficiencies of users, genderscript studies draw attention to the design of technologies (Oudshoorn 1996; Oudshoorn et al., forthcoming; Rommes et al. 1999; Rommes 2002). These studies make visible how specific practices of configuring the user may lead to the exclusion of specific users.

At first glance, the script approach seems to be very similar to Woolgar's approach of configuring the user, since both approaches are concerned with understanding how designers inscribe their views of users and use in technological objects. A closer look, however, reveals important differences. Although both approaches deal with technological objects and designers, the script approach makes users more visible as active participants in technological development. Akrich in particular is very much aware that a focus on how technological objects constrain the ways in which people relate to things and to one another can be easily misunderstood as a technological determinist view that represents designers as active and users as passive. To avoid this misreading, she emphasizes the reciprocal relationship between objects and subjects and explicitly addresses the question of the agency of users (Akrich 1992: 207). Although technological objects can define the relationships between human and nonhuman actors, Akrich suggests that "this geography is open to question and may be resisted" (ibid.). To avoid technological determinism, Akrich urges us to analyze the negotiations between

designers and users and concludes that "we cannot be satisfied methodologically with the designer's or user's point of view alone. Instead we have to go back and forth continually between the designer and the user, between the designer's projected users and the real users, between the world inscribed in the object and the world described by its displacement" (ibid.: 209).

To further capture the active role of users in shaping their relationships to technical objects, Akrich and Latour have introduced the concepts of subscription, de-inscription, and antiprogram. "Antiprogram" refers to the users' program of action that is in conflict with the designers' program (or vice versa). "Subscription" or "de-inscription" is used to describe the reactions of human (and nonhuman) actors to "what is prescribed and proscribed to them" and refers respectively to the extent to which they underwrite or reject and renegotiate the prescriptions (Akrich and Latour 1992: 261). In contrast to Woolgar's work on configuring the user, script analyses thus conceptualize both designers and users as active agents in the development of technology. However, compared to domestication theory, the script approach gives more weight to the world of designers and technological objects. The world of users, particularly the cultural and social processes that facilitate or constrain the emergence of users' antiprograms, remains largely unexplored by actor-network approaches. More recently, this imbalance has been repaired to an extent by the work of scholars who have extended actor-network theory to include the study of "subject networks." These studies aim to understand the "attachment" between people and things, particularly but not exclusively between disabled people and assistive technologies, and to explore how technologies work to articulate subjectivities (Callon and Rabehariso 1999; Moser 2000; Moser and Law 1998, 2001). This scholarship conceptualizes subjects in the same way as actor-network theorists previously approached objects. Subject positions such as disability and ability are constituted as effects of actor networks and hybrid collectives.

Cultural and Media Studies: Consumption and Domestication

In contrast to the approaches to user-technology relations we have discussed thus far, scholars in the fields of cultural and media studies acknowledged the importance of studying users from the very beginning.

Whereas historians and sociologists of technology have chosen technology as their major topic of analysis, those who do cultural and media studies have focused primarily on users and consumers. Their central thesis is that technologies must be culturally appropriated to become functional. This scholarship draws inspiration from Bourdieu's (1984) suggestion that consumption has become more important in the political economy of late modernity. Consequently, human relations and identities are increasingly defined in relation to consumption rather than production. In his study of differences in consumption patterns among social classes, Bourdieu defined consumption as a cultural and material activity and argued that the cultural appropriation of consumer goods depends on the "cultural capital" of people (ibid.). This view can be traced back to the tradition of the anthropological study of material culture, most notably the work of Mary Douglas and Baron Isherwood (1979). Among the first to criticize the view (then dominant among consumption theorists) that consumption is merely an economic activity, they suggested that consumption is always a cultural as well as an economic phenomenon (Lury 1996: 10). Describing the use of consumer goods in ritual processes, they defined consumer culture as a specific form of material culture, and they conceptualized the circulation of material things as a system of symbolic exchange. This scholarship articulates the importance of the sign value rather than the utility value of things. From this perspective, material things can act as sources and markers of social relations and can shape and create social identities (Lury 1996: 10, 12, 14; Douglas and Isherwood 1979; McKracken 1988; Appadurai 1986).

Feminist historians have also been important actors in signaling the relevance of studying consumption rather than production (McGaw 1982). Feminists have long been aware of the conventional association and structural relations of women with consumption as a consequence of their role in the household and as objects in the commodity-exchange system (de Grazia 1996: 7). Whereas early feminist studies focused on the (negative) consequences of mass consumption for women, more recent studies address the question of whether women have been empowered by access to consumer goods. They conceptualize consumption as a site for the performance of gender and other identities. The notion of consumption as a status and identity project was elaborated further by Jean Baudrillard (1988), who criticizes the view that the needs of consumers are dictated, manipulated, and fully controlled by the modern capitalist marketplace and by producers. Theodor Adorno, Herbert Marcuse, and Max Horkheimer of the Frankfurt School had argued that the expansion of the production of consumer goods throughout the twentieth century had resulted in an increase in ideological control and manipulation by the "culture industries" (Adorno 1991; Horkheimer and Adorno 1979; Marcuse 1964). Since the 1970s, this view of consumption as manipulation had resulted in a literature dominated by studies oriented toward production and marketing—studies that highlighted big companies and advertising agencies as the forces driving consumption. In these studies, consumption was characterized as a passive and adaptive process and consumers are represented as the anonymous buyers and victims of mass production. In contrast, Baudrillard emphasized the mutual dependencies between production and consumption and suggested that consumers are not passive victims but active agents in shaping consumption, social relations, and identities.

Cultural and media studies also emphasize the creative freedom of users to "make culture" in the practice of consumption as well as their dependence on the cultural industries, not because they control consumers but because they provide the means and the conditions of cultural creativity (Storey 1999: xi). This scholarship portrays consumers as "cultural experts" who appropriate consumer goods to perform identities, which may transgress established social divisions (du Gay et al. 1997: 104; Chambers 1985).

Semiotic approaches to analyzing user-technology relations also came to the fore in cultural and media studies. Stuart Hall, one of the leading scholars in this field, introduced the "encoding/decoding" model of media consumption (Hall 1973), which aims to capture both the structuring role of the media in "setting agendas and providing cultural categories and frameworks" and the notion of the "active viewer, who makes meaning from signs and symbols that the media provide" (Morley 1995: 300). Since the 1980s, the symbolic and communicative character of consumption has been studied extensively by scholars in the fields of cultural and media studies. Consumption fulfills a wide range of social and personal aims and serves to articulate who we are or who we would like to be; it may provide symbolic means of creating and establishing friendship and celebrating success; it may serve to produce certain lifestyles; it may provide the material for daydreams; it may be used to articulate social difference and social distinctions

(Bocock 1993; du Gay et al. 1997; Lie and Sørensen 1996; Mackay 1997; Miller 1995; Storey 1999). Cultural and media studies thus articulate a perspective on user-technology relations that emphasizes the role of technological objects in creating and shaping social identities, social life, and culture at large.

Roger Silverstone coined the term "domestication" to describe how the integration of technological objects into daily life involves "a taming of the wild and a cultivation of the tame." New technologies have to be transformed from "unfamiliar, exciting, and possible threatening things" into familiar objects embedded in the culture of society and the practices and routines of everyday life (Silverstone and Hirsch 1992; Lie and Sørensen 1996). Domestication processes include symbolic work, in which people create symbolic meanings of artifacts and adopt or transform the meanings inscribed in the technology; practical work, in which users develop a pattern of use to integrate artifacts into their daily routines; and cognitive work, which includes learning about artifacts (Lie and Sørensen 1996: 10; Sørensen et al. 2000). In this approach, domestication is defined as a dual process in which both technical objects and people may change. The use of technological objects may change the form and the practical and symbolic functions of artifacts, and it may enable or constrain performances of identities and negotiations of status and social positions (Silverstone et al. 1989; Lie and Sørensen 1996). The notion of domestication also reflects a preference for studying the use of technology in a specific location: the home. British scholars in this tradition have largely restricted their analyses to the household and the politics of family life (Silverstone 1989, 1992). In their work, processes of domestication are understood in terms of the "dynamics of the household's moral economy" (Silverstone, Hirsch, and Morley 1992). More recently, Norwegian scholars have extended the scope of research to other domains. Merete Lie and Knut Sørensen (1996: 13, 17) argue that the domestication of technical objects has been too easily associated with the "private sector" (meaning the home). Various chapters in the volume edited by Lie and Sørensen show how similar processes are taking-place in work, in leisure, and within subcultures.

Domestication approaches have enriched our understanding of user-technology relations by elaborating the processes involved in consumption. In *Consuming Technologies,* Roger Silverstone and his colleagues specify four phases of domestication: appropriation, objectification, incorporation, and conversion. Appropriation occurs when a technical product or service is sold and individuals or households become its owners (Silverstone et al. 1992: 21). In objectification, processes of display reveal the norms and principles of the "household's sense of itself and its place in the world" (ibid.: 22). Incorporation occurs when technological objects are used in and incorporated into the routines of daily life. "Conversion" is used to describe the processes in which the use of technological objects shape relationships between users and people outside the household (ibid.: 25). In this process, artifacts become tools for making status claims and for expressing a specific lifestyle to neighbors, colleagues, family, and friends (Silverstone and Haddon 1996: 46).

Although at first sight "domestication" and "decoding" or "de-inscription" may seem synonymous, there is an important difference. By specifying the processes involved in the diffusion and the use of technology, domestication approaches take the dynamics of the world of users as their point of departure. The concepts of decoding and de-inscription, on the other hand, give priority to the design context in order to understand the emergence of user-technology relations. Domestication approaches thus emphasize the complex cultural dynamics in which users appropriate technologies (ibid.: 52). This contrasts with semiotic approaches that tend to define the user as an isolated individual whose relationship to technology is restricted to technical interactions with artifacts. As Silverstone and Haddon suggest, a focus on how designers configure the user runs the risk of reifying the innovator's conceptions of users. In contrast, domestication approaches conceptualize the user as a part of a much broader set of relations than user-machine interactions, including social, cultural, and economic aspects. By employing cultural approaches to understand user-technology relations, this scholarship aims to go beyond a rhetoric of designers' being in control. Semiotic approaches tend to reinforce the view that technological innovation and diffusion are successful only if designers are able to control the future actions of users. Although semiotic approaches have introduced notions that are useful in understanding the worlds of designers and users, "script" and "configuring the user" conceptualize the successes and failures of technologies mainly in terms of the extent to which designers adequately anticipate users' skills and behavior. In this view, users tend to be degraded to objects of innovators' strategies. The semiotic approaches have therefore been criticized for staying too close to the old linear model of technological innovation and diffusion, which prioritizes the agency of designers and producers over the agency of users and other actors involved in technological innovation (Oudshoorn 1999). Even the concept of antiprogram, introduced by Akrich and Latour to describe how users may try to counter the original intentions of the design of the artifact, remains within the rhetoric of designer's control (Sørensen 1994: 5). The only option available to the user seems to be to adopt or to reject the designers' intended use and meaning of technological objects. These approaches are inadequate to understand the full dynamics of technological innovation where users invent completely new uses and meanings of technologies or where users are actively involved in the design of technologies.

Most important, cultural and media studies inspire us to transcend the artificial divide between design and use. This scholarship has drastically reconceptualized the traditional distinction between production and consumption by re-introducing Karl Marx's claim that the process of production is not complete until users have defined the uses, meanings, and significance of the technology: "Consumption is production." They describe design and domestication as "the two sides of the innovation coin" (Lie and Sørensen 1996: 10).

Reading 17

Touch Someone: The Telephone Industry Discovers Sociability

Claude S. Fischer

The familiar refrain, "Reach out, reach out and touch someone," has been part of American Telephone and Telegraph's (AT&T's) campaign urging use of the telephone for personal conversations. Yet, the telephone industry did not always promote such sociability; for decades it was more likely to discourage it. The industry's "discovery" of sociability illustrates how structural and cultural constraints interact with public demand to shape the diffusion of a technology. While historians have corrected simplistic notions of "autonomous technology" in showing how technologies are produced, we know much less about how consumers use technologies. We too often take those uses (especially of consumer products) for granted, as if they were straightforwardly derived from the nature of the technology or dictated by its creators.[1]

In the case of the telephone, the initial uses suggested by its promoters were determined by—in addition to technical and economic considerations— its cultural heritage: specifically, practical uses in common with the telegraph. Subscribers nevertheless persisted in using the telephone for "trivial gossip." In the 1920s, the telephone industry shifted from resisting to endorsing such sociability, responding, at least partly, to consumers' insistent and innovative uses of the technology for personal conversation. After summarizing telephone history to 1940, this article will describe the changes in the uses that telephone promoters advertised and the changes in their attitudes toward sociability; it will then explore explanations for these changes.[2]

---

[1] See C. S. Fischer, "Studying Technology and Social Life," pp. 284–301 in *High Technology, Space, and Society: Emerging Trends,* ed. M. Castells (Beverly Hills, Calif., 1985). For a recent example of a study looking at consumers and sales, see M. Rose, "Urban Environments and Technological Innovation: Energy Choices in Denver and Kansas City, 1900–1940," *Technology and Culture* 25 (July 1984): 503–39.

[2] The primary sources used here include telephone and advertising industry journals; internal telephone company reports, correspondence, collections of advertisements, and other documents, primarily from

A Brief History of the Telephone

Within about two years of A. G. Bell's patent award in 1876, there were roughly 10,000 Bell telephones in the United States and fierce patent disputes over them, battles from which the Bell Company (later to be AT&T) emerged a victorious monopoly. Its local

---

AT&T and Pacific Telephone (PT&T); privately published memoirs and corporate histories; government censuses, investigations, and research studies; and several interviews, conducted by John Chan, with retired telephone company employees who had worked in marketing. The archives used most are the AT&T Historical Archives, New York (abbreviated hereafter as AT&T ARCH), and the Pioneer Telephone Museum, San Francisco (SF PION MU), with some material from the Museum of Independent Telephony, Abilene (MU IND TEL); Bell Canada Historical, Montreal (BELL CAN HIST); Illinois Bell Information Center, Chicago (ILL BELL INFO); and the N. W. Ayer Collection of Advertisements and the Warshaw Collection of Business Americana, National Museum of American History, Smithsonian Institution, Washington, D.C. A bibliography on the social history of the telephone is unusually short, especially in comparison with those on later technologies such as the automobile and television. There are industrial and corporate histories, but the consumer side is largely untouched. For some basic sources, see J. W. Stehman, *The Financial History of the American Telephone and Telegraph Company* (Boston, 1925); A. N. Holcombe, *Public Ownership of Telephones on the Continent of Europe* (Cambridge, Mass., 1911); H. B. MacMeal, *The Story of Independent Telephony* (Chicago: Independent Pioneer Telephone Association, 1934); J. L. Walsh, *Connecticut Pioneers in Telephony* (New Haven, Conn.: Morris F. Tyler Chapter of the Telephone Pioneers of America, 1950); J. Brooks, *Telephone: The First Hundred Years* (New York, 1976); A. Hibbard, *Hello-Goodbye: My Story of Telephone Pioneering* (Chicago, 1941); Robert Collins, *A Voice from Afar: The History of Telecommunications in Canada* (Toronto, 1977); R. L. Mahon, "The Telephone in Chicago," ILL BELL INFO, MS, ca. 1955; J. C. Rippey, *Goodbye, Central; Hello, World: A Centennial History of Northwestern Bell* (Omaha, Nebr.: Northwestern Bell, 1975); G. W. Brock, *The Telecommunications Industry: The Dynamics of Market Structure* (Cambridge, Mass, 1981); I. de S. Pool, *Forecasting the Telephone* (Norwood, N.J., 1983); R. W. Garnet, *The Telephone Enterprise: The Evolution of the Bell System's Horizontal Structure, 1876–1909* (Baltimore, 1985); R. A. Atwood, "Telephony and Its Cultural Meanings in Southeastern Iowa, 1900–1917" (Ph.D. diss., University of Iowa, 1984); Lana Fay Rakow, "Gender, Communication, and the Technology: A Case Study of Women and the Telephone" (Ph.D. diss., University of Illinois at Urbana-Champaign, 1987); and I. de S. Pool, ed., *The Social Impact of the Telephone* (Cambridge, Mass., 1977). (Note that AT&T, Bell, and similar corporate names refer, of course, to these companies—or their direct ancestors—up to the U.S. industry reorganization of January 1, 1984.)

Table 1
Telephone Development, 1880–1940

	Number of Telephones	Telephones per 1,000 People	Percentage in Bell System	Percentage Independent, Connected to Bell	Percentage Residential, Connected to Bell
1880	54,000	1	100	0	. . .
1885	156,000	3	100	0	. . .
1890	228,000	4	100	0	. . .
1895	340,000	5	91	0	. . .
1900	1,356,000	18	62	1	. . .
1905	4,127,000	49	55	6	. . .
1910	7,635,000	82	52	26	. . .
1915	10,524,000	104	57	30	. . .
1920	13,273,000	123	66	29	68
1925	16,875,000	145	75	24	67
1930	20,103,000	163	80	20	65
1935	17,424,000	136	82	18	63
1940	21,928,000	165	84	16	65
1980	180,000,000	790	81	19	74

Sources—U.S. Bureau of the Census, *Historical Statistics of the United States,* Bicentennial Ed., pt. 2 (Washington, D.C., 1975), pp. 783–84; and U.S. Bureau of the Census, *Statistical Abstract of the United States 1982–83* (Washington, D.C., 1984), p. 557.

franchisees' subscriber lists grew rapidly and the number of telephones tripled between 1880 and 1884. Growth slowed during the next several years, but the number of instruments totaled 266,000 by 1893.[3] (See table 1.)

As long-distance communication, telephony quickly threatened telegraphy. Indeed, in settling its early patent battle with Western Union, Bell gave financial concessions to Western Union as compensation for loss of business. As local communication, telephony quickly overwhelmed nascent efforts to establish signaling exchange systems (except for stock tickers).

During Bell's monopoly, before 1894, telephone service consisted basically of an individual line for which a customer paid an annual flat fee allowing unlimited calls within the exchange area. Fees varied widely, particularly by size of exchange. Bell rates dropped in the mid-1890s, perhaps in anticipation of forthcoming competition. In 1895, Bell's average residential rate was $4.66 a month (13 percent of an average worker's monthly wages). Rates remained

high, especially in the larger cities (the 1894 Manhattan rate for a two-party line was $10.41 a month).[4]

On expiration of the original patents in 1893–94, thousands of new telephone vendors, ranging from commercial operations to small cooperative systems, sprang up. Although they typically served areas that Bell had ignored, occasional head-to-head competition drove costs down and spurred rapid diffusion: almost a nine-fold increase in telephones per capita between 1893 and 1902, as compared to less than a twofold increase in the prior nine years.[5]

Bell responded fiercely to the competition, engaging in price wars, political confrontations, and other aggressive tactics. It also tried to reach less affluent customers with cheaper party lines, coin-box telephones, and "measured service" (charging by the call). Still, Bell lost at least half the market by 1907. Then, a new management under Theodore N. Vail, the most influential figure in telephone history,

[3]Statistics from AT&T, *Events in Telecommunications History* (New York: AT&T, 1979), p. 6; U.S. Bureau of the Census (BOC), *Historical Statistics of the United States*, Bicentennial Ed., pt. 2 (Washington, D.C., 1975), pp. 783–84.

[4]Rates are reported in scattered places. For these figures, see BOC, *Telephones and Telegraphs 1902*, Special Reports, Department of Commerce and Labor (Washington, D.C., 1906), p. 53; and *1909 Annual Report of AT&T* (New York, 1910), p. 28. Wage data are from *Historical Statistics* (n. 3 above), tables D735–38.

[5]BOC, *Telephones, 1902* (n. 4 above); Federal Communications Commission (FCC), *Proposed Report: Telephone Investigation* (Washington, D.C., 1938), p. 147. AT&T has always officially challenged this interpretation; see, e.g., *1909 Annual Report of AT&T*, pp. 26–28.

changed strategies. Instead of reckless, preemptive expansion and price competition, AT&T bought out competitors where it could and ceded territories where it was losing. With tighter fiscal control, and facing capital uncertainties as well, AT&T's rate of expansion declined.[6] Meanwhile, the "independents" could not expand much beyond their small-town bases, partly because they were unable to build their own long-distance lines and were cut off from Bell-controlled New York City. Many were not competitive because they were poorly financed and provided poor service. Others accepted or even solicited buyouts from AT&T or its allies. By 1912, the Bell System had regained an additional 6 percent of the market.

During this competitive era, the industry offered residential customers a variety of economical party-line plans. Bell's average residential rate in 1909 was just under two dollars a month (about 4 percent of average wages).[7] How much territory the local exchange covered and what services were provided—for example, nighttime operators—varied greatly, but costs dropped and subscriber lists grew considerably. These basic rates changed little until World War II (although long-distance charges dropped).

In the face of impending federal antitrust moves, AT&T agreed in late 1913 to formalize its budding accommodation with the independents. Over several years, local telephone service was divided into regulated geographic monopolies. The modern U.S. telephone system—predominandy Bell local service and exclusively Bell long-distance service—was essentially fixed from the early 1920s to 1984.

The astronomical growth in the number of telephones during the pre-Vail era (a compound annual rate of 23 percent per capita from 1893 to 1907) became simply healthy growth (4 percent between 1907 and 1929). The system was consolidated and technically improved, and, by 1929, 42 percent of all households had telephones. That figure shrank during the Depression to 31 percent in 1933 but rebounded to 37 percent of all households in 1940.

---

[6]See, e.g., *Annual Report of AT&T,* 1907–10; and FCC, *Proposed Report* (n. 5 above), pp. 153–154. On making deals with competitors, see, e.g., Rippey (n. 2 above), pp. 143ff.

[7]*1909 Annual Report of AT&T*, p. 28. Charges for minimal, urban, four-party lines ranged from $3.00 a month in New York (about 6 percent of the average manufacturing employee's monthly wages) to $1.50 in Los Angeles (about 3 percent of wages) and much less in small places with mutual systems; see BOC, *Telephones and Telegraphs and Municipal Electric Fire-Alarm and Police-Patrol Signaling Systems, 1912* (Washington, D.C., 1915); and *Historical Statistics* (n. 3 above), table D740.

Sales Strategies

The telephone industry believed, as President Vail testified in 1909, that the "public had to be educated . . . to the necessity and advantage of the telephone."[8] And Bell saluted itself on its success in an advertisement entitled "Blazing the Way": Bell "had to invent the business uses of the telephone and convince people that they were uses. . . . [Bell] built up the telephone habit in cities like New York and Chicago. . . . It has from the start created the need of the telephone and then supplied it."[9]

"Educating the public" typically meant advertising, face-to-face solicitations, and public relations. In the early years, these efforts included informational campaigns, such as publicizing the existence of the telephone, showing people how to use it, and encouraging courteous conversation on the line.[10] Once the threat of nationalization became serious, "institutional" advertising and publicity encouraged voters to feel warmly toward the industry.[11]

As to getting paying customers, the first question vendors had to ask was, Of what use is this machine? The answer was not self-evident.

For roughly the first twenty-five years, sales campaigns largely employed flyers, simple informational notices in newspapers, "news" stories supplied to friendly editors (many of whom received free service

---

[8]Testimony on December 9, 1909, in State of New York, *Report of the Committee of the Seriate and Assembly Appointed to Investigate Telephone and Telegraph Companies* (Albany, 1910), p. 398.

[9]Ayer Collection of AT&T Advertisements, Collection of Business Americana, National Museum of American History, Smithsonian Institution.

[10]See, e.g., *Pacific Telephone Magazine* (PT&T employee magazine, hereafter PAC TEL MAG), 1907–40, passim; 1914 advertisements in SF PION MU folder labeled "Advertising"; MU IND TEL "Scrapbook" of Southern Indiana Telephone Company clippings; advertisements in directories of the day; "Educating the Public to the Proper Use of the Telephone," *Telephony* 64 (June 21, 1913): 32–33; "Swearing over the Telephone," *Telephony* 9 (1905): 418; and "Advertising and Publicity—1906–1910," box 1317, AT&T ARCH.

[11]On AT&T's institutional advertising, see R. Marchand, "Creating the Corporate Soul: The Origins of Corporate Image Advertising in America" (paper presented to the Organization of American Historians, 1980), and N. L. Griese, "AT&T: 1908 Origins of the Nation's Oldest Continuous Institutional Advertising Campaign," *Journal of Advertising* 6 (Summer 1977): 18–24. FCC, *Proposed Report* (n. 5 above), has a chapter on "Public Relations"; see also N. R. Danielian, *AT&T: The Story of Industrial Conquest* (New York, 1939), chap. 13. For a defense of AT&T public relations, see A. W. Page, *The Bell Telephone System* (New York, 1941). Among the publicity efforts along these lines were "free" stories, subsidies of the press, and courting of reporters and politicians (documented in AT&T ARCH). In one comical case, AT&T frantically and apparently unsuccessfully tried in 1920 to pressure Hal Roach to cut out from a Harold Lloyd film he was producing a burlesque scene of central exchange hysteria (see folder "Correspondence—E. S. Wilson, V.P., AT&T," SF PION MU).

or were partners in telephony), public demonstrations, and personal solicitations of businessmen. As to uses, salesmen typically stressed those that extended applications of telegraph signaling. For example, an 1878 circular in New Haven—where the first exchange was set up—stated that "your wife may order your dinner, a hack, your family physician, etc., all by Telephone without leaving the house or trusting servants or messengers to do it." (It got almost no response.)[12] In these uses, the telephone directly competed with—and decisively defeated—attempts to create telegraph exchanges that enabled subscribers to signal for services and also efforts to employ printing telegraphs as a sort of "electronic mail" system.[13]

In this era and for some years later, the telephone marketers sought new uses to add to these telegraphic applications. They offered special services over the telephone, such as weather reports, concerts, sports results, and train arrivals. For decades, vendors cast about for novel applications: broadcasting news, sports, and music, night watchman call-in services, and the like. Industry magazines eagerly printed stories about the telephone being used to sell products, alert firefighters about forest blazes, lullaby a baby to sleep, and get out voters on election day. And yet, industry men often attributed weak demand to not having taught the customer "what to do with his telephone."[14]

In the first two decades of the 20th century, telephone advertising became more professionally "modern."[15] AT&T employed a Boston agency to dispense "free publicity" and later brought its chief, J. D. Ellsworth, into the company. It began national advertising campaigns and supplied local Bell companies with copy for their regional presses. Some of the advertising was implicitly competitive (e.g., stressing that Bell had long-distance service), and much of it was institutional, directed toward shaping a favorable public opinion about the Bell System. Advertisements for selling service employed drawings, slogans, and texts designed to make the uses of the telephone—not just the technology—attractive. (The amount and kind of advertising fluctuated, especially in the Bell System, in response to competition, available supplies, and political concerns.)[16]

From roughly 1900 to World War I, Bell's publicity agency advertised uses of the telephone by planting newspaper "stories" on telephones in farm life, in the church, in hotels, and the like.[17] The national advertisements, beginning around 1910, addressed mostly businessmen. They stressed that the telephone was impressive to customers and saved time, both at work and at home, and often noted the telephone's convenience for planning and for keeping in touch with the office during vacations.

A second major theme was household management. A 1910 series, for example, presented detailed suggestions: Subscribers could telephone dressmakers, florists, theaters, inns, rental agents, coal dealers,

[12]Walsh (n. 2 above), p. 47.
[13]S. Schmidt, "The Telephone Comes to Pittsburgh" (master's thesis, University of Pittsburgh, 1948); Pool, *Forecasting* (n. 2 above), p. 30; D. Goodman, "Early Electrical Communications and the City: Applications of the Telegraph in Nineteenth-Century Urban America" (unpub. paper, Department of Social Sciences, Carnegie-Mellon University, n.d., courtesy of Joel Tarr); and "Telephone History of Dundee, Ontario," City File, BELL CAN HIST.
[14]On special services and broadcasting, see Walsh (n. 2 above), p. 206; S. H. Aronson, "Bell's Electrical Toy: What's the Use? The Sociology of Early Telephone Usage," pp. 15–39, and I. de S. Pool et al., "Foresight and Hindsight: The Case of the Telephone," pp. 127–58, both in Pool, ed., *Social Impact* (n. 2 above); "Broadening the Possible Market," *Printers' Ink* 74 (March 9, 1911): 20; G. O. Steel, "Advertising the Telephone," *Printers' Ink* 51 (April 12, 1905): 14–17; and F. P. Valentine, "Some Phases of the Commercial Job," *Bell Telephone Quarterly* 5 (January 1926): 34–43. For illustrations of uses, see, e.g., PAC TEL MAG (October 1907), p. 6, (January 1910), p. 9, (December 1912), p. 23, and (October 1920), p. 44; and the independent magazine, *Telephony*. E.g., the index to vol. 71 (1916) of *Telephony* lists the following under "Telephone, novel uses of": "degree conferred by telephone, dispatching tugs in harbor service, gauging water by telephone, telephoning in an aeroplane." On complaints about not having taught the public, see the quotation from H. B. Young, ca. 1929, pp. 91, 100 in "Publicity Conferences—Bell System—1921–34," box 1310, AT&T ARCH, but similar comments appear in earlier years, as well as positive claims, such as Vail's in 1909.

[15]The following discussion draws largely from examination of advertisement collections at the archives listed in n. 2. Space does not permit more than a few examples of hundreds of advertisements in the sources. See esp. at AT&T ARCH, files labeled "Advertising and Publicity"; at SF PION MU, folders labeled "Advertising" and "Publicity Bureau"; at BELL CAN HIST, "Scrapbooks"; at ILL BELL INFO, "AT&T Advertising" and microfilm 384B, "Adver."; and at the Ayer Collection (n. 9 above), the AT&T series.
[16]For explicit discussions, see Mahon (n. 2 above), e.g., pp. 79, 89; Publicity Vice-President A. W. Page's comments in "Bell System General Commercial Conference, 1930," microfilm 368B, ILL BELL INFO; and comments by Commercial Engineer K. S. McHugh in "Bell System General Commercial Conference on Sales Matters, 1931," microfilm 368B, ILL BELL INFO. On the origins of in-house advertising, see N. L. Griese, "1908 Origins" (n. 11 above).
[17]See correspondence in "Advertising and Publicity—Bell System—1906–1910, Folder 1," box 1317, AT&T ARCH. Some reports claimed that thousands of stories were placed in hundreds of publications. Apparently no national advertising campaigns were conducted prior to these years; Bell marketing strategy seemed largely confined to price and service competition. See N. C. Kingsbury, "Results from the American Telephone's National Campaign," *Printers' Ink* (June 29, 1916): 182–84.

schools, and the like. Other uses were suggested, too, such as conveying messages of moderate urgency (a businessman calling home to say that he will be late, calling a plumber), and conveying invitations (to an impromptu party, for a fourth at bridge).

Sociability themes ("visiting" kin by telephone, calling home from a business trip, and keeping "In Touch with Friends and Relatives") appeared, but they were relatively rare and almost always suggested sending a message such as an invitation or news of safe arrival rather than having a conversation. A few advertisements also pointed out the modernity of the telephone ("It's up to the times!"). But the major uses suggested in early telephone advertising were for business and household management; sociability was rarely advised.[18]

With the decline of competition and the increase in regulation during the 1910s, Bell stressed public relations even more and pressed local companies to follow suit. AT&T increasingly left advertising basic services and uses to its subsidiaries, although much of the copy still originated in New York, and the volume of such advertising declined. Material from Pacific Telephone and Telegraph (PT&T), apparently a major advertiser among the Bell companies, indicates the substance of "use" advertising during that era.[19]

PT&T advertisements for 1914 and 1915 include, aside from informational notices and general paeans to the telephone, a few suggestions for businessmen (e.g., "You fishermen who feel these warm days of Spring luring you to your favorite stream. . . . You can adjust affairs before leaving, ascertain the condition of streams, secure accommodations, and always be in touch with business and home"). Several advertisements mention the home or women, such as those suggesting that extension telephones add to safety and those encouraging shopping by telephone. Just one advertisement in this set explicitly suggests an amiable conversation: A grandmotherly woman is speaking on the telephone, a country vista visible through the window behind her, and says: "My! How sweet and clear my daughter's voice sounds! She seems to be right here with me!" The text reads:

"Let us suggest a long distance visit home today." But this sort of advertisement was unusual.

During and immediately after World War I, there was no occasion to promote telephone use, since the industry struggled to meet demand pent up by wartime diversions. Much publicity tried to ease customer irritation at delays.

Only in the mid-1920s did AT&T and the Bell companies refocus their attention, for the first time in years, to sales efforts.[20] The system was a major advertiser, and Bell leaders actively discussed advertising during the 1920s. Copy focused on high-profit services, such as long distance and extension sets; modern "psychology," so to speak, influenced advertising themes; and Bell leaders became more sensitive to the competition from other consumer goods. Sociability suggestions increased, largely in the context of long-distance marketing.

In the United States, long-distance advertisements still overwhelmingly targeted business uses, but "visiting" with kin now appeared as a frequent suggestion. Bell Canada, for some reason, stressed family ties much more. Typical of the next two decades of Bell Canada's long-distance advertisements are these, both from 1921: "Why night calls are popular. How good it would sound to hear mother's voice tonight, he thought—for there were times when he was lonely—mighty lonely in the big city"; and "it's a weekly affair now, those fond intimate talks. Distance rolls away and for a few minutes every Thursday night the familiar voices tell the little family gossip that both are so eager to hear." Sales pointers to employees during this era often suggested providing customers with lists of their out-of-town contacts' telephone numbers.

In the 1920s, the advertising industry developed "atmosphere" techniques, focusing less on the product and more on its consequences for the consumer.[21] A similar shift may have begun in Bell's advertising, as well: "The Southwestern Bell Telephone Company has decided [in 1923] that it is selling something more vital than distance, speed or accuracy. . . . [T]he telephone . . . almost brings [people] face to

[18]In addition to the advertising collections, see A. P. Reynolds, "Selling a Telephone" (to a businessman), *Telephony* 12 (1906): 280–81; id., "The Telephone in Retail Business," *Printers' Ink* 61 (November 27, 1907): 3–8; and "Bell Encourages Shopping by Telephone," ibid., vol. 70 (January 19, 1910).

[19]Letter from AT&T Vice-President Reagan to PT&T President H. D. Pillsbury, March 4, 1929, in "Advertising," SF PION MU; W. J. Phillips, "The How, What, When and Why of Telephone Advertising," talk given July 7, 1926, in ibid.; and "Advertising Conference—Bell System—1916," box 1310, AT&T ARCH, p. 44.

[20]See n. 16 above.

[21]D. Pope, *The Making of Modern Advertising* (New York, 1983); S. Fox, *The Mirror Makers: A History of American Advertising and Its Creators* (New York, 1984); M. Schudson, *Advertising: The Uneasy Persuasion* (New York, 1985), pp. 60ff; R. Marchand, *Advertising the American Dream: Making Way for Modernity, 1920–1940* (Berkeley, Calif., 1985); and R. Pollay, "The Subsiding Sizzle: A Descriptive History of Print Advertising, 1900–1980," *Journal of Marketing* 49 (Summer 1985): 24–37.

face. It is the next best thing to personal contact. So the fundamental purpose of the current advertising is to sell the company's subscribers their voices at their true worth—to help them realize that 'Your Voice is You.'. . . to make subscribers think of the telephone whenever they think of distant friends or relatives. . . ."[22] This attitude was apparently only a harbinger, because during most of the 1920s the sociability theme was largely restricted to long distance and did not appear in many basic service advertisements.

Bell System salesmen spent the 1920s largely selling ancillary services, such as extension telephones, upgrading from party lines, and long distance, to current subscribers, rather than finding new customers. Basic residential rates averaged two to three dollars a month (about 2 percent of average manufacturing wages), not much different from a decade earlier, and Bell leaders did not consider seeking new subscribers to be sufficiently profitable to pursue seriously.[23] The limited new subscriber advertising continued the largely practical themes of earlier years. PT&T contended that residential telephones, especially extensions, were useful for emergencies, for social convenience (don't miss a call about an invitation, call your wife to set an extra place for dinner), and for avoiding the embarrassment of borrowing a telephone, as well as for its familiar business uses. A 1928 Bell Canada sales manual stressed household practicality first and social invitations second as tactics for selling basic service.[24]

Then, in the late 1920s, Bell System leaders—prodded perhaps by the embarrassment that, for the first time, more American families owned automobiles, gas service, and electrical appliances than subscribed to telephones—pressed a more aggressive strategy. They built up a full-fledged sales force. And they sought to market the telephone as a "comfort and convenience"—that is, as more than a practical device—drawing somewhat on the psychological, sensualist themes in automobile advertising. They focused not only on upgrading the service of current subscribers but also on reaching those car owners and electricity

users who lacked telephones. And the *social* character of the telephone was to be a key ingredient in the new sales strategies.[25]

Before "comfort and convenience" could go far, however, the Depression drew the industry's attention to basic service once again. Subscribers were disconnecting. Bell companies mounted campaigns to save residential connections by mobilizing *all* employees to sell or save telephone hookups on their own time (a program that had started before the Crash), expanding sales forces, advertising to current subscribers, and mounting door-to-door "save" and "nonuser" campaigns in some communities.[26] The "pitches" PT&T suggested to its employees included convenience (e.g., saving a trip to market), avoiding the humiliation of borrowing a neighbor's telephone, and simply being "modern." Salesmen actually seemed to rely more on pointing out the emergency uses of the telephone—an appeal especially telling to parents of young children—and suggesting that job offers might come via the telephone. Having a telephone so as to be available to friends and relatives was a lesser sales point. By now, a half-century since A. G. Bell's invention, salespeople did not have to sell telephone service itself but had to convince potential customers that they needed a telephone in their own homes.[27]

During the Depression, long-distance advertising continued, employing both business themes and the themes of family and friendship. But basic service advertising, addressed to both nonusers and would-be disconnectors, became much more common than it had been for twenty years.

The first line of argument in print ads for basic service was practicality—emergency uses, in particular—but suggestions for sociable conversations were more prominent than they had been

[22]W. B. Edwards, "Tearing Down Old Copy Gods," *Printers' Ink* 123 (April 26, 1923): 65–66.

[23]On rates, see W. F. Gray, "Typical Schedules for Rates of Exchange Service," and related discussion, in "Bell System General Commercial Engineers' Conference, 1924," microfilm 364B, ILL BELL INFO.

[24]Bell Telephone Company of Canada, "Selling Service on the Job," ca. 1928, cat. 12223, BELL CAN HIST.

[25]Comments, esp. by AT&T vice-presidents Page and Gherardi, during "General Commercial Conference, 1928," and "Bell System General Commercial Conference, 1930," both microfilm 368B, ILL BELL INFO, expressed a view that telephones should be part of consumers' "life-styles," not simply their practical instruments. One hears many echoes of "comfort and convenience" at lower Bell levels during this period.

[26]See A. Fancher, "Every Employee Is a Salesman for American Telephone and Telegraph," *Sales Management* 28 (February 26, 1931): 45–51, 472; "Bell Conferences," 1928 and 1930 (n. 25 above), esp. L. J. Billingsley, "Presention of Disconnections," in 1930 conference; *Pacemaker*, a sales magazine for PT&T, ca. 1928–31, SF PION MU; and *Telephony*, passim, 1931—36.

[27]PT&T *Pacemaker;* interviews by John Chan with retired industry executives in northern California; see also J. E. Harrel, "Residential Exchange Sales in New England Southern Area," in "Bell Conference, 1931" (n. 16 above), pp. 67ff.

before. A 1932 advertisement shows four people sitting around a woman who is speaking on the telephone. "Do Come Over!" the text reads, "Friends who are linked by telephone have good times." A 1934 Bell Canada advertisement features a couple who have just re-subscribed and who testify, "We got out of touch with all of our friends and missed the good times we have now." A 1935 advertisement asks, "Have you ever watched a person telephoning to a friend? Have you noticed how readily the lips part into smiles . . . ?" And 1939 copy states, "Some one thinks of some one, reaches for the telephone, and all is well." A 1937 AT&T advertisement reminds us that "the telephone is vital in emergencies, but that is not the whole of its service. . . . Friendship's path often follows the trail of the telephone wire." These family-and-friend motifs, more frequent and frank in the 1930s, forecast the jingles of today, such as ". . . a friendly voice, like chicken soup/is good for your health/Reach out, reach out and touch someone."[28]

This brief chronology draws largely from prepared copy in industry archives, not from actual printed advertisements. A systematic survey, however, of two newspapers in northern California confirms the impression of increasing sociability themes. Aside from one 1911 advertisement referring to farm wives' isolation, the first sociability message in the *Antioch Ledger* appeared in 1929, addressed to parents: "No girl wants to be a wallflower." It was followed in the 1930s with notices for basic service such as "Give your friends straight access to your home," and "Call the folks now!" In 1911, advertisements in the *Marin* (County) *Journal* stressed the convenience of the telephone for automotive tourists. Sociability became prominent in both basic and long-distance advertisements in the late 1920s and the 1930s with suggestions that people "broaden the circle of friendly contact" (1927), "Voice visit with friends in nearby cities" (1930), and call grandmother (1935), and with

the line, "I got my telephone for convenience. I never thought it would be such fun!" (1940).[29]

The emergence of sociability also appears in guides to telephone salesmen. A 1904 instruction booklet for sales representatives presents many selling points, but only one paragraph addresses residential service. That paragraph describes ways that the telephone saves time and labor, makes the household run smoothly, and rescues users in emergencies, but the only barely social use it notes is that the telephone "invites one's friends, asks them to stay away, asks them to hurry and enables them to invite in return." Conversation—telephone "visiting"—per se is not mentioned.

A 1931 memorandum to sales representatives, entitled "Your Telephone," is, on the other hand, full of tips on selling residential service and encouraging its use. Its first and longest subsection begins: *"Fosters friendships.* Your telephone will keep your personal friendships alive and active. Real friendships are too rare and valuable to be broken when you or your friends move out of town. Correspondence will help for a time, but friendships do not flourish for long on letters alone. When you can't visit in person, telephone periodically. Telephone calls will keep up the whole intimacy remarkably well. There is no need for newly-made friends to drop out of your life when they return to distant homes." A 1935 manual puts practicality and emergency uses first as sales arguments but explicitly discusses the telephone's "social importance," such as saving users from being "left high and dry by friends who can't reach [them] conveniently."[30]

This account, so far, covers the advertising of the Bell System. There is less known and perhaps less to know about the independent companies' advertising. Independents' appeals seem much like those of the Bell System, stressing business, emergencies, and practicality, except perhaps for showing an earlier sensitivity to sociability among their rural clientele.[31]

---

[28]There is some variation among the advertising collections I examined. Illinois Bell's basic service advertisements used during the Depression are, for the most part, similar to basic service ads used a generation earlier. The Pacific Bell and Bell Canada advertisements feature sociable conversations much more. On the other hand, the Bell Canada ads are distinctive in that sociability is almost exclusively a family matter. Friendship, featured in U.S. ads all along, emerges clearly in the Canadian ads only in the 1930s. The 1932 ad cited in the text appears in the August 17 issue of the *Antioch* (Calif.) *Ledger*. The "chicken soup" jingle, sung by Roger Miller, was a Bell System ad in 1981. On the "Touch Someone" campaigns, see M. J. Arlen, *Thirty Seconds* (New York, 1980). See also "New Pitch to Spur Phone Use," *New York Times,* October 23, 1985, p. 44.

[29]These particular newspapers were examined as part of a larger study on the social history of the telephone that will include case studies of three northern California communities from 1890 to 1940.
[30]Central Union Telephone Company Contracts Department, *Instructions and Information for Solicitors,* 1904, ILL BELL INFO. Note that Central Union had been, at least through 1903, one of Bell's most aggressive solicitors of business. Illinois Bell Commercial Department, *Sales Manual* 1931, microfilm, ILL BELL INFO. Ohio Bell Telephone Company, "How You Can Sell Telephones," 1935, file "Salesmanship," BELL CAN HIST.
[31]Until 1894, independent companies did not exist. For years afterward, they largely tried to meet unfilled demand in the small cities and towns Bell had underserved. In other places, they advertised competitively

Table 2

Counts of Dominant Advertising Themes by Period

Sources and Types of Advertisements	Prewar		1919–29		1930–40	
Antioch (Calif.) *Ledger:*						
Social, sociability....................................	1	(1)	1	(1)	6	(4)
Business, businessmen.............................	6	(5)	1	(1)	2	(1)
Household, convenience, etc......................	5	(5)	3	(3)	4	(3)
Public relations, other..............................	0	(0)	4	(3)	1	(1)
Total......................................	12	(11)	9	(8)	13	(9)
Approximate ratio of social to others......	1:11	(1:10)	1:8	(1:7)	1:1	(1:1)
Marin (Calif.) *Journal:*.....................................						
Social, sociability....................................	1	(1)	5	(2)	43	(20)
Business, businessmen.............................	2	(2)	8	(2)	10	(3)
Household, convenience, etc......................	12	(12)	3	(3)	20	(20)
Public relations, other..............................	0	(0)	19	(13)	25	(16)
Total......................................	15	(15)	35	(20)	98	(59)
Approximate ratio of social to others......	1:14	(1:14)	1:6	(1:9)	1:1	(1:2)
Bell Canada:..						
Social, sociability....................................	5	(2)	25	(1)	59*	(9)
Business, businessmen.............................	20*	(20)	15	(2)	24*	(4)
Household, convenience, etc......................	28	(28)	3	(3)	23*	(6)
Public relations, other..............................	30*	(30)	25	(40)	2	(2)
Total......................................	83*	(80)	68	(46)	108*	(21)
Approximate ratio of social to others......	1:16	(1:39)	1:2	(1:45)	1:1	(1:1)
Pacific Telephone, 1914–15:						
Social, sociability....................................	2	(1)	. . .		. . .	
Business, businessmen.............................	7	(6)	. . .		. . .	
Household, convenience, etc......................	18	(16)	. . .		. . .	
Public relations, other..............................	16	(9)	. . .		. . .	
Total......................................	43	(32)	. . .		. . .	
Approximate ratio of social to others......	1:21	(1:31)	. . .		. . .	
Assorted Bell ads, 1906–10:						
Social, sociability....................................	4	(4)	. . .		. . .	
Business, businessmen.............................	13	(12)	. . .		. . .	
Household, convenience, etc......................	11	(11)	. . .		. . .	
Public relations, other..............................	9	(9)	. . .		. . .	
Total......................................	37	(36)	. . .		. . .	
Approximate ratio of social to others......	1:8	(1:8)	. . .		. . .	

Sources—Advertisements in the *Antioch Ledger* were sampled from 1906 to 1940 by Barbara Loomis; those in the *Marin Journal* were sampled from 1900 to 1940 by John Chan. The Bell Canada collection appears in scrapbooks at Bell Canada Historical; the Pacific collection is in the San Francisco Pioneer Telephone Museum. The AT&T advertisements are from AT&T ARCH, box 1317. Other, spotty collections were used for the study but not counted here because they were not as systematic. All coding was done by the author.

Note—Counts in parentheses exclude explicitly long-distance advertisements. Usually each ad had one dominant theme. When more than one seemed equal in weight, the ad was counted in both categories. "Social, sociability" refers to the use of the telephone for personal contact, including season's greetings, invitations, and conversation between friends and family. (Note that the inclusion of brief messages in this category makes the analysis a conservative test of the argument that there was a shift toward sociability themes.) "Business, businessmen" refers to the explicit use of the telephone for business purposes or general appeals to businessmen—e.g., that the telephone will make one a more forceful entrepreneur. "Household, convenience, etc." includes the use of the telephone for household management, personal convenience (e.g., don't get wet, order play tickets), and for emergencies, such as illness or burglary. "Public relations, other" includes general institutional advertising, informational notices (such as how to use the telephone), and other miscellaneous. Perhaps the most conservative index is the ratio of non-long-distance social ads to non-long-distance household ads. (Business ads move to speciality magazines over the years; public information ads fluctuate with political events; and long-distance ads may be "inherently" social.) In the *Antioch Ledger*, this ratio changes from 1:5 to 4:3; in the *Marin Journal*, from 1:12 to 1:1; and in Bell Canada's ads, from 1:14 to 1.5:1. Even these ratios understate the shift, for several reasons. One, I was much more alert to social than to other ads and was more thorough with early social ads than any other category. Two, the household category is increased in the later years by numerous ads for extension telephones. Three, the nature of the social ads counted here changes. The earlier ones overwhelmingly suggest using the telephone for greetings and invitations, not conversation. With rare exception, only the later ones discuss friendliness and "warm human relationships" and suggest chats.

*Estimated.

In sum, the variety of sales materials portray a similar shift. From the beginning to roughly the mid-1920s, the industry sold service as a practical business and household tool, with only occasional mention of social uses and those largely consisting of brief messages. Later sales arguments, for both long-distance and basic service, featured social uses prominently, including the suggestion that the telephone be used for converations ("voice visiting") among friends and family. While it would be helpful to confirm this impressionistic account with firm statistics, for various reasons it is difficult to draw an accurate sample of advertising copy and salesmen's pitches for over sixty years. (For one, we have no easily defined "universe" of advertisements. Are the appropriate units specific printed ads, or ad campaigns? How are duplicates to be handled? Or ads in neighboring towns? Do they include planted stories, inserts in telephone bills, billboards, and the like? Should locally generated ads be included? And what of nationally prepared ads not used by the locals? For another, we have no clear "population" of ads. The available collections are fragmentary, often preselected for various reasons.) An effort in that direction appears, however, in table 2, in which the numbers of "social" advertisements show a clear increase, both absolutely and relatively.

Industry Attitudes toward Sociability

This change in advertising themes apparently reflected a change in the actual beliefs industry men held about the telephone. Alexander Graham Bell himself forecast social chitchats using his invention. He predicted that eventually Mrs. Smith would spend an hour on the telephone with Mrs. Brown "very enjoyably . . . cutting up Mrs. Robinson."[32] But for decades few of his successors saw it that way.

Instead, the early telephone vendors often battled their residential customers over social conversations, labeling such calls "frivolous" and "unnecessary." For example, an 1881 announcement complained, "The fact that subscribers have been free to use the wires as they pleased without incurring additional expense [i.e., flat rates] has led to the transmission of large numbers of communications of the most trivial character."[33] In 1909, a local telephone manager in Seattle listened in on a sample of conversations coming through a residential exchange and determined that 20 percent of the calls were orders to stores and other businesses, 20 percent were from subscribers' homes to their own businesses, 15 percent were social invitations, and 30 percent were "purely idle gossip"—a rate that he claimed was matched in other cities. The manager's concern was to reduce this last, "unnecessary use." One tactic for doing so, in addition to "education" campaigns on proper use of the telephone, was to place time limits on calls (in his survey the average call had lasted over seven minutes). Time limits were often an explicit effort to stop people who insisted on chatting when there was "business" to be conducted.[34]

An exceptional few in the industry, believing in a more "populist" telephony, did, however, try to encourage such uses. E. J. Hall, Yale-educated and originally manager of his family's firebrick business, initiated the first "measured service" in Buffalo in 1880 and later became an AT&T vice-president. A pleader for lower rates, Hall also defended "trivial" calls, arguing that they added to the total use-value of the system. But the evident isolation of men like Hall underlines the dominant antisociability view of the pre–World War I era.[35]

---

against Bell. Nevertheless, advertising men often exhorted the independents to use "salesmanship in print" to encourage basic service and extensive use. See, e.g., J. A. Schoell, "Advertising and Other Thoughts of the Small Town Man," *Telephony* 70 (June 10, 1916): 40–41; R. D. Mock, "Fundamental Principles of the Telephone Business: Part V, Telephone Advertising," series in ibid., vol. 71 (July 22–November 21, 1916); D. Hughes, "Right Now Is the Time to Sell Service," ibid., 104 (June 10, 1933): 14–15; and L. M. Berry, "Helpful Hints for Selling Service," ibid., 108 (February 2, 1935): 7–10. See also Kellogg Company, "A New Business Campaign for _____" (Chicago: Kellogg, 1929), MU IND TEL.

[32]Quoted in Aronson, "Electrical Toy" (n. 14 above).

[33]Proposed announcement by National Capitol Telephone Company, in letter to Bell headquarters, January 20, 1881, box 1213, AT&T ARCH. In a similar vein, the president of Bell Canada confessed, ca. 1890, to being unable to stop "trivial conversations"; see Collins, *A Voice* (n. 2 above), p. 124. The French authorities were also exasperated by nonserious uses; see C. Bertho, *Télégraphes et téléphones* (Paris, 1980), pp. 244–45.

[34]C. H. Judson, "Unprofitable Traffic—What Shall Be Done with It?" *Telephony* 18 (December 11, 1909): 644–47, and PAC TEL MAG 3 (January, 1910): 7. He also writes, "the telephone is going beyond its original design, and it is a positive fact that a large percentage of telephones in use today on a flat rental basis are used more in entertainment, diversion, social intercourse and accommodation to others, than in actual cases of business or household necessity" (p. 645). MacMeal, *Independent* (n. 2 above), p. 240, reports on a successful campaign in 1922 to discourage gossipers through letters and advertisements. Typically, calls were—at least officially—limited to five minutes in many places, although it is unclear how well limits were enforced.

[35]Hall's philosophy is evident in the correspondence over measured service before 1900, box 1127, AT&T ARCH. Decades later, he pushed it in a letter to E. M. Burgess, Colorado Telephone Company, March 30, 1905, box 1309, AT&T ARCH, even arguing that operators should stop turning away calls made by children and should instead encourage such "trivial uses." The biographical information comes from an obituary in AT&T ARCH. Another, more extreme populist was John L. Sabin, of PT&T and the Chicago Telephone Co.; see Mahon (n. 2 above), pp. 29ff.

Official AT&T opinions came closer to Hall's in the later 1920s when executives announced that, whereas the industry had previously thought of telephone service as a practical necessity, they now realized that it was more: it was a "convenience, comfort, luxury"; its value included its "trivial" social uses. In 1928, Publicity Vice-President A. W. Page, who had entered AT&T from the publishing industry the year before, was most explicit when he criticized earlier views: "There had also been the point of view [in the Bell System and among the public] about not using the telephone for frivolous conversation. This is about as commercial as if the automobile people should advertise. 'Please do not take out this car unless you are going on a serious errand. . . .' We are faced, I think, with a state of public consciousness that the telephone is a necessity and not to be trifled with, certainly in the home." Bell sales officials were told to sell telephone service as a "comfort and convenience," including as a conversational tool.[36]

Although this change in opinion is most visible for the Bell System, similar trends can be seen in the pages of the journal of the independent companies, *Telephony*, especially in regard to rural customers. Indeed, early conflict about telephone sociability was most acute in rural areas. During the monopoly era, Bell companies largely neglected rural demand. The depth and breadth of that demand became evident in the first two decades of this century, when proportionally more farm than urban households obtained telephones, the former largely from small commercial or cooperative local companies. Sociability both spurred telephone subscription and irritated the largely non-Bell vendors.

The 1907 Census of Telephones argued that in areas of isolated farmhouses "a sense of community life is impossible without this ready means of communication. . . . The sense of loneliness and insecurity felt by farmers' wives under former conditions disappears, and an approach is made toward the solidarity of a small country town." Other official investigations bore similar witness.[37] Rural telephone men also dwelt on sociability. One independent company official stated: "When we started the farmers thought they could get along without telephones. . . . Now you couldn't take them out. The women wouldn't let you even if the men would. Socially, they have been a godsend. The women of the county keep in touch with each other, and with their social duties, which are largely in the nature of church work."[38]

Although the episodic sales campaigns to farmers stressed the practical advantages of the telephone, such as receiving market prices, weather reports, and emergency aid, the industry addressed the social theme more often to them than to the general public. A PT&T series in 1911, for example, focused on the telephone in emergencies, staying informed, and saving money. But one additional advertisement said it was: "A Blessing to the Farmer's Wife. . . . It relieves the monotony of life. She CANNOT be lonesome with the Bell Service. . . ."[39] For all that, telephone professionals who dealt with farmers often fought the use of the line for nonbusiness conversations, at least in the early years. The pages of *Telephony* overflow with complaints about farmers on many grounds, not the least that they tied up the lines for chats.

More explicit appreciation of the value of telephone sociability to farmers emerged later. A 1931 account of Bell's rural advertising activities stressed business uses, but noted that "only within recent years [has] emphasis been given to [the telephone's] usefulness in everyday activities . . . the commonplaces of rural life." A 1932 article in the *Bell Telephone Quarterly* notes that "telephone usage

---

[36]A. W. Page, "Public Relations and Sales," "General Commercial Conference, 1928," p. 5, microfilm 368B, ILL BELL INFO. See also comments by Vice-President Gherardi and others in same conference and related ones of the period. On Page and the changes he instituted, see G. J. Griswold, "How AT&T Public Relations Policies Developed," *Public Relations Quarterly* 12 (Fall 1967): 7–16; and Marchand, *Advertising* (n. 21 above), pp. 117–20.

[37]BOC, *Special Reports: Telephones: 1907* (Washington, D.C., 1910), pp. 77–78; see also U.S. Congress, Senate, Country Life Commission, 60th Cong., 2d sess., 1909, S. Doc. 705; and F. E. Ward, *The Farm Woman's Problems,* USDA Circular 148 (Washington, D.C., 1920). See also C. S. Fischer, "The Revolution in Rural Telephony," *Journal of Social History* (in press).

[38]Quoted in R. F. Kemp, "Telephones in Country Homes," *Telephony* 9 (June 1905): 433. A 1909 article claims that "[t]he principle use of farm line telephones has been their social use. . . . The telephones are more often and for longer times held for neighborly conversations than for any other purpose." It goes on to stress that subscribers valued conversation with anyone on the line; see G. R. Johnston, "Some Aspects of Rural Telephony," *Telephony* 17 (May 8, 1909): 542. See also R. L. Tomblen, "Recent Changes in Agriculture as Revealed by the Census," *Bell Telephone Quarterly* 9 (October 1932): 334–50; and J. West (C. Withers), *Plainville, U.S.A.* (New York, 1945), p. 10.

[39]The PT&T series appeared in the *Antioch* (Calif.) *Ledger* in 1911. For some examples and discussions of sales strategies to farmers, see Western Electric, "How to Build Rural Lines," n.d., "Rural Telephone Service, 1944–46," box 1310, AT&T ARCH; Stromberg-Carlson Telephone Manufacturing Company, *Telephone Facts for Farmers* (Rochester, N.Y., 1903), Warshaw Collection, Smithsonian Institution; "Facts regarding the Rural Telephone," *Telephony* 9 (April 1905): 303. In *Printers' Ink*, "The Western Electric," 65 (December 23, 1908): 3–7; F. X. Cleary, "Selling to the Rural District," 70 (February 23, 1910): 11–12; "Western Electric Getting Farmers to Install Phones," 76 (July 27, 1911): 20–25; and H. C. Slemin, "Papers to Meet 'Trust' Competition," 78 (January 18, 1912): 28.

for social purposes in rural areas is fundamentally important." Ironically, in 1938, an independent telephone man claimed that the social theme *had been* but was *no longer* an effective sales point because the automobile and other technologies had already reduced farmers' isolation![40]

As some passages suggest, the issue of sociability was also tied up with gender. When telephone vendors before World War I addressed women's needs for the telephone, they usually meant household management, security, and emergencies. There is evidence, however, that urban, as well as rural, women found the telephone to be useful for sociability.[41] When industry men criticized chatting on the telephone, they almost always referred to the speaker as "she." Later, in the 1930s, the explicit appeals to sociability also emphasized women; the figures in such advertisements, for example, were overwhelmingly women.

In rough parallel with the shift in manifest advertising appeals toward sociability, there was a shift in industry attitudes from irritation with to approval of sociable conversations as part of the telephone's "comfort, convenience, and luxury."

Economic Explanations

Why were the telephone companies late and reluctant to suggest sociable conversations as a use? There are several, not mutually exclusive, possible answers. The clearest is that there was no profit in sociability at first but profit in it later.

Telephone companies, especially Bell, argued that residential service had been a marginal or losing proposition, as measured by the revenues and expenses accounted to each instrument, and that business service had subsidized local residential service. Whether this argument is valid remains a matter of debate. Nevertheless, the belief that residential customers were unprofitable was common, especially among line workers, and no doubt discouraged intensive sales efforts to householders.[42] At times, Bell lacked the capital to construct lines needed to meet residential demand. These constraints seemed to motivate occasional orders from New York not to advertise basic service or to do so only to people near existing and unsaturated lines.[43] And, at times, there was a technical incompatibility between the quality of service Bell had accustomed its business subscribers to expect and the quality residential customers were willing to pay for. Given these considerations, Bell preferred to focus on the business class, who paid higher rates, bought additional equipment, and made long-distance calls.[44]

Still, when they did address residential customers, why did telephone vendors not employ the sociability theme until the 1920s, relying for so long only on practical uses? Perhaps social calls were an untouched and elastic market of consumer demand. Having sold the service to those who might respond to practical appeals—and perhaps by World War I everyone knew those practical uses—vendors might

[40]R. T. Barrett, "Selling Telephones to Farmers by Talking about Tomatoes," *Printers' Ink* (November 5, 1931): 49–50; Tomblen (n. 38 above); and J. D. Holland, "Telephone Service Essential to Progressive Farm Home," *Telephony* 114 (February 19, 1938): 17–20. See also C. S. Fischer, "Technology's Retreat: The Decline of Rural Telephones, 1920–1940," *Social Science History* (in press).

[41]A 1925 survey of women's attitudes toward home appliances by the General Federation of Women's Clubs showed that respondents preferred automobiles and telephones above indoor plumbing; see M. Sherman, "What Women Want in Their Homes," *Woman's Home Companion* 52 (November 1925): 28, 97–98. A census survey of 500,000 homes in the mid-1920s reportedly found that the telephone was considered a primary household appliance because it, with the automobile and radio, "offer[s] the homemaker the escape from monotony which drove many of her predecessors insane"; reported in *Voice Telephone Magazine,* in-house organ of United Communications, December 1925, p. 3, MU IND TEL. One of our interviewees who conducted door-to-door telephone sales in the 1930s said that women were attracted to the service first in order to talk to kin and friends, second for appointments and shopping, and third for emergencies, while, for men, employment and business reasons ranked first. See also Rakow, "Gender" (n. 2 above), and C. S. Fischer, "Women and the Telephone, 1890–1940," paper presented to the American Sociological Association, 1987.

[42]See, e.g., J. W. Sichter, "Separations Procedures in the Telephone Industry," paper P-77-2, Harvard University Program on Information Resources (Cambridge, Mass., 1977); *Public Utilities Digest,* 1930s–1940s, passim; "Will Your Phone Rates Double?" *Consumer Reports* (March 1984): 154–56. Chan's industry interviewees believed this cross subsidy to be true, as, apparently, did AT&T's commercial engineers; see various "Conferences" cited above, AT&T ARCH and ILL BELL INFO.

[43]E.g., commercial engineer C. P. Morrill wrote in 1914 that "we are not actively seeking new subscribers except in a few places where active competition makes this necessary. Active selling is impossible due to rapid growth on the Pacific Coast." He encouraged sales of party lines in congested areas, individual lines in place of party lines elsewhere, extensions, more calling, directory advertisements, etc., rather than expanding basic service into new territories; see PAC TEL MAG 7 (1914): 13–16. And, in 1924, the Bell System's commercial managers decided to avoid canvassing in areas that would require plant expansion and to stress instead long-distance calls and services, especially for large business users; see correspondence from B. Gherardi, vice-president, AT&T, to G. E. McFarland president, PT&T, July 14, 1924, and November 26, 1924, folder "282—Conferences," SF PION MU, and exchage with McFarland, May 10 and May 20, 1924, folder "Correspondence—B. Gherardi," SF PION MU.

[44]The story of the Chicago exchange under John L. Sabin illustrates the point. See R. Garnet, "The Central Union Telephone Company," box 1080, AT&T ARCH.

have thought that further expansion depended on selling "new" social uses of the telephone.[45] Similarly, vendors may have thought they had already enrolled all the subscribers they could—42 percent of American households in 1930—and shifted attention to encouraging use, especially of toll lines. We have seen how sales efforts for intercity calls invoked friends and family. But this explanation does not suffice. It leaves as a puzzle why the sociability themes continued in the Depression when the industry focused again on simply ensuring subscribers and also why the industry's internal attitudes shifted as well.

Perhaps the answer is in the rate structures. Initially, telephone companies charged a flat rate for unlimited local use of the service. In such a system, extra calls and lengthy calls cost users nothing but are unprofitable to providers because they take operator time and, by occupying lines, antagonize other would-be callers. Some industry men explicitly blamed "trivial" calls on flat rates.[46] Discouraging "visiting" on the telephone then made sense.

Although flat-rate charges continued in many telephone exchanges, especially smaller ones, throughout the period, Bell and others instituted "measured service" in full or in part—charging additionally per call—in most large places during the era of competition. In St. Louis in 1898, for example, a four-party telephone cost forty-five dollars a year for 600 calls a year, plus eight cents a call in excess.[47] This system allowed companies to reduce basic subscription fees and thus attract customers who wanted the service only for occasional use.

Company officials had conflicting motives for pressing measured service. Some saw it simply as economically rational, charging according to use. Others saw it as a means of reducing "trivial" calls and the borrowing of telephones by nonsubscribers. A few others, such as E. J. Hall, saw it as a vehicle for bringing in masses of small users.

The industry might have welcomed social conversations, if it could charge enough to make up for uncompleted calls and for the frustrated subscribers busy lines produced. In principle, under measured service, it could. (As it could with long distance, where each minute was charged.) Although mechanical time metering was apparently not available for most or all of this period, rough time charges for local calls existed in principle, since "messages" were typically defined as five minutes long or any fraction thereof. Thus, "visiting" for twenty minutes should have cost callers four "messages." In such systems, the companies would have earned income from sociability and might have encouraged it.[48]

However, changes from flat rates to measured rates do *not* seem to explain the shift toward sociability around the 1920s. Determining the extent that measured service was actually used for urban residential customers is difficult because rate schedules varied widely from town to town even within the same states. But the timing does not fit. The big exchanges with measured residential rates had them early on. For example, in 1904, 96 percent of Denver's residential subscribers were on at least a partial measured system, and, in 1905, 90 percent of those in Brooklyn, New York, were as well. (Yet, Los Angeles residential customers continued to have flat rates.)[49] There is little sign that these rate systems altered significantly in the next twenty-five years while sociability themes emerged.

Conversely, flat rates persisted in small exchanges beyond the 1930s. Moreover, sociability themes appeared more often in rural sales campaigns than in urban ones, despite the fact the rural areas remained on flat-rate schedules.

Although concern that long social calls occupied lines and operators—with financial losses to the

[45]This point was suggested by John Chan from the interviews.

[46]See n. 33, 34. This is also the logic of a recent New York Telephone Co. campaign to encourage social calls: The advertising will not run in upstate New York "since the upstaters tend to have flat rates and there would be no profit in having them make unnecessary calls" (see "New Pitch," n. 28 above).

[47]Letter to AT&T President Hudson, December 27, 1898, box 1284, AT&T ARCH. On measured service in general, see "Measured Service Rates," boxes 1127, 1213, 1287, 1309, AT&T ARCH; F. H. Bethell, "The Message Rate," repr. 1913, AT&T ARCH; H. B. Stroud, "Measured Telephone Service," *Telephony* 6 (September 1903): 153–56, and (October 1903): 236–38; and J. E. Kingsbury, *The Telephone and Telephone Exchanges* (London, 1915), pp. 469–80.

[48]Theodore Vail claimed in 1909 that mechanical time metering was impossible (in testimony to a New York State commission, see n. 8 above, p. 470). See also Judson (n. 34 above), p. 647. In 1928, an operating engineer suggested overtime charges on five-minute calls and stated that equipment for monitoring overtime was now available; see L. B. Wilson, "Report on Commercial Operations, 1927," in "General Commercial Conference, 1928," p. 28, microfilm 368B, ILL BELL INFO. On the five-minute limit, see "Measured Service," box 1127, AT&T ARCH, passim; and Bell Canada, *The First Century of Service* (Montreal, 1980), p. 4. There is no confirmation on how strict operators in fact were in charging overtime. The Bell System, at least, was never known for its laxness in such matters.

[49]Denver: letter from E. J. Hall to E. W. Burgess, 1905, box 1309, AT&T ARCH; Brooklyn: BOC, *Telephones, 1902* (n. 4 above); Los Angeles: "Telephone on the Pacific Coast, 1878–1923," box 1045, AT&T ARCH.

companies—no doubt contributed to the industry's resistance to sociability, it is not a sufficient explanation of those attitudes or, especially, of the timing of their change.

Technical Explanations

Industry spokesmen early in the era would probably have claimed that technical considerations limited "visiting" by telephone. Extended conversations monopolized party lines. That is why companies, often claiming customer pressure, encouraged, set—or sought legal permission to set—time limits on calls. Yet, this would not explain the shift toward explicit sociability, because as late as 1930, 40–50 percent of Bell's main telephones in almost all major cities were still on party lines, a proportion not much changed from 1915.[50]

A related problem was the tying up of toll lines among exchanges, especially those among villages and small towns. Rural cooperatives complained that the commercial companies provided them with only single lines between towns. The companies resisted setting up more, claiming they were underpaid for that service. This single-line connection would create an incentive to suppress social conversations, at least in rural areas. But this does not explain the shift toward sociability either. The bottleneck was resolved much later than the sales shift when it became possible to have several calls on a single line.[51]

The development of long distance might also explain increased sociability selling. Over the period covered here, the technology improved rapidly, AT&T's long-distance charges dropped, and its costs dropped even more. The major motive for residential subscribers to use long distance was to greet kin or friends. Additionally, overtime was well monitored and charged. Again, while probably contributing to the overall frequency of the sociability theme, long-distance development seems insufficient to explain the change. Toll calls as a proportion of all calls increased from 2.5 percent in 1900 to 3.2 percent in 1920 and 4.1 percent in 1930, then dropped to 3.3 percent in 1940. They did not reach even 5 percent of all calls until the 1960s.[52] More important, the shift toward sociability appears in campaigns to sell basic service and to encourage local use, as well as in long-distance ads. (See table 2.)

Cultural Explanations

While both economic and technical considerations no doubt framed the industry's attitude toward sociability, neither seems sufficient to explain the historical change. Part of the explanation probably lies in the cultural "mind-set" of the telephone men.

In many ways, the telephone industry descended directly from the telegraph industry. The instruments are functionally very similar; technical developments sometimes applied to both. The people who developed, built, and marketed telephone systems were predominantly telegraph men. Theodore Vail himself came from a family involved in telegraphy and started his career as a telegrapher. (In contrast, E. J. Hall and A. W. Page, among the supporters of "triviality," had no connections to telegraphy. J. L. Sabin, a man of the same bent, did have roots in telegraphy.) Many telephone companies had started as telegraph operations. Indeed, in 1880, Western Union almost displaced Bell as the telephone company. And the organization of Western Union served in some ways as a model for Bell. Telephone use often directly substituted for telegraph use. Even the language used to talk about the telephone revealed its ancestry. For example, an early advertisement claimed that the telephone system was the "cheapest telegraph service ever."

[50]On company claims, see, e.g., "Limiting Party Line Conversations," *Telephony* 66 (May 2, 1914): 21; and MacMeal (n. 2 above), p. 224. On party-line data, compare the statistics in the letter from J. P. Davis to A. Cochrane, April 2, 1901, box 1312, AT&T ARCH, to those in B. Gherardi and F. B. Jewett, "Telephone Communications System of the United States," *Bell System Technical Journal* 1 (January 1930): 1–100. The former show, e.g., that, in 1901, in the five cities with the most subscribers, an average of 31 percent of telephones were on party lines. For those five cities in 1929, the percentage was 36. Smaller exchanges tended to have even higher proportions. See also "Supplemental Telephone Statistics, PT&T," "Correspondence—Du Bois," SF PION MU. The case of Bell Canada also fails to support a party-line explanation. Virtually all telephones in Montreal and Toronto were on individual lines until 1920.

[51]"Carrier currents" allowed multiple conversations on the same line. The first one was developed in 1918, but for many years they were limited to use on long-distance trunk lines, not local toll lines. See, e.g., R. Coe, "Some Distinguishing Characteristics of the Telephone Business," *Bell Telephone Quarterly* 6 (January 1927): 47—51, esp. pp. 49–50; and R. C. Boyd, J. D. Howard, Jr., and L. Pederson, "A New Carrier System for Rural Service," *Bell System Technical Journal* 26 (March 1957): 349–90. The first long-distance carrier line was established in Canada in 1928, after the long-distance sociability theme had emerged; see Bell Canada, *First Century,* no. 46, p. 28.

[52]BOC, *Historical Statistics* (n. 3 above), p. 783.

Telephone calls were long referred to as "messages." American telegraphy, finally, was rarely used even for brief social messages.[53]

No wonder, then, that the uses proposed first and for decades to follow largely replicated those of a printing telegraph: business communiqués, orders, alarms, and calls for services. In this context, industry men reasonably considered telephone "visiting" to be an abuse or trivialization of the service. Internal documents suggest that most telephone leaders typically saw the technology as a business instrument and a convenience for the middle class, claimed that people had to be sold vigorously on these marginal advantages, and believed that people had no "natural" need for the telephone—indeed, that most (the rural and working class) would never need it. Customers would have to be "educated" to it.[54] AT&T Vice-President Page was reacting precisely against this telegraphy perspective in his 1928 defense of "frivolous" conversation. At the same conference, he also decried the psychological effect of telephone advertisements that explicitly compared the instrument to the telegraph.[55]

Industry leaders long ignored or repressed telephone sociability—for the most part, I suggest, because such conversations did not fit their understandings of what the technology was supposed to be for. Only after decades of customer insistence on making such calls—and perhaps prodded by the popularity of competing technologies, such as the automobile and radio—did the industry come to adopt sociability as a means of exploiting the technology.

This argument posits a generation-long lag, a mismatch, between how subscribers used the telephone and how industry men thought it would be used. A variant of the argument (posed by several auditors of this article) suggests that there was no mismatch, that the industry's attitudes and advertising accurately reflected public practice. Sales strategies changed toward sociability around the mid-19208 because, in fact, people began using the telephone that way more. This increase in telephone visiting occurred for perhaps one or more reasons—a drop in real costs, an increase in the number of subscribers available to call, clearer voice transmission, more comfortable instruments (from wall sets to the "French" handsets), measured rates, increased privacy with the coming of automatic dial switching, and so on—and the industry's marketing followed usage.

To address this argument fully would require detailed evidence on the use of the telephone over time, which we do not have. Recollections by some elderly people suggest that they visited by telephone less often and more quickly in the "old days," but they cannot specify exact rates or in what era practices changed.[56] On the other hand, anecdotes, comments by contemporaries, and fragments of numerical data (e.g., the 1909 Seattle "study") suggest that residential users regularly visited by telephone before the mid-1920s, whatever the etiquette was supposed to be, and that such calls at least equaled calls regarding household management. Yet, telephone advertising in the period overwhelmingly stressed practical use and ignored or suppressed sociability use.

Changes in customers' practices may have helped spur a change in advertising—although there is no direct evidence of this in the industry archives—but some sort of mismatch existed for a long time between actual use and marketing. Its source appears to be, in large measure, cultural.

[53]On the telegraph background of early telephone leaders, see, e.g., A. B. Paine, *Theodore N. Vail* (New York, 1929); Rippey (n. 2 above); and W. Patten, *Pioneering the Telephone in Canada* (Montreal: Telephone Pioneers, 1926). Interestingly, this was true of Bell and the major operations. But the leaders of small-town companies were typically businessmen and farmers; see, e.g., *On the Line* (Madison: Wisconsin State Telephone Association, 1985). On Western Union and Bell, see G. D. Smith, *The Anatomy of a Business Strategy: Bell, Western Electric, and the Origins of the American Telephone Industry* (Baltimore, 1985). The "cheapest telegraph" appears in a Buffalo flier of November 13, 1880, box 1127, AT&T ARCH. On the infrequent use of the telegraph for social messages, see R. B. DuBoff, "Business Demand and Development of the Telegraph in the United States, 1844–1860," *Business History Review* 54 (Winter 1980): 459–79.

[54]In the very earliest days, Vail had expected that the highest level of development would be one telephone per 100 people; by 1880, development had reached four per 100 in some places; see Garnet (n. 2 above), p. 133, n. 3. It reached one per 100 Americans before 1900 (see table 1). In 1905, a Bell estimate assumed that twenty telephones per 100 Americans was the saturation point and even that "may appear beyond reason"; see "Estimated Telephone Development, 1905–1920," letter from S. H. Mildram, AT&T, to W. S. Allen, AT&T, May 22, 1905, box 1364, AT&T ARCH. The saturation date was forecast for 1920. This estimate was optimistic in its projected *rate* of diffusion—twenty per 100 was reached only in 1945—but very pessimistic in its projected *level* of diffusion. That level was doubled by 1960 and tripled by 1980. One reads in Bell documents of the late 1920s of concern that the automobile and other new technologies were far outstripping telephone diffusion. Yet, even then, there seemed to be no assumption that the telephone would reach the near universality in American homes of, say, electricity or the radio.

[55]Page 53 in L. B. Wilson (chair), "Promoting Greater Toll Service," "General Commercial Conference, 1928," microfilm 368B, ILL BELL INFO.

[56]This comment is based on the oral histories reported by Rakow (n. 2 above) and by several interviews conducted in San Rafael, Calif., by John Chan for this project. See also Fischer, "Women" (n. 41).

This explanation gains additional plausibility from the parallel case of the automobile, about which space permits only brief mention. The early producers of automobiles were commonly former bicycle manufacturers who learned their production techniques and marketing strategies (e.g., the dealership system, annual models) during the bicycle craze of the 1890s. As the bicycle was then, so was the automobile initially a plaything of the wealthy. The early sales campaigns touted the automobile as a leisure device for touring, joyriding, and racing. One advertising man wondered as late as 1906 whether "the automobile is to prove a fad like the bicycle or a lasting factor in the industry of the country."[57]

That the automobile had practical uses dawned on the industry quickly. Especially after the success of the Ford Model T, advertisements began stressing themes such as utility and sociability—in particular, that families could be strengthened by touring together. Publicists and independent observers alike praised the automobile's role in breaking isolation and increasing community life.[58] As with the telephone, automobile vendors largely followed a marketing strategy based on the experience of their "parent" technology; they stressed a limited and familiar set of uses; and they had to be awakened, it

seems, to wider and more popular uses. The automobile producers learned faster.

No doubt other social changes also contributed to what I have called the discovery of sociability, and other explanations can be offered. An important one concerns shifts in advertising. Advertising tactics, as noted earlier, moved toward "softer" themes, with greater emphasis on emotional appeals and on pleasurable rather than practical uses of the product. They also focused increasingly on women as primary consumers, and women were later associated with telephone sociability.[59] AT&T executives may have been late to adopt these new tactics, in part because their advertising agency, N. W. Ayer, was particularly conservative. But in this analysis, telephone advertising eventually followed general advertising, perhaps in part because AT&T executives attributed the success of the automobile and other technologies to this form of marketing.[60]

Still, there is circumstantial and direct evidence to suggest that the key change was the loosening, under the influence of public practices with the telephone, of the telegraph tradition's hold on the telephone industry.

Conclusion

Today, most residential calls are made to friends and family, often for sociable conversations. That may well have been true two or three generations ago, too.[61] Today, the telephone industry encourages such calls; seventy-five years ago it did not. Telephone salesmen then claimed the residential telephone was good for emergencies; that function is now taken for granted. Telephone salesmen then claimed the telephone was good for marketing; that function persists ("Let your fingers do the walking. . . .") but never seemed to be too important

[57]Among the basic sources on the history of the automobile drawn from are: J. B. Rae, *The American Automobile: A Brief History* (Chicago, 1965); id., *The Road and Car in American Life* (Cambridge, Mass., 1971); J. J. Flink, *America Adopts the Automobile, 1895–1910* (Cambridge, Mass., 1970); id., *The Car Culture* (Cambridge, Mass., 1976); and J.-P. Bardou, J.-J. Chanaron, P. Fridenson, and J. M. Laux, *The Automobile Revolution*, trans. J. M. Laux (Chapel Hill, N.C., 1982). The advertising man was J. H. Newmark, "Have Automobiles Been Wrongly Advertised?" *Printers' Ink* 86 (February 5, 1914): 70–72. See also id., "The Line of Progress in Automobile Advertising," ibid., 105 (December 26, 1918): 97–102.

[58]G. L. Sullivan, "Forces That Are Reshaping a Big Market," *Printers' Ink* 92 (July 29, 1915): 26–28. Newmark (n. 57 above, p. 97) wrote in 1918 that it "has taken a quarter century for manufacturers to discover that they are making a utility." A 1930s study suggested that 80 percent of household automobile expenditures was for "family living"; see D. Monroe et al., *Family Income and Expenditures. Five Regions*, Part 2. *Family Expenditures*, Consumer Purchases Study, Farm Series, Bureau of Home Economics, Misc. Pub. 465 (Washington, D.C., 1941), pp. 34–36. Recall the 1925 survey of women's attitudes toward appliances (n. 41 above). The author of the report, Federation President Mary Sherman, concluded that "Before toilets are installed or washbasins put into homes, automobiles are purchased and telephones are connected . . . [b]ecause the housewife for generations has sought escape from the monotony rather than the drudgery of her lot" (p. 98). See also *Country Life* and Ward (n. 37 above); E. de S. Brunner and J. H. Kolb, *Rural Social Trends* (New York, 1933); and F. R. Allen, "The Automobile," pp. 107–32 in F. R. Allen et al., *Technology and Social Change* (New York, 1957).

[59]Recall that, early on, women were associated in telephone advertising with emergencies, security, and shopping.

[60]On changes in advertising, see sources cited in n. 21 above. The comment on N. W. Ayer's conservatism comes from Roland Marchand (personal communication).

[61]It is difficult to establish for what purpose people actually use the telephone. A few studies suggest that most calls by far are made for social reasons, to friends and family. (This does not mean, however, that people subscribe to telephone service for such purposes.) See Field Research Corporation, *Residence Customer Usage and Demographic Characteristics Study: Summary*, conducted for Pacific Bell, 1985 (courtesy R. Somer, Pacific Bell); B. D. Singer, *Social Functions of the Telephone* (Palo Alto, Calif.: R&E Associates, 1981), esp. p. 20; M. Mayer, "The Telephone and the Uses of Time," in Pool, *Social Impact* (n. 2 above), pp. 225–45; and A. H. Wurtzel and C. Turner, "Latent Functions of the Telephone," ibid., pp. 246–61.

to residential subscribers.[62] The sociability function seems so obviously important today, and yet was ignored or resisted by the industry for almost the first half of its history.

The story of how and why the telephone industry discovered sociability provides a few lessons for understanding the nature of technological diffusion. It suggests that promoters of a technology do not necessarily know or determine its final uses; that they seek problems or "needs" for which their technology is the answer (cf. the home computer business); but that consumers may ultimately determine those uses for the promoters. And the story suggests that, in promoting a technology, vendors are constrained not only by its technical and economic attributes but also by an interpretation of its uses shaped by its and their own histories, a cultural constraint that can be enduring and powerful.

---

[62]A 1934 survey found that up to 50 percent of women respondents with telephones were "favorable" to shopping by telephone. Presumably, fewer actually did so; see J. M. Shaw, "Buying by Telephone at Department Stores," *Bell Telephone Quarterly* 13 (July 1934): 267–88. This is true despite major emphases on telephone shopping in industry advertising. See also Fischer, "Women" (n. 41 above).

Reading 18

Shaping the Early Development of Television

Jan van den Ende, Wim Ravesteijn, and Dirk De Wit

This article is an adapted and translated version of a piece entitled "Waarom geen Nipkowschijf in elke huiskamer? De sociale constructie van televisie" which appeared in Het Jaarboek Mediageschiedenis, 1993 [1].

Few of the billions of today's television viewers are aware that their early predecessors were served by a mechanical form of television. The central element in this early television was a rotating disc with a spiral series of perforations, called the Nipkow disc, after its Russian inventor Paul Nipkow. Two such discs were needed, one in the camera and one in the television itself. Rotating Nipkow discs, lamps, and light-sensitive elements together formed a means of recording and displaying television images.

Today the mechanical television is usually regarded as a historical curiosity; a cumbersome machine with a large, motor-driven wheel, a window, and flickering, blurred images: at best, merely a technological precursor to electronic television. However, we believe that there were once important incentives supporting mechanical television in its rivalry to electronic television. The electronic television gained preeminence not because of purely technical considerations, as is so often assumed, but also because of social ones. The programming that the television public wanted to see (the "software") influenced the choice in television technology (the "hardware"), and the institutional incorporation of television also played an important role.

Received wisdom on the early history of television originates from writers who have largely focused on technical aspects. This material conforms to the traditional approaches to the history of technology, which explained the development of technology mainly referring to technical factors and the ideas of inventors. We will address these approaches with the usual term "internalist," although it has been demonstrated that this concept has never been defined accurately, in contrast to its opposite "externalist" [2]. The internalist kind of historical study resulted in good narratives but offered little explanation for the direction of technological developments. The internalists

assumed that the new technology was always superior to the old and took this as the self-evident reason for the spread of new technologies within society. In the case of television, according to this view mechanical television may well have enjoyed some success, but did not stand a chance against the newer, and therefore better, electronic television. Naturally, in a short while electronic television had gained complete superiority and mechanical television could be regarded as a historical curiosity. In hindsight, those involved could also be appraised accordingly: conservatives had held on to the mechanical system, those who had embraced electronic television early on were progressives.

> Today, few viewers are aware that a mechanical form of television served their predecessors.

It has since become clear that the internalist approach provides an insufficient explanation of technological development. For example, it gives no explanation of the fact that new technologies sometimes lie idle for years before coming into use, while others quickly displace the old. With regard to the history of television, an internalist approach fails to answer two questions. The first is, why was mechanical television so successful? Around 1930 the system was in widespread use in both Great Britain and the United States. In America, numerous broadcasting stations were in operation; in England, the BBC broadcasted regular programs. For many years the system underwent continuous development and in 1936, when an extensive comparison between the mechanical and the electronic systems was organized in England, it was still not clear which was better. The second question is, why did the development of television technology run such different courses in the United States and in Great Britain? In the U.S., development was discontinuous; around 1930 a "television boom" saw the use of mechanical television grow strongly, then stagnate around 1933. Electronic television subsequently came slowly into more widespread use. In the U.K., the development was more gradual; the use of mechanical systems continued right up to a national comparison in 1936, after which they were replaced by electronic systems.

Besides the internalist approach, a second type of historical perspective on the growth of television is provided by media history. In this, the social infrastructure necessary for television, such as broadcasting organizations, actors, and press offices, is held to be central. Media historians also pay attention to programming and to public evaluation. However, the television itself is kept out of the discussion; in practice it is regarded as a "given" technology, the implementation of which (in the form of television programs) is socially determined.[1] Media historians therefore regard the choice between mechanical and electronic television systems as a technical issue. Nevertheless, media historians would throw more light onto the subject were it not for the fact that they are primarily concerned with the later, rather than the earliest, stages in the development and spread of television.

Despite all criticism on internalist and other traditional approaches of technology development, they persist in recent literature, while new social-constructivist approaches, until now, have not led to a satisfactory alternative theoretical framework.[2] Basically, the new approaches stress the interaction between internal and external (or contextual) factors in technology development. We support this notion in regard to the early development of television. We do so on the basis of materials largely extracted from writers who focused on the technical history of television.[3] We see that it may be difficult to demonstrate the role of contextual influences in the history of television, but hope our analysis contributes in indicating that such influences indeed existed, even on the basis of existing materials.[4]

We focus on the choice between mechanical and electronic television systems. We analyze this choice from a broad perspective, with attention to social and technical factors and their interaction. We employ social-constructivist notions, especially the insight that several choices are always present in the historical development of a technology (see [6], [26]). Furthermore we take upon the idea that social groups, according to the meanings they attribute to the technology, can influence the course of this development.[5] We understand social groups, not on the social-constructivist basis of the meanings attached to a given technology, but in the more traditional sociological sense. We also include individuals, organizations, and the relationships between them in our deliberations (see also, [29]).

In contrast to social-constructivism, we accord a specific role to technological factors. In our approach, a given technology has a certain objective reality that circumscribes the meanings that can be attributed to it (see also [30]). This is one of the reasons that agreement often exists between different parties on the given attributes of a new technology, even when they have different preferences with regard to the technology itself. We shall see that both mechanical and electronic television had strong and weak points that were recognized by most of those involved. Moreover, we hold that to a certain extent technological innovations do indeed have their own dynamic. Physical reality puts constraints and offers opportunities for technological development. For instance, technical limitations in 1930s Britain made it very difficult to develop a mechanical camera for outside broadcasts. Moreover, innovations developed for a specific application create opportunities in other fields. The invention of the vacuum tube, which made the development of electronic television possible, forms an example.

We concentrate our attention on developments as they took place in two of the leading countries in this field, Great Britain and the United States, with only a passing comment on a third front runner, Germany. The Netherlands provide us with an illustration of developments in a "subsequent" country; although the choice between the two television systems was strongly influenced by events abroad, the contest between mechanical and electronic television was also fought on Dutch soil. In fact, the unique social situation in the Netherlands gave developments a specific character.

---

[1]This standpoint is explicitly embodied in [3].

[2]Even one of the main protagonists of social constructivist approaches, Wiebe Bijker, states this in. [4].

[3]There are some exceptions of publications that focus on the social context, for instance J. Udelson's, *The Great Television Race, a History of the American Television Industry 1925–1941* [5].

[4]Citing M.J. Mulkay, Wiebe Bijker *et al.* explicitly state that it is difficult to apply contextual approaches on the history of television. See [6]. See also [7]–[10].

---

[5]Our approach shows similarities to Patrick C. Carbonara who also focuses on the circle of interdependence between players that affected the dispersion of monochrome and color television in [28].

Early History

The nineteenth and first half of the twentieth century saw a large number of new ideas and inventions having to do with sending images over long distances. In hindsight, many of these can be seen as having formed part of the history of television; however, the inventors often had other applications in mind, such as a "picture telegraph," film, or "picture telephone."

A number of historical moments come immediately to mind [6]. Firstly, the invention of the selenium cell, whose photo-electric characteristics were accidentally discovered by an English telegrapher in 1873. In 1875, the American G.R. Carey was the first to suggest that it might be possible to send images by electricity: he talked of "seeing by electricity", in analogy to "hearing by electricity" with a telephone. The telephone inventor, Alexander Graham Bell, also thought that an image could be added to the sound.

At that time, researchers saw the main problem as one of being unable to send an entire image over just one line. Between 1880 and 1925, therefore, various scanning devices were proposed to allow the image to be sent, piece by piece, over one line. One of the most important of these was devised by Paul Nipkow, whose invention formed the basis for many subsequent mechanical television systems. While still a student at the University of Berlin in 1884, Nipkow had come up with the idea of a television that employed rotating perforated discs. The camera would consist of such a disc, a number of lenses, and a selenium cell. It would work as follows: a light source illuminated the object; lenses projected the reflected light onto the rotating disc; the disc allowed one point of the projected image at a time onto the selenium cell, so that for each complete rotation of the disc the entire image had been projected, a bit at a time, onto the selenium cell; the cell transformed this varying light intensity into an electrical current, and thereby turned the image into a series of consecutive electrical impulses. In the television receiver, the process would be reversed: the electrical signal would be amplified and connected to a lamp situated behind a synchronously rotating Nipkow disc, so that the lamp flickered with an intensity corresponding to the electrical signal generated by the selenium cell. The image would then be visible on the face of the disc. Nipkow never built a prototype: the first fully operational model based on his principles was hot produced until 1925.

> Received wisdom on the early history of television focuses on technical aspects.

Operation of Mechanical Television

Around 1910, the Russian Boris Rosing developed a television system in which the function of the Nipkow disc was performed by rotating mirrors. His television receiver, however, employed a cathode ray tube, developed in 1897 by Karl Ferdinand Braun at the University of Straatsburg. In this tube a glowing cathode radiated electrons which were then guided towards a light-emitting screen by means of an electromagnetic field. In 1911 Rosing succeeded in producing an image consisting of four lines using this method. Rosing called his system the "electric telescope." Rapid developmental progress was prevented by problems in directing the cathode rays. Tragically, Rosing himself was arrested in the turmoil of the Russian revolution, exiled, and never heard of again.

At about the same time as Rosing, the Englishman A.A. Campbell Swinton devised a completely electronic television system. In recording, the image was projected onto a photo-electric surface within a vacuum tube, and an electron beam scanned this surface. Campbell Swinton never succeeded in building a working model, but his work was to form the basis of the first patent application for a fully electronic television in 1929.

The following years also saw improvements in many of the elements of the mechanical system. In 1913 the German physicists Julius Elster and Hans Geitel made progress with the selenium cell and in 1917 D. McFarlan Moore developed a neon lamp capable of the rapid flickering required of the light source in the television receiver.

So by 1920 various ideas existed for television systems. Designs existed for a system using mechanical cameras and receivers (using Nipkow discs), a system employing a mechanical camera (with rotating mirrors) and an electronic receiver, and a system employing both an electronic camera and an electronic receiver. None of these systems could yet boast of a working prototype; but this was to change fast.

Mechanical Television in Great Britain: Baird

Thanks to the work of John Logie Baird, Great Britain can be called the birthplace of mechanical television.[6] Baird was trained as an electrical engineer at the Royal Technical College in Scotland and was a born innovator. After a number of failures, which included a jam factory in Trinidad, in 1923 he began developing a mechanical television system based on Nipkow's principles [10]. Within a year

---

[6]For the history of television in Great Britain, we have made extensive use of [9].

he had succeeded in sending shadowy images from a camera to a television receiver along an electrical wire. In April 1925, Baird gave the first public demonstration of his television system. The television produced eight image lines. It was a primitive affair; the Nipkow discs were made of hatbox lids.

Nevertheless, the demonstration was a success. A London *Times* journalist described the images as being faint and blurred but nevertheless sharp enough to be able to distinguish facial expressions. Baird was able to obtain financial support from a number of sources which enabled him to continue his research. He formed Baird Television Ltd. and in 1926 was able to present his new "televisor" to members of the Royal Institute and *The Times*. The pictures were still of a person's head and shoulders, but Baird was now working with thirty lines; a neon lamp illuminated the disc in the television, and the discs in the camera and the receiver were electrically synchronized [11]. Moreover, he now used a new lighting method, the "spotlight" method, in which he changed the positions of the light source and photocells. The light source was behind the disc and the photocells in front, trained on the object. It was the object that was illuminated piece by piece. By employing more photocells, the total effective light intensity could be raised and this method also prevented the object from being excessively strongly lit.

There were three main technical problems. Baird himself complained constantly about the quality of the lenses and the photocells, his employees pointed to the imperfect signal amplification, and according to spectators the mechanical aspects, the cardboard discs, the wires and strings were shoddy. In an attempt to improve the working of the photocell he even considered using a human eye; a surgeon obtained one for him but after a crude dissection with a razor he gave up and threw it into a canal.

Baird wanted to provide an integrated system, that is, he wanted to provide programs as well as equipment. He was not interested in collaborating directly with the national broadcasting company, the BBC. He wanted to finance his research out of the sale of television sets and his programs out of advertising. However, the organization responsible for controlling broadcasting, the General Post Office (GPO), objecting to the low image quality, refused Baird permission to use his transmitters. In the meantime, he devised a number of remarkable firsts, for example, the *Noctovisor* (1926), an invention employing infrared light which ensured, as had the "spotlight" method, that actors were not subjected to excessive lighting intensities. With transatlantic transmission (1923) and color television (1929) Baird succeeded in attracting publicity. He needed it badly; his good contacts with supportive patrons were vital, since as long as he could not transmit and sold no television sets, Baird had no other income.

In 1929, Baird finally received permission to employ a medium-wave radio frequency for his transmissions. He worked with thirty lines and 12.5 images/s. The reason for the GPO's change of mind was that television had now become an issue of national prestige, and Baird had begun to establish international contacts (he also worked, for instance, with the German Fernseh AG) and had threatened to move to the United States if he did not receive a British transmission license. Though he marketed his television sets, or *televisors*, himself, they were manufactured by a number of other companies and were not immediately available; it would be six months before they would be on the market. In this first period the transmissions were mostly intended for amateur enthusiasts. Two radios were needed, one tuned to the audio signal, the other to the image signal; the loudspeaker output of the second formed the input signal for a homemade television receiver.

Baird's television sets cost £25, an average monthly wage. The DIY kit versions that he also marketed cost between £12 and £16 pounds. All models had a screen measuring 5 cm by 10 cm. Estimates of the number of television sets in use varied widely. In January 1931 Baird reported to the GPO that he had sold 1000, but at the end of 1932 he stated that 500 were in use. An unknown number of homemade sets were also sold. In addition, other companies, such as Electrical Music Industries (EMI), began to develop mechanical television systems. In 1933 the BBC put the figure of television sets in use at 8000. Slowly but surely, mechanical television was becoming a success.

Baird was now working on improving the image quality by increasing the number of image lines, but this was no simple matter. It meant increasing the diameter of the Nipkow disc, increasing the number of its perforations and reducing their size. In 1933, Baird managed to increase the number of lines to 120, and in 1936 to 240. By this time the perforations were so small that dust was blocking them and so Baird switched to a camera employing revolving mirrors. He also brought larger, more expensive television sets onto the market, with a screen measuring 10 cm by 22 cm; these varied from £50 to £75 in price. Few if any of these sets were ever sold.

Electronic Television in Great Britain: A Gradual Takeover

The development of electronic television in the U.K. made use of American inventions and patents. One organization, in particular, was involved: the specially-formed company Marconi-EMI, of which the aforementioned EMI was one of the founders. The new company was given access to diverse American patents, including one for an electronic camera, the *iconoscope*, developed for the American company RCA by the Russian Vladimir K. Zworykin. He had studied under Rosing in Russia. The heart of the iconoscope was a vacuum tube, in which the image was projected onto a light-sensitive surface. Each point on this surface formed a small capacitor together with the metal plate situated immediately behind it. An electron beam scanned the surface vertically and horizontally, and the discharge current, which varied with the degree of illumination at each scanned point, provided the television signal.

However, the range of the transmitters was smaller for electronic television. The reason for this was that an increased amount of information per second could be transmitted only on higher radio frequencies, particularly in the VHF band, whose range was only about 50 km. Mechanical television mostly employed short- and medium-wave frequencies which had much greater ranges. Of course, it was not necessary for electronic television to have larger number of image lines and a higher amount of information per second than mechanical systems. Electronic television could also be employed with similar numbers of image lines, and then the same frequencies could be used and the same geographic range be reached. However, in that case the potentials and advantages of the electronic system compared with the mechanical one would not exist, whereas the electronic system still had the disadvantage of a higher price. So in practice, the amount of information of electronic was indeed higher. The eventual solution for electronic TV broadcasts was the construction of relay stations that allowed the VHF signal to be heard further afield.

In England, Marconi-EMI developed a camera, the *emitron*, based on Zworykin's principles, and which the company displayed at the 1935 annual radio exhibition. The *emitron* had an improved image quality and was also mobile. A number of other television systems were also on show. The electronic television sets were more expensive than the mechanical ones; at about £100 they represented an average Englishman's three months' wages, but the price did nothing to dampen the public's enthusiasm. Only for films (where the object was relatively small and could be perfectly lit) was Baird's mechanical system still considered more attractive. Baird tried to improve his system further, and even started to do research into electronic television.

In 1936, the Television Advisory Committee, which comprised representatives of the GPO and the BBC, created a two-year period to decide which system was better. Numerous television system manufacturers put themselves forward but only Baird and Marconi-EMI made it to the final round. Baird only participated with mechanical systems in the contest. He was now working with 240 lines and had two recording systems. For close-ups and studio work he used a conventional mechanical camera. For outside broadcasts he had developed a special process: the object was first filmed, the film automatically ran on through a developing machine, and the developed film was then broadcast using a special scanner. The whole process took only forty seconds. The purpose was to circumvent the mechanical camera's inadequacy in weak light. Baird's mechanical outside-broadcast camera had important disadvantages; it was large and difficult to transport, and the fragility of the film only increased the chances of an accident. Moreover, the whole process was extremely expensive. Baird's troubles took a turn for the worse when Crystal Palace burnt down in 1936; his company was housed there and much of his equipment was lost.

EMI was already demonstrating the emitron when the two-year trial period was introduced. The machine used 405 lines, and the television screen was now 30 cm by 30 cm. The screen tube was so long that it was mounted vertically, with a 45° mirror providing the final viewing screen. Even before the trial period was over, the machine was declared the best available; the image quality was better, the system was relatively reliable, and the emitron camera, being truly portable, was better suited to outside broadcasts than were Baird's mechanical cameras.

In January 1937 the BBC — which the GPO had accorded a television broadcasting monopoly, as it had in radio broadcasting — began transmitting regular broadcasts using an electronic television system with 405 lines. At first, few television sets were sold but a month later EMI dropped the price from £100 to £60. Six months after broadcasts had begun, 1500 sets (of all makes combined) had been sold. The spread of electronic television had begun.

Social Shaping

The development of mechanical television in the U.K. involved a number of different social groups and organizations: the manufacturers (Baird and his company), the supervising governmental body (the GPO), the broadcasting company (the BBC), actors, amateur television enthusiasts, and the viewing public. Each group attributed different meanings to television.

For Baird, mechanical television was a means to set up and run a company. Television, for him, was a marketable product; his desire to produce his own programs was part of his marketing strategy. This was made possible in 1929 when the BBC first allowed him to use a transmitter. His broadcasts were mostly made up of studio entertainment, educational programs, and films. In this period, while he was broadcasting on a small scale and sold no television receivers, he depended on financiers to sponsor his activities. This makes clear why he put so much effort into coming up with new inventions; they would convince his patrons of the viability of the medium.

To start with, the BBC showed extreme caution; the broadcasting company would only be interested in talks about transmission, let alone participation, when an acceptable image quality could be guaranteed. Cinema film quality was informally taken as a standard. The difficulty of making outside broadcasts with the mechanical system also formed a problem for the BBC. Baird's solution to this problem, using an intermediary film stage, had its own problems; the camera was practically immobile, and the time delay that the film development produced was another possible problem.

At first the GPO was reluctant to give Baird a broadcasting license. Their argument was the same as the BBC's: Baird's inability to guarantee a satisfactory image quality. Only when Baird threatened to leave the U.K. and set up in business abroad was he finally given a license—for England could not afford to lose Baird's company; television had become an issue of national importance.

Actors, too, had problems with mechanical television, for it depended on intolerably intense studio lighting. Baird's "spotlight" method was only a partial solution to this problem; the process put the actors into flickering light with the rest of the studio in utter darkness, so that they could read their parts only with difficulty and could not see the director. The system also left the actors precious little freedom of movement. At one point the BBC feared that actors would refuse to extend their cooperation to any mechanical television broadcasts.

Electronic television promised to solve most of these problems, but only by introducing a new problem, that of the extremely limited range of the existing VHF transmitters; many viewers would be left without a signal. To tackle this problem, an entire network of new VHF transmitters had to be set up. In 1934, when the BBC was considering replacing the mechanical system with the electronic, letters arrived from listeners in Scotland and the north of England, protesting that the change would leave them with no television signal at all. Electronic television, they argued, would divide British television viewers into two geographical halves: those within reach of VHF transmissions, and those without. The price of electronic television sets was probably another obstacle to the consuming public; they were much more expensive than the mechanical model.

Amateur television enthusiasts also found difficulties with the electronic version. VHF frequencies could not be picked up on ordinary radios, and the electronic television was too complex to build at home. And finally, electronic television of course brought Baird a big problem: it formed a threat to the continuation of his company.

Mechanical Television in the United States: A "Boom"

Baird had an American counterpart. It was another inventor, Charles Francis Jenkins, who gave America its first mechanical television.[7] In June 1925 he was able to show pictures of a Dutch windmill turning; the images had been taken from a film screen. Jenkins' camera used two ground glass discs, that formed prisms, which deflected the beam of light in one direction as the discs rotated. He called his system *radiovision* because the system was wireless, preferring to reserve the term *television* for transmissions that took place over electrical cables, in analogy with the telephone and the telegraph. He envisaged numerous applications, such as film shows, theater shows, and sporting contests [5]. Like Baird he formed a company, but unlike Baird's it was supported by the banks.

It was not long before other large concerns began to show interest. AT&T's Bell Laboratories started

---

[7]For the history of the mechanical television in the United States we have made extensive use of [7] and [12].

work in this field in 1925 and General Electric also carried out research into mechanical television around this time. AT&T saw the device's potential as a "picture telephone," and between 1925 and 1929 the company poured over $300 000 into research into mechanical television. Two hundred engineers and scientists were put to work. By 1927 Bell was able to demonstrate two systems. One employed a large screen measuring 60 cm by 75 cm, the other a small screen measuring 5 cm by 6 cm. The latter was intended for installation in a "picture phone." In June 1929, exactly four years after Jenkins' original television, AT&T demonstrated a mechanical color television, which worked using three color filters and three photocells, each photocell being connected via one of three transmission cables to a separate neon lamp in the television receiver.

From 1928 to 1932, mechanical television in the USA prospered — these were the years of a "television boom." Dozens of television stations started broadcasting, having been granted "experimental broadcasting licenses" by the Federal Radio Commission (FRC), the body that had been set up to regulate the use of radio waves. In 1931, 31 television stations were in active service and Chicago alone had an estimated 8000 television sets. Thereafter most of these stations disappeared, except for three from universities in the Chicago area that continued to broadcast educational programs using the mechanical system until 1937. Although the depression may also have been of influence, after this initial period it gradually became clear that the mechanical television system was not a viable option in the United States. We shall shortly examine the reasons below. Jenkins' company went bankrupt, and other manufacturers had it no easier.

Electronic Television in the U.S.: A Slow Start

The first patent application for an entirely electronic television system came from the above-mentioned Zworykin. In the 1920s he worked at Westinghouse in the U.S., where at first little interest was shown in his activities but where he was later given a free hand. In 1923, Zworykin applied for a patent on his device, although the machine had yet to produce more than the shadowy image of a cross.

Six years later Zworykin was able to present a working cathode ray tube television set. Around this time he went to work at RCA, a company that had set up a central radio research laboratory in which

researchers from various RCA companies were at work. In 1930 Zworykin was made head of the laboratory, and under his direction in 1933 it produced the iconoscope, the electronic camera.

The higher image quality of electronic television was due, in Zworykin's terms, to the "storing of electrical charges." In the mechanical camera, an increase in the number of screen image lines reduced the amount of light falling onto the selenium cell at a given moment; since the cell's low-light sensitivity was limited, at a certain point this began to create problems. In the electronic camera, the light sensitive plate was permanently illuminated, and in raising the number of screen lines, only the scanning speed was increased [13]. This allowed the electronic camera to operate in lower light levels than the mechanical camera.

Zworykin's successes encouraged RCA to invest a million dollars in television in 1935 alone. The concern built a new television transmitter on the Empire State Building in New York City and a special TV studio was built at the National Broadcasting Company (NBC), the RCA broadcasting organization. Although the company still had no official license to transmit TV commercially, in 1936 RCA began a campaign to attract potential TV advertisers. RCA's efforts to gain advantage over its competitors by building a patent portfolio became obstructed by the efforts of one man, who started with a single invention, Philo T. Farnsworth.

Farnsworth was responsible for the invention of another electronic television system. Farnsworth started working on what he called his television system in the mid 1920s. He managed to get several Californian businessmen, especially George Everson, interested in his ideas. In the fall of 1927 Farnsworth succeeded in a laboratory demonstration. Within a year his television system was made public with an article in the *San Francisco Chronicle* describing Farnsworth' camera, the *image dissector*, as ordinary jars that housewives used to preserve fruit. Farnsworth' achievements did not go unnoticed. In 1930 RCA tried to acquire "Television Laboratories Inc.," an offer that was declined by George Everson. After a brief liaison with Philco, Farnsworth continued on his own account with the Farnsworth Television Inc. The company soon became entangled in a patent conflict with RCA, which argued that Zworykin's iconoscope was the first camera.

Commercialization of television in the U.S. was delayed by the fight over patents. Farnsworth could not license his patents in the U.S. as long as his patents were under contest. In 1934 Farnsworth received an invitation from Baird Television for a patent license in England. Baird was at that stage forced by his financial backer, British Gaumont, to turn his mechanical television into an electronic system, because BBC had advised him to stop working on his experiments. With the money from this license Farnsworth could continue the perfection of his television system.

Electronic TV broadcasts took place in the U.S. on a limited scale after 1932. A number of stations that had employed mechanical equipment continued with electronic. RCA, which had broadcast with mechanical and shortly with a hybrid mechanical/electronic system, in 1933 set up a network of transmitters for all-electronic television. In this way the company offset the limited VHF signal range. In 1935 AT&T produced the coaxial cable, the first of which was laid between New York and Philadelphia. Nevertheless, electronic television made slow progress over the next few years. This was partly caused by the fact that companies such as Philco and Dumont, which were not yet able to produce a complete system themselves, convinced the Federal Communications Commission (FCC), the successor of the FRC, that no consensus yet existed within the industry on an "ideal" television system. It was not until Farnsworth entered into a cross-licensing agreement with AT&T in 1937 that could have left RCA out of commercial television altogether, that RCA and Farnsworth began the negotiation of cross-licensing agreements. In 1939 RCA was obliged to reach a licensing agreement with Farnsworth before they could market their television. In the meantime the agreement between AT&T and Farnsworth convinced the Columbia Broadcasting System (CBS) that commercial television had become viable.

In 1939 television became "official" when President F.D. Roosevelt initiated the first television broadcast at the New York World's Fair. In the next week the first television sets went on sale. Commercial television was within reach, but the World's Fair only started a new phase of experimentation in which the public was to be included through the sale of a handful of receivers. Not even RCA had permission to sell commercial time to advertisers until the FCC decided on standardization. However, the FCC delayed standardization since it demanded unanimity within the industry. FCC's concern for consensus became the catalyst for the Federal Government to establish the National Television System Committee (NTSC), which operated under the auspices of the branch organization of the industry, the Radio Manufacturer's Association. The NTSC reached consensus soon after the outbreak of World War II, and in 1941 standards were established. But now the war would prevent the commercialization of television.

Social Shaping of Television in the United States

The U.S. media infrastructure differed greatly from the British, a situation dating from the radio age. In the early years of radio development, various companies energetically competed for and contested patent and broadcasting rights. In setting up NBC in 1926, RCA brought a number of these companies together; the radio stations belonging to the RCA, General Electric, Westinghouse, and AT&T were now operated by a single organization. AT&T was now responsible only for the transmission itself.

As had also been the case for radio, mechanical television equipment was produced by a number of different manufacturers, including Jenkins, AT&T, and General Electric. A number of these companies also undertook broadcasting activities; one of these was the RCA's own NBC. Most of the manufacturers saw mechanical television as a broadcasting medium, but AT&T concentrated on the potential for a "picture phone."

During the radio age, government and industry had reached an agreement embodied in the Radio Act of 1927. Government's primary aim had been to prevent the formation of monopolies in the broadcasting, programming, and television manufacturing industries. Secondly, it sought standardization, so that every radio could receive every station. It also expressed the view that the development of radio should remain a matter of private initiative; radio manufacturers and broadcasting companies should finance themselves. The governmental Federal Radio Commission was formed to confer licenses and to allocate radio frequencies, but was not to concern itself with the canvassing of advertising, nor with most aspects of programming. With the subsequent introduction of television, the FRC — later to become the FCC — was also the regulatory body.

Television broadcasts were dominated by singing, cabaret, educational programs (which included, for instance, a course on bookbinding) provided by the universities, and the occasional film or sporting contest. In Chicago a television station was run by a newspaper together with a television manufacturer.

It televised little news, though; most of the programming consisted of cartoons.

In the U.S., too, amateur television receivers formed a significant social group. Special DIY televisions, which were considerably cheaper than the electronic sets, were brought onto the market.

What specific problems were faced by the various social groups using mechanical television in the U.S.? The FRC was faced with three problems; first, it was unsatisfied with the image quality and therefore only conferred temporary licenses. Secondly, there was still no transmission standard. Different stations employed different standards with regard to the number of lines and the number of images per second. In choosing between different standards, television buyers ended up effectively choosing between stations. Later the FCC therefore advocated standardization. Thirdly, station frequencies regularly overlapped. Here, too, the FRC wanted to sort things out.

For the broadcasting companies, the main problem was one of income. A temporary license did not allow a station to broadcast advertisements, and this denied them an important source of income. A third social group, the actors, had the same difficulties with the lighting as they did in the U.K. Finally, for the consumer public, the absence of technical standards and the overlapping of broadcast frequencies were the greatest problems.

How did this compare with the situation for electronic television? For the most important manufacturers of electronic television (RCA, Farnsworth, and Philco) patent rights formed a problem; as we have seen, RCA was eventually obliged to come to a licensing agreement with Farnsworth. The FCC, on the other hand, had to decide on certain technical standards that would unavoidably hinder the introduction of later technical improvements. Nevertheless, after the establishment of the NTSC, in 1941 a set of standards was agreed to. In other countries, such standards were only drawn up much later, and this explains why to this day the U.S. has a technically less sophisticated television system.

> Social groups, according to the meaning they attribute to the technology, can influence the course of its development.

The broadcasters employing electronic television were by and large the same companies that had employed mechanical television. The main problem was the limited VHF transmission range. To begin with, reception was limited to the larger cities, and only after 1933 was a solution provided by a wider network of transmitters. By the end of the 1930s broadcasters and manufacturers joined forces after the FCC, in order to promote technical improvements, decided to confer broadcasting licenses only to those companies actively engaged in research into such improvements.

Since from the mid-1930s onwards broadcasters were allowed to carry advertisements, advertisers gradually became another important social group involved in electronic television. Initially, however, low viewing figures made these advertisers cautious.

The viewing public, meanwhile, was pleased that electronic television made it possible to broadcast events that took place outside the studio: sport and news. However, the absence of transmission standards was a problem for consumers, as in all probability was the expense of an electronic television set: in 1939 a set cost between $200 and $600. Again, part of the mechanical television viewing public was not reached by electronic television broadcasts. The conviction that areas outside the large cities should also have the benefit of television played a large part in the FCC's decision to allow the three universities that employed mechanical television to continue doing so for many years.

Finally, American amateur enthusiasts had the same problems with electronic television as did their British counterparts: its overwhelming technical complexity and the greater technical differences with ordinary radio reception.

Mechanical Television in the Netherlands: Hat Shows

For a short time, mechanical television enjoyed a certain popularity in the Netherlands, although this was less the case than in either the U.K. or the U.S.[8] Just as had been the case in these two countries, it was an individual pioneer, the Eindhoven radio amateur Freek Kerkhof, who got Dutch television off the ground. In 1924 he decided to develop a television system using Nipkow discs, but it was 1927 before he could demonstrate his ten-line system, and little more than shadowy images could be discerned. Philips followed developments in image transmission from 1925 onward. Between 1928 and 1931, radio and newspapers showed considerable interest in mechanical television.

---

[8]A small number of Dutch publications have appeared on the development of mechanical and electronic television. Sources for the study of mechanical television are [14]–[16]; see also [17].

Philip's first experiments date from this time. Although its research laboratory's directors had resisted mechanical television from the start, under commercial pressure they presented, in 1928, a television image made up of 48 lines, employing Nipkow discs, a photocell, and a neon lamp. The device was able to radiotransmit an image of static objects a distance of 400 m. Philips had two aims in mind for these demonstrations of mechanical television: first to show that it was interested in the medium, and second to demonstrate its shortcomings. For all its sophistication, the Nipkow television was a large, cumbersome and susceptible machine and Philips saw little future for it in Dutch households.

In the first half of the 1930s, Dutch broadcasters showed little interest in the medium of television. One, the Vereniging van Arbeiders en Radio Amateurs (VARA), was an exception; in 1931 it transmitted an experimental broadcast using a thirty-line machine obtained through the German television manufacturer Telehor AG. The 3 cm by 4 cm viewing screen was enlarged by means of an enormous lens. The broadcast lasted 15 min and was accompanied by widespread publicity. Radio magazines and newspapers described the limitations that thousands had witnessed for themselves. The portrait of a woman, they reported, had been indistinct and shaky; "one moment she was visible, and the next moment it looked more like a film of railway tracks shaken by the passage of a highspeed train" [18]. Not surprisingly, the public was warned not to expect too much of the new medium.

Kerkhof, in the meantime, continued his experiments. By 1935 he was able to demonstrate a thirty-line system, comprising a mechanical mirrordisc camera and three receivers, to the Dutch National Society of Radio Amateurs. The program showed Mrs. Kerkhof, busily displaying one hat after another. This demonstration was followed up in 1936 when Kerkhof started regular Sunday morning transmissions for Eindhoven amateurs using homemade 30-line receivers. A year later Kerkhof had built a second transmitter for the accompanying sound signal. He had also begun using cathode ray tubes in his cameras, which resulted in a considerably improved image contrast. After other improvements were made in the studios, he was able to start transmissions employing well-known artists, and these broadcasts continued until his transmission license was withdrawn as the result of Dutch war mobilization in 1939.

Electronic Television in the Netherlands: Philips and the Government

Philips began building an experimental electronic television system in the 1930s.[9] Their research laboratory experimented with camera recordings in the studio, outside broadcasts, and with film shows. In 1937 Philips began to give public demonstrations; systems were available using 405 and 567 lines. In the Utrecht trade fair held in the spring of 1938, Philips demonstrated a number of 405-line television sets and also a large-screen set called the *protelgram*. The yellow-reddish image of the mechanical TV set had now given way to the sepia-greenish image of the electronic ones.

For all this, Philips was not actually very interested in television, a view the company retained right up to the 1940s. Philips directors saw the limited range of VHF transmitters and the high cost of television sets as serious obstacles to their successful marketing. Neither did they expect the public to be satisfied with a blurred image measuring only three by seven centimeters. And Philips did not expect the glass industry to suddenly come up with a large enough image tube.

Philips concentrated on two alternatives to television: the *huiscineac* and the above-mentioned *protelgram*, which was first shown at the Radiolympia in England in 1937. With the *huiscineac* (literally, "home cinema") viewers could project films at home, but the process was ultimately never exploited commercially. With the *protelgram*, Philips was reacting to the fact that many people found the picture of the first generation receiver disappointingly small: the *protelgram* projected a small, intense electronic screen image, by means of a large spherical mirror and a correcting lens, onto a screen or wall [22]. The image was then 40 cm by 50 cm, but its clarity left much to be desired. In 1937 Philips sold several sets, but shortly afterwards was proposing to take them back because the tube life was unacceptably low.

Philips also suffered an unpleasant and unexpected side effect from its experiments with television; the "picture radio" led to a stagnation in demand for ordinary radios [17]! In 1938, Philips therefore embarked on a rather unusual campaign; staff traveled throughout Europe ostensibly demonstrating televisions but actually concentrating on promoting the sale of radios. For Philips too, the start of the World War II, put a lengthy stop to further activities [23].

The radio broadcasting companies had not ignored these developments. In 1935, three of these

---

[9]The section on Dutch electronic television is primarily based on [19]–[21].

companies submitted an application to jointly run a television station. This application led to the formation of the Television Commission, which was to seek information and to advise on the technical, legal, and economic aspects of mechanical and electronic television systems. The existing administrative structure with regard to radio served as a guideline. The Commission was composed of broadcasters and the PTT, the Dutch Post Office, which had a role in the provision of communication cables.

The Commission examined the situation in various countries, and it watched experimental transmissions of the German Olympic Games. Its interim report expressed reservations about costs and program content, and the Commission felt that the difficulties were serious enough to warrant deferring a regular television service for the time being. However, the rapid course of developments abroad made an experiment with electronic television imperative, a conclusion which was repeated in the Commission's definitive report published in December 1937. Broadcasters and the PTT would take care of the technical side of such an experimental transmission; television receivers would be situated in public places, with the expectation that they would later appear in private houses. Philips declared itself prepared to supply apparatus, and after the broadcasters had acknowledged themselves prepared to make the programs and foot the bill for the transmissions, the Commission announced that everything was ready for a trial television service. At this point, World War II intervened and the experiment was postponed.

After the war, the various groups involved took matters up once more. Philips still had the American marketing of its own *protelgram* high on its agenda; the company was still convinced that television would spread only slowly in the Netherlands. Nevertheless by 1947 Philip's experimental research laboratory Natlab had produced a television transmitter, and its receiver was almost ready for mass manufacture. When a Philips manager in America sent a telegram strongly urging his colleagues to increase their television activities (the *protelgram* was all very well, but it looked like the future lay in television), Philips changed course [24].

The company was given a license to transmit within a radius of 40 km, and began experimental broadcasts to a select public. Three times per week Philips employees, local worthies, and radio dealers were treated to ninety minutes of television. Philips television receivers operated at a standard 625 lines, in contrast to France's 819 lines and Britain's 405 [25, 26].

Together with the PTT, Philips began to lobby government for permission to start more frequent broadcasts. One of its arguments was that a good domestic market was crucial to the sale of television sets abroad. However, party political and broadcasting companies' reservations held back the introduction of an experimental television service until 1951.

Social Shaping of Television in the Netherlands

In the Netherlands too, various social groups and organizations influenced the development of both mechanical and electronic television: television amateurs, broadcasting companies (especially the VARA), Philips, government, and the PTT. As elsewhere, mechanical television was most important to the amateur enthusiasts. Kerkhof called for their comments on the programming. Ultimately, however, they formed no important pressure group in defense of mechanical television, and this had much to do with Kerkhof's attitudes. Unlike Baird, Kerkhof did not try to wring commercial profit out of television but aimed only to provide the best possible quality given the means available to amateurs.

> Electronic television was more suited to the broadcasting of live outdoor events, where good image quality was vital.

Broadcasting companies were quick to lay claim to television, perhaps in order to prevent the medium from being stolen by other organizations. The VARA's 1931 experimental broadcasts gave an important boost to the development of mechanical television, but as a whole the broadcasting companies' involvement in these experiments was short-lived. The same can be said of Philips; though the company began experiments with mechanical television in the 1920s, it appears that these experiments were not considered of great importance. Government involvement was represented by the formation of the first Television Commission, representing the broadcasting companies and the PTT, and which appears to have been equally unenthusiastic about mechanical television.

By and large, the same groups were involved in the growth of electronic television. Philips was still barely interested; when the company began demonstrating this equipment in the late 1930s the main purpose was to stimulate the sale of ordinary radios. Moreover, Philips was principally interested in a medium for the domestic reproduction of large-screen films, and was therefore more preoccupied with its *protelgram*. Still, its apparent disinterest in television ought not to be exaggerated; a significant portion of Philips' patent applications in 1934 and 1935 had to do with television.

In an important sense radio was the medium of the 1930s, and since the same parties were involved this was reflected in the development of both mechanical and electronic television. Practically everyone was content to wait and see what television had to offer; this also had much to do with the limited television programming of the time. The widespread opinion was that if television had anything to offer beyond adding images to a radio signal, it was the showing of films. Neither did Philips' attitude do anything to stimulate the further development of mechanical television.

After the Second World War the Netherlands opened an intense debate on the introduction of electronic television. Most of the parties who had been involved before the war reappeared and it was remarkable how many of them had changed their minds completely in the meantime. Philips' vacillation had been replaced by an almost aggressive approach, after the company had been put onto the scent by developments abroad. Having also gained the support of the PTT, Philips argued in terms of employment opportunities, export opportunities, and national prestige, and that these would first require a secure position for Philips in the Dutch market. The broadcasting companies, too, had modified their views; they were prepared, primarily from a protective standpoint, to work with experimental television broadcasts.

Government adopted no very clear position but did raise a number of financial and cultural objections. Political and broadcasting representatives (who in the Dutch social system were already strongly aligned) shared a certain hesitancy with respect to television which yielded to the pressure being generated by Philips, the PTT, and Nozema, a regulatory government installed office. Administrative structures that had evolved in the age of radio went on to determine the way in which television was used; indeed, in programming terms, television was initially little more than radio with pictures.

Social Factors an Influence

In considering the developments recounted here, the choice between mechanical and electronic television can be perceived as having been the result of a combination of technical and social factors. Technical factors had to do with the advantages and disadvantages of each system. Mechanical television was cheap and TV sets could be constructed by amateurs themselves, all sorts of technical extensions (such as the use of color, image recordings, and large screens) were easy to put into practice, and as a consequence of low information density, the transmitters could have a long range. However, the image quality was poor and the system was best suited to studio work and film shows; it proved impossible to develop a practical camera for outside broadcasts. Electronic television on the other hand had a high image quality and was well suited to outside broadcasts, but the technique was expensive and not within easy reach of amateurs. Moreover, electronic television could only exploit its advantages in image quality if a higher number of image lines was applied. This required transmission on a VHF bandwidth with a low geographical range.

The choice between these two systems cannot, however, be attributed to these technical factors alone. Social factors also played a role. Different meanings were attributed to television by the various agents in its development: some wanted to see it used for public broadcasts, others for "picture phones." Its use in broadcasting was itself subject to differing appraisals: some saw it as a medium for film shows, others saw its potential as a news broadcaster, and still others noted its educational possibilities. In a social process involving a number of different social groups and organizations, the use of television for broadcasting purposes became predominant. With regard to programming, live events, and sports contests in particular, seem to be considered important. This appears, for instance, in the use of the 1936 Olympic Games in Berlin as a test between the two television systems in Germany. Electronic television was more suited to the broadcasting of such live events, as good image quality was vital and these events normally took place outdoors.

However, had another set of connotations predominated, for instance if television had been a medium of studio recordings and video telephony alone, then mechanical television would have stood a better chance. Certainly, in such applications its image quality would not have reached that of the electronic television, but the difference would have been smaller in these applications, and its advantages in other respects would probably have improved its overall acceptance. Although mechanical television had the advantage of a head start on electronic television, its early success largely resulted from the fact that studio recordings and video telephone were originally perceived as important purposes. Had there been a need for transmissions with worldwide reception (a need which is now met by satellites), mechanical television would also have had advantages. It is possible that other interpretations also played a part in the eventual choice for electronic television; actors, we may recall, had many difficulties with the strong lighting required by mechanical television.

In other words: internalist historical accounts take the importance of the image quality criterion for granted. In contrast to this, our approach to the early history of television reveals that in the course of a social process, image quality and the possibility of making outside broadcasts became dominant criteria.

The difference between historical developments in Great Britain and the United States must be attributed to the financial structures of the broadcasting organizations in each country. In the U.S., broadcasting companies had to meet their costs by selling advertising. Since they were given only experimental transmission licenses which stipulated that no advertising could be carried, they were deprived of this income and many eventually folded. In Britain, the BBC had a more secure future and the development of mechanical television became part of the national interest. Another explanation for the difference may lie in the fact that American mechanical television was faced with more problems, thanks to the large number of broadcasting companies, the lack of standardization, and limited radio frequencies. The reason that American universities continued to use mechanical television for so long is that it suited their purposes reasonably well; they broadcast mainly studio programs.

While the Netherlands was no television pioneer, it was no slow starter either. Early on, a mechanical system was developed that remained in use until 1939. Later, Dutch television fell behind the field. Philips, the most important Dutch electrical goods manufacturer, was only moderately interested in electronic television, and until 1947 it was more concerned with the alternatives to television. Developments abroad finally persuaded Philips to change tack.

A note should be made of the relationship between television and its predecessor, radio. We have seen that the organizations, set up by government and industry in all three countries, responsible for supervising radio broadcasts, were also given (or assumed) responsibility for television. Television did not develop in a political or industrial vacuum. Significantly, the organizational structure of radio was simply grafted onto television, though the technical system chosen was one less compatible with radio as far as frequency and type of technology was concerned.

Overall, our broad, technical-social perspective gives a better understanding of the early history of television than earlier one-sided approaches. Our analysis reveals, for example, that screen image quality and the possibility of making outside broadcasts became dominant criteria in the course of a social process. However, our analysis also shows that we cannot rub out technical factors, as some new approaches tend to do, thus in fact introducing a new type of bias. Perhaps, researchers busy in constructing new approaches should not polish off the older ones too soon, but try to find more embracing perspectives.

References

1. J. Van den Ende, W. Ravesteijn, D. de Wit, "Waarom geen Nipkowschijf in elk huiskamer? De sociale constructie van televisie," *Het Jaarboek Mediageschiedenis*, vol. 5, pp. 131–161, 1993. (Published by Stichting Film en Wetenschap, Zeeburgerkade 8, 1019 HA Amsterdam, the Netherlands).

2. S. Shapin, "Discipline and bounding: The history and sociology of science as seen through the externalism-internalism debate," *History of Science*, vol. 30, pp. 333–369, 1992.

3. R. Williams, *Television, Technology and Cultural Form*. London: Fontana, Collins, 1974.

4. W.E. Bijker, "Sociohistorical technology studies," in *Handbook of Science and Technology Studies*, S. Jasanoff, G.E. Markle, J.C. Peterson, and T. Pinch, Eds. Thousand Oaks/London/New Delhi, 1995. p. 255.

5. J. Udelson's, *The Great Television Race, A History of the American Television Industry 1925–1941*. AL: Univ. of Alabama Press, 1982.

6. W.E. Bijker, T.P. Hughes, and T. Pinch, Eds., *The Social Construction of Technological Systems*. Cambridge, MA: M.I.T. Press, 1987, pp. 41–42.

7. A. Abramson, *The History of Television, 1880–1941*. Jefferson, 1987.

8. *The History of Television from the Early Days to the Present*. London, IEE Conf. Pub. 271, 1986.

9. R.W. Burns, *British Television, The Formative Years*. London: Peregrinus, 1986.

10. T. McArthur and P. Waddell, *The Secret Life of John Logie Baird*. London: Hutchinson, 1986.

11. J. Swift, *Adventure in Vision, the First Twenty-five Years of Television*. London: Lehmann, 1950.
12. L.W. Lichty and M.C. Topping, Eds., *American Broadcasting*. New York: Hastings House, 1975.
13. G. Shiers, Ed., *Technical Development of Television*. New York: Amo, 1977.
14. W.J. de Gooijer, *Beheersing van technologische vernieuwing*. Alphen a/d Rijn, 1976.
15. J. Libbenga, "Tv or not tv, veertig jaar Nederlands televisie," in *Intermediair*, vol. 39, pp. 38–43, 1991.
16. G. Bekooy, *Philips honderd, een industriële onderneming*. Europese bibliotheek, 1991.
17. Jan Wieten, "Vuistslagen op peluws, de valse start van de televisie in Nederland," in *Het Jaarboek Mediageschiedenis* vol. 5, pp. 163–197, 1993.
18. E. Smulders, "Het Wonder van Morgen, de televisierage in Nederland 1928–1931," Rotterdam, Master's thesis, 1993, p. 56.
19. P. Gros, *Televisie, Parlement, Pers, Publiek*. Assen: Van Gorcum, 1960.
20. H. Wijfjes, *Hallo hier Hilversum, driekwart eeuw radio en televisie*. Weesp, 1985.
21. W. Knulst, *Van vaudeville tot video*. Alphen a/d Rijn, 1989.
22. K. Geddes and G. Bussey, *The Setmakers, A History of the Radio and Television Industry*. London: Brema, 1991.
23. L. van der Linden, "Vijftig jaar kastje kijken," in *De Ingenieur*, vol. 9, pp. 11–19, 1985.
24. H. Beunders, "De prehistorie van de televisie," in *NRC Handelsblad*, Sept. 28, 1991.
25. F. Kerkhof and W. Werner, *Televisie*. Deventer, 1951.
26. N.Tj. Swierstra, *Tien jaar televisie in Nederland*. Eindhoven: Philips, 1958.
27. W.E. Bijker, *Of Bicycles, Bakelites and Bulbs. Towards a Theory of Sociotechnical Change*. Cambridge, MA, 1995.
28. P.C. Carbonara, "A historical perspective of management, technology and innovation in the American television industry," Ph.D. diss., Univ. of Texas, Austin, 1989.
29. D. de Wit, *The Shaping of Automation, a Historical Analysis of the Interaction Between Technology and Organization, 1950–1985*. Hilversum, pp. 53–56, 1994.
30. J. van den Ende, *The Turn of the Tide. Computerization in Dutch Society, 1900–1965*. Delft, 1994.

The Personal Computer

Paul Ceruzzi

Ready or not, computers are coming to the people.

That's good news, maybe the best since psychedelics.

Those words introduced a story in the fifth anniversary issue of *Rolling Stone* (December 7, 1972). "Spacewar: Fanatic Life and Symbolic Death Among the Computer Bums" was written by Stewart Brand, a lanky Californian who had already made a name for himself as the publisher of the *Whole Earth Catalog*. Brand's resumé was unique, even for an acknowledged hero of the counterculture. At Stanford in the 1960s, he had participated in Defense Department-sponsored experiments with hallucinogenic drugs. In 1968 he had helped Doug Engelbart demonstrate his work on interactive computing at a now-legendary session of the Fall Joint Computer Conference in San Francisco. Brand was no stranger to computers or to the novel ways one might employ them as interactive tools.

Brand was right. Computers did come to the people. The spread of computing to a mass market probably had a greater effect on society than the spread of mind-altering drugs. Personal computing, however, did not arrive in the way that Brand—or almost anyone else—thought it would. The development of personal computing followed a trajectory that is difficult to explain as rational. When trying to describe those years, from 1972 through 1977, one is reminded of Mark Twain's words: "Very few things happen at the right time, and the rest do not happen at all. The conscientious historian will correct these defects." This chapter will examine how computers came "to the people," not as Twain's historian would have written it, but as it really occurred.

What triggered Brand's insight was watching people at the Stanford Artificial Intelligence Laboratory playing a computer game, Spacewar. Spacewar revealed computing as far from the do-not-fold-spindle-or-mutilate punched-card environment as one could possibly find. The hardware they were using was not "personal," but the way it was being used was personal: for fun, interactively, with no concern for how many ticks of the processor one was using. That was what people wanted when, two years later, personal computers burst into the market.

Spacewar was running on a PDP-10. In terms of its hardware, a PDP-10 had nothing in common with the personal computers of the next decades. It was large—even DEC's own literature called it a mainframe. It had a 36-bit word length. A full system cost around a half million dollars and easily took up a room of its own. It used discrete transistors and magnetic cores, not integrated circuits, for logic and memory. Still, one can think of the PDP-10 as an ancestor of the personal computer. It was designed from the start to support interactive use. Although its time-sharing abilities were not as ambitious as those of MIT's Project MAC, it worked well. Of all the early time-sharing systems, the PDP-10 best created an illusion that each user was being given the full attention and resources of the computer. That illusion, in turn, created a mental model of what computing could be—a mental model that would later be realized in genuine personal computers.

Chapter 5 discussed the early development of time-sharing and the selection of a General Electric computer for Project MAC at MIT. While that was going on, the MIT Artificial Intelligence Laboratory obtained a DEC PDP-6, the PDP-10's immediate predecessor, for its research. According to the folklore, MIT students, especially members of the Tech Model Railroad Club, worked closely with DEC on the PDP-6, especially in developing an operating system for it, which would later have an influence on the PDP-10's system software. As a pun on the Compatible Time Sharing System that was running on an IBM mainframe nearby, the students called their

"The Personal Computer, 1972–1977," from *A History of Modern Computing*, by Paul E. Ceruzzi, pp. 207–241, text only, © 1998 Massachusetts Institute of Technology, by permission of The MIT Press.

PDP-6 system ITS—Incompatible Time Sharing System. The PDP-6 did not have the disk storage necessary to make it a viable time-sharing system and only about twenty were sold. The PDP-10 did have a random-access disk system, which allowed its users direct access to their own personal files. Like other DEC computers, the PDP-10 also allowed users to load personal files and programs onto inexpensive reels of DECtape, which fitted easily into a briefcase.

The feeling that a PDP-10 was one's own personal computer came from its operating system—especially from the way it managed the flow of information to and from the disks or tapes. With MIT's help, DEC supplied a system called "TOPS-10," beginning in 1972. In the introduction to the TOPS-10 manual, the authors stated, "Our goal has always been that in a properly configured system, each user has the feeling that he owns his portion of the machine for the time he needs to use it." Users could easily create, modify, store, and recall blocks of data from a terminal. The system called these blocks by the already-familiar term, "files." Files were named by one to six characters, followed by a period, then a three-character extension (which typically told what type of file it was, e.g.: xxxxxx.BAS for a program written in BASIC). By typing DIR at a terminal users could obtain a directory of all the files residing on a disk. They could easily send the contents of a file to a desired output device, which typically consisted of a three-letter code, for example, LPT for line printer, or TTY for Teletype.

A small portion of TOPS-10 was always present in core memory. Other programs were stored on the disk and could be called up as necessary. One, called PIP (Peripheral Interchange Program), allowed users to move files in a variety of ways to and from input/output equipment. Another program, TECO (Text Editor and Corrector), allowed users to edit and manipulate text from a terminal. DDT (Dynamic Debugging Tool) allowed users to analyze programs and correct errors without going through the long turnaround times that plagued batch processing.

For PDP-10 users, TOPS-10 was a marvel of simplicity and elegance and gave them the illusion that they were in personal control. TOPS-10 was like a Volkswagen Beetle: basic, simple, and easy to understand and work with. Using a PDP-10 was not only fun but addictive. It was no accident that Brand saw people playing Spacewar on one, or that it was also the computer on which Adventure—perhaps the most long-lasting of all computer games—was written.

On the West Coast another system appeared with similar capabilities, the SDS-940, offered by Scientific Data Systems (SDS) of southern California. The 940 was an extension of a conventional computer, the SDS 930, modified by researchers at Berkeley with support from the Defense Department's Advanced Research Projects Agency. The 940 was more polished than the PDP-10, and it performed well. Still, the PDP-10 seemed to be preferred. At the Xerox Palo Alto Research Center, the legendary lab where so much of personal computing would be created, the staff was encouraged to use SDS machines, since Xerox had just purchased SDS. But the researchers there resisted and instead built a clone of a PDP-10, which they called MAXC—Multiple Access Xerox Computer—the name a pun on Max Palevsky, the founder of SDS. (Palevsky, after becoming very wealthy from the sale of SDS to Xerox, dabbled in Hollywood movies, politics, and culture—and joined the board of *Rolling Stone*. Palevsky also became a venture capitalist with that money, helping to fund Intel, among other companies.)

For a while, when Wall Street was enamored of anything connected with computers, it was easy to raise money to buy or lease a PDP-10 or SDS-940, and then sell computer time to engineering companies or other customers. Most of these firms were undercapitalized and did not understand the complexities of what they were selling. Like their counterparts in the electric utility industry, they had to have enough capacity to handle peak loads, in order not to discourage customers. But that meant that during off-peak times they would be wasting unused and expensive computing equipment. The capital requirements necessary to manage the cycles of the business were as large as they were in the electric power business, which had gone through decades of chaos and turmoil before settling down. Only a few survived, and even fewer, like Tymshare of Cupertino, California, did well (although it was sold to McDonnell-Douglas in the early 1980s). Among those many companies, one is worth mentioning, Computer Center Corporation, or C-Cubed, which installed one of the first PDP-1Os in the Seattle area in 1968. While it was getting started, it offered a local teenager, Bill Gates, free time on the computer in exchange for helping find and rid the system of bugs. C-Cubed folded in 1970, having given Gates a taste of the potential of interactive computing.

Many of those who had access to these systems saw the future of computing. But the financial troubles of time-sharing companies also showed that it would be difficult to make personal, interactive use widely available. There were attempts to make terminals accessible to the public for free or at low cost—the most famous being the Resource One project in the San Francisco Bay area (partially funded by the *Whole Earth Catalog*). But it did not last, either.

Calculators and Corporate Personal Computer Projects

Economics prevented the spread of computing to the public from the top down—from large mainframes through time-shared terminals. But while those attempts were underway, the underlying technology was advancing rapidly. Could personal computing arrive from the bottom up—from advances in semiconductor electronics?

Many engineers believe that a mental model of the personal computer was irrelevant. They believe that no one invented the personal computer, it simply flowed from advances in semiconductors. Chuck House, an engineer involved with the early Hewlett-Packard calculators, said, "One could uncharitably say that we invented essentially nothing; we simply took all the ideas that were out there and figured out how to implement them cost-effectively." Gordon Bell stated, "The semiconductor density has really been the driving force, and as you reach different density levels, different machines pop out of that in time." To them, inventions are like a piece of fruit that falls to the ground when it is ripe, and the inventor is given credit for doing little more than picking it up. If that were true, one would find a steady progression of machines offering personal, interactive use, as advances in semiconductors made them viable. And these would have come from established firms who had the engineering and manufacturing resources to translate those advances into products.

Products that took advantage of advances in semiconductors did appear on the market. It is worth looking at them to see whether they validate or refute the bottom-up explanation of the PC's invention.

The first electronic computers were of course operated as if they were personal computers. Once a person was granted access to a machine (after literally waiting in a queue), he or she had the whole computer to use, for whatever purpose. That gave way to more restricted access, but those at MIT and Lincoln Labs who used the Whirlwind, TX-0, and TX-2 that way never forgot its advantages.

In 1962 some of them developed a computer called the LINC, made of Digital Equipment Corporation logic modules and intended for use by a researcher as a personal tool. A demonstration project, funded by the NIH, made sixteen LINCs available to biomedical researchers. DEC produced commercial versions, and by the late 1960s, about 1,200 were in use as personal computers. A key feature of the LINC was its compact tape drive and tapes that one could easily carry around: the forerunner of DECtape. The ease of getting at data on the tape was radically different from the clumsy access of tape in mainframes, and this ease would be repeated with the introduction of floppy-disk systems on personal computers. DEC also marketed a computer that was a combination of a LINC and a PDP-8, for $43,000. Although DECtape socn was offered on nearly all DEC's products, the LINC did not achieve the same kind of commercial success as the PDP-8 and PDP-11 lines of minicomputers.

Advances in chip density first made an impact on personal devices in calculators. For decades there had been a small market for machines that could perform the four functions of arithmetic, plus square root. In the 1950s and 1960s the calculator industry was dominated by firms such as Friden and Marchant in the United States, and Odhner in Europe. Their products were complex, heavy, and expensive. In 1964 Wang Laboratories, a company founded by An Wang, a Chinese immigrant who had worked with Howard Aiken at Harvard, came out with an electronic calculator. The Wang LOCI offered more functions, at a lower cost, than the best mechanical machines. Its successor, the Wang 300, was even easier to use and cheaper, partly because Wang deliberately set the price of the 300 to undercut the competitive mechanical calculators from Friden and others. (Only one or two of the mechanical calculator firms survived the transition to electronics.) A few years later Hewlett-Packard, known for its oscilloscopes and electronic test equipment, came out with the HP-9100A, a calculator selling for just under $5,000. And the Italian firm Olivetti came out with the Programma 201, a $3,500 calculator intended primarily for accounting and statistical work. Besides direct calculation, these machines could also execute a short sequence of steps recorded on magnetic cards. Like the LINC, these calculators used discrete circuits. To display digits, the Wang used "Nixie" tubes, an ingenious tube invented by Burroughs in 1957. HP used a small cathode-ray tube, as might be expected from a company that made oscilloscopes.

By 1970 the first of a line of dramatically cheaper and smaller calculators appeared that used integrated circuits. They were about the size of a paperback book and cost as little as $400. A number of wealthy consumers bought them immediately, but it wasn't until Bowmar advertised a Bowmar Brain for less than $250 for the 1971 Christmas season that the calculator burst into public consciousness. Prices plummeted: under $150 in 1972; under $100 by 1973; under $50 by 1976; finally they became cheap enough to be given away as promotional trinkets. Meanwhile Hewlett-Packard stunned the market in early 1972 with the HP-35, a $400 pocket calculator that performed all the logarithmic and trigonometric functions required by engineers and scientists. Within a few years the slide rule joined the mechanical calculator on the shelves of museums.

Like processed foods, whose cost is mostly in the packaging and marketing, so with calculators: technology no longer determined commercial success. Two Japanese firms with consumer marketing skills, Casio and Sharp, soon dominated. Thirty years after the completion of the half-million dollar ENIAC, digital devices became throw-away commodities. The pioneering calculator companies either stopped making calculators, as did Wang, or went bankrupt, as did Bowmar. Hewlett-Packard survived by concentrating on more advanced and expensive models; Texas Instruments survived by cutting costs.

The commodity prices make it easy to forget that these calculators were ingenious pieces of engineering. Some of them could store sequences of keystrokes in their memory and thus execute short programs. The first of the programmable pocket calculators was Hewlett-Packard's HP-65, introduced in early 1974 for $795. Texas Instruments and others soon followed. As powerful as they were, the trade press was hesitant to call them computers, even if Hewlett-Packard introduced the HP-65 as a "personal computer" (possibly the first use of that term in print). Their limited programming was offset by their built-in ability to compute logarithms and trigonometric functions, and to use floating-point arithmetic to ten decimal digits of precision. Few mainframes could do that without custom-written software.

The introduction of pocket programmable calculators had several profound effects on the direction of computing technology. The first was that the calculator, like the Minuteman and Apollo programs of the 1960s, created a market where suppliers could count on a long production run, and thereby gain economies of scale and a low price. As chip density, and therefore capability, increased, chip manufacturers faced the same problem that Henry Ford had faced with his Model T: only long production runs of the same product led to low prices, but markets did not stay static. That was especially true of integrated circuits, which by nature became ever more specialized in their function as the levels of integration increased. (The only exception was in memory chips, which is one reason why Intel was founded to focus on memories.) The calculator offered the first consumer market for logic chips that allowed companies to amortize the high costs of designing complex integrated circuits. The dramatic drop in prices of calculators between 1971 and 1976 showed just how potent this force was.

The second effect was just as important. Pocket calculators, especially those that were programmable, unleashed the force of personal creativity and energy of masses of individuals. This force had already created the hacker culture at MIT and Stanford (observed with trepidation by at least one MIT professor). Their story is one of the more colorful among the dry technical narratives of hardware and software design. They and their accomplishments, suitably embellished, have become favorite topics of the popular press. Of course their strange personal habits made a good story, but were they true? Developing system software was hard work, not likely to be done well by a salaried employee, working normal hours and with a family to go home to in the evening. Time-sharing freed all users from the tyranny of submitting decks of cards and waiting for a printout, but it forced some users to work late at night, when the time-shared systems were lightly loaded and thus more responsive.

The assertion that hackers created modern interactive computing is about half-right. In sheer numbers there may never have been more than a few hundred people fortunate enough to be allowed to "hack" (that is, not do a programming job specified by one's employer) on a computer like the PDP-10. By 1975, there were over 25,000 HP-65 programmable calculators in use, each one owned by an individual who could do whatever he or she wished to with it. Who were these people? HP-65 users were not "strange". Nearly all were adult professional men, including civil and electrical engineers, lawyers, financial people, pilots, and so on. Only a few were students (or professors), because an HP-65 cost $795. Most purchased the HP-65 because they had a practical need for calculation in their jobs. But this was a *personal* machine—one could take it home at night. These users—perhaps 5 or 10 percent of those who owned machines—did not fit the popular notion of hackers as kids with "[t]heir rumpled clothes, their unwashed and unshaven faces, and their uncombed hair." But their passion for programming made them the intellectual cousins of the students in the Tech Model Railroad Club. And their numbers—only to increase as the prices of calculators dropped—were the first indication that personal computing was truly a mass phenomenon.

Hewlett-Packard and Texas Instruments were unprepared for these events. They sold the machines as commodities; they could ill-afford a sales force that could walk a customer through the complex learning process needed to get the most out of one. That was what IBM salesmen were known for—but they sold multimillion dollar mainframes. Calculators were designed to be easy enough to use to make that unnecessary, at least for basic tasks. What was unexpected was how much more some of those customers wanted to do. Finding little help from the supplier, they turned to one another. Users groups, clubs, newsletters, and publications proliferated.

This supporting infrastructure was critical to the success of personal computing; in the following decade it would become an industry all its own. Many histories of the personal computer emphasize this point; they often cite the role of the Homebrew Computer Club, which met near the Stanford campus in the mid-1970s, as especially important. The calculator users groups were also important, though for different reasons. As the primitive first personal computers like the Altair gave way to more complete systems, a number of calculator owners purchased one of them as well. In the club newsletters there were continuous discussions of the advantages and drawbacks of each—the one machine having the ability to evaluate complex mathematical expressions with ease, the other more primitive but *potentially* capable of doing all that and more. There was no such thing as a typical member of the Homebrew Computer Club, although calculator owners tended to be professionals whose jobs required calculation during the day, and who thought of other uses at night. Many of them were bitten by the PC bug; at the same time they took a show-me attitude toward the computer. Could you rely on one? Could you use one to design a radar antenna? Could it handle a medium-sized mailing list? Was the personal computer a serious machine? At first the answers were, "not yet," but gradually, with some firm prodding by this community, the balance shifted. Groups like the Homebrew

Computer Club emphasized the "personal" in personal computer; calculator users emphasized the word computer.

Ever since time-sharing and minicomputers revealed an alternative to mainframe computing, there have been prophets and evangelists who raged against the world of punched cards and computer rooms, promising a digital paradise of truly interactive tools. The most famous was Ted Nelson, whose self-published book *Computer Lib* proclaimed (with a raised fist on the cover): "You can and must understand computers *now*." By 1974 enough of these dreams had become real that the specific abilities—and limits—of actual "dream machines" (the alternate title to Nelson's book) had to be faced. Some of the dreamers, including Nelson, were unable to make the transition. They dismissed the pocket calculator. They thought it was puny, too cheap, couldn't do graphics, wasn't a "von Neumann machine," and so on. For them, the dream machine was better, even if (or because) it was unbuilt. By 1985 there would be millions of IBM Personal Computers and their copies in the offices and homes of ordinary people. These computers would use a processor that was developed for other purposes, and adapted for the personal computer almost by accident. But they would be real and a constant source of inspiration and creativity to many who used them, as well as an equal source of frustration for those who knew how much better they could be.

The Microprocessor

Calculators showed what integrated circuits could do, but they did not open up a direct avenue to personal interactive computing. The chips used in them were too specialized for numerical calculation to form a basis for a general-purpose computer. Their architecture was ad-hoc and closely guarded by each manufacturer. What was needed was a set of integrated circuits—or even a single integrated circuit—that incorporated the basic architecture of a general-purpose, stored-program computer. Such a chip, called a "microprocessor," did appear.

In 1964 Gordon Moore, then of Fairchild and soon a cofounder of Intel, noted that from the time of the invention of integrated circuits in 1958, the number of circuits that one could place on a single integrated circuit was doubling every year. By simply plotting this rate on a piece of semi-log graph paper, "Moore's Law" predicted that by the mid 1970s one could buy a chip containing logic circuits equivalent to those used in a 1950s-era mainframe. (Recall that the UNIVAC I had about 3,000 tubes, about the same number of active elements contained in the first microprocessor discussed below.) By the late 1960s transistor-transistor logic (TTL) was well established, but a new type of semiconductor called metal-oxide semiconductor (MOS), emerged as a way to place even more logic elements on a chip. MOS was used by Intel to produce its pioneering 1103 memory chip, and it was a key to the success of pocket calculators. The chip density permitted by MOS brought the concept of a computer-on-a-chip into focus among engineers at Intel, Texas Instruments, and other semiconductor firms. That did not mean that such a device was perceived as useful. If it was generally known that enough transistors could be placed on a chip to make a computer, it was also generally believed that the market for such a chip was so low that its sales would never recoup the large development costs required.

By 1971 the idea was realized in silicon. Several engineers deserve credit for the invention. Ted Hoff, an engineer at Intel, was responsible for the initial concept, Federico Faggin of Intel deserves credit for its realization in silicon, and Gary Boone of Texas Instruments designed similar circuits around that time. In 1990, years after the microprocessor became a household commodity and after years of litigation, Gil Hyatt, an independent inventor from La Palma, California, received a patent on it. Outside the courts he has few supporters, and recent court rulings may have invalidated his claim entirely.

The story of the microprocessor's invention at Intel has been told many times. In essence, it is a story encountered before: Intel was asked to design a special-purpose system for a customer. It found that by designing a general-purpose computer and using software to tailor it to the customer's needs, the product would have a larger market.

Intel's customer for this circuit was Busicom, a Japanese company that was a top seller of hand-held calculators. Busicom sought to produce a line of products with different capabilities, each aimed at a different market segment. It envisioned a set of custom-designed chips that incorporated the logic for the advanced mathematical functions. Intel's management assigned Martian E. ("Ted") Hoff, who had joined the company in 1968 (Intel's twelfth employee), to work with Busicom.

Intel's focus had always been on semiconductor memory chips. It had shied away from logic chips like those suggested by Busicom, since it felt that markets for them were limited. Hoff's insight was to recognize that by designing fewer logic chips with more general capabilities, one could satisfy Busicom's needs elegantly. Hoff was inspired by the PDP-8, which had a very small set of instructions, but which its thousands of users had programmed to do a variety of things. He also recalled using an IBM 1620, a small scientific computer with an extremely limited instruction set that nevertheless could be programmed to do a lot of useful work.

Hoff proposed a logic chip that incorporated more of the concepts of a general-purpose computer. A critical feature was the ability to call up a subroutine, execute it, and return to the main program as needed. He proposed to do that with a register that kept track of where a program was in its execution and saved that status when interrupted to perform a subroutine. Subroutines themselves could be interrupted, with return addresses stored on a "stack": an arrangement of registers that automatically retrieved data on a last-in-first-out basis.

With this ability, the chip could carry out complex operations stored as subroutines in memory, and avoid having those functions permanently wired onto the chip. Doing it Hoff's way would be slower, but in a calculator that did not matter, since a person could not press keys that fast anyway. The complexity of the logic would now reside in software stored in the memory chips, so one was not getting something for nothing. But Intel was a memory company, and it knew that it could provide memory chips with enough capacity. As an added inducement, sales of the logic chips would mean more sales of its bread-and-butter memories.

That flexibility meant that the set of chips could be used for many other applications besides calculators. Busicom was in a highly competitive and volatile market, and Intel recognized that. (Busicom eventually went bankrupt.) Robert Noyce negotiated a deal with Busicom to provide it with chips at a lower cost, giving Intel in return the right to market the chips to other customers for noncalculator applications. From these unsophisticated negotiations with Busicom, in Noyce's words, came a pivotal moment in the history of computing.

The result was a set of four chips, first advertised in a trade journal in late 1971, which included "a microprogrammable computer on a chip!" That was the 4004, on which one found all the basic registers and control functions of a tiny, general-purpose stored-program computer. The other chips contained a read-only memory (ROM), random-access memory

(RAM), and a chip to handle output functions. The 4004 became the historical milestone, but the other chips were important as well, especially the ROM chip that supplied the code that turned a general-purpose processor into something that could meet a customer's needs. (Also at Intel, a team led by Dov Frohman developed a ROM chip that could be easily reprogrammed and erased by exposure to ultraviolet light. Called an EPROM (erasable programmable read-only memory) and introduced in 1971, it made the concept of system design using a microprocessor practical.)

The detailed design of the 4004 was done by Stan Mazor. Federico Faggin was also crucial in making the concept practical. Masatoshi Shima, a representative from Busicom, also contributed. Many histories of the invention give Hoff sole credit; all players, including Hoff, now agree that that is not accurate. Faggin left Intel in 1974 to found a rival company, Zilog. Intel, in competition with Zilog, felt no need to advertise Faggin's talents in its promotional literature, although Intel never showed any outward hostility to its ex-employee. The issue of whom to credit reveals the way many people think of invention: Hoff had the idea of putting a general-purpose computer on a chip. Faggin and the others "merely" implemented that idea in silicon. At the time, Intel was not sure what it had invented either: Intel's patent attorney resisted Hoff's desire at the time to patent the work as a "computer." Intel obtained two patents on the 4004, covering its architecture and implementation; Hoff's name appears on only one of them. (That opened the door to rival claims for patent royalties from TI, and eventually Gil Hyatt.)

The 4004 worked with groups of four bits at a time—enough to code decimal digits but no more. At almost the same time as the work with Busicom, Intel entered into a similar agreement with Computer Terminal Corporation (later called Datapoint) of San Antonio, Texas, to produce a set of chips for a terminal to be attached to mainframe computers. Again, Mazor and Hoff proposed a microprocessor to handle the terminal's logic. Their proposed chip would handle data in 8-bit chunks, enough to process a full byte at a time. By the time Intel had completed its design, Datapoint had decided to go with conventional TTL chips. Intel offered the chip, which they called the 8008, as a commercial product in April 1972.

In late 1972, a 4-bit microprocessor was offered by Rockwell, an automotive company that had merged with North American Aviation, maker of the Minuteman Guidance System. In 1973 a half dozen other companies began offering microprocessors as well. Intel responded to the competition in April 1974 by announcing the 8080, an 8-bit chip that could address much more memory and required fewer support chips than the 8008. The company set the price at $360—a somewhat arbitrary figure, as Intel had no experience selling chips like these one at a time. (Folklore has it that tire $360 price was set to suggest a comparison with the IBM System/360.) A significant advance over the 8008, the 8080 could execute programs written for the other chip, a compatibility that would prove crucial to Intel's dominance of the market. The 8080 was the first of the microprocessors whose instruction set and memory addressing capability approached those of the minicomputers of the day.

From Microprocessor to Personal Computer

There were now, in early 1974, two converging forces at work. From one direction were the semiconductor engineers with their ever-more-powerful microprocessors and ever-more-capacious memory chips. From the other direction were users of time-sharing systems, who saw a PDP-10 or XDS 940 as a basis for public access to computing. When these forces met in the middle, they would bring about a revolution in personal computing.

They almost did not meet. For the two years between Brand's observation and the appearance of tire Altair, the two forces were rushing past one another. The time-sharing systems had trouble making money even from industrial clients, and the public systems like community Memory were also struggling. At the other end, semiconductor companies did not think of their products as a possible basis for a personal computer.

A general-purpose computer based on a microprocessor did appear in 1973. In May of that year Thi T. Truong, an immigrant to France from Viet Nam, had his electronics company design and build a computer based on the Intel 8008 microprocessor. The MICRAL was a rugged and well-designed computer, with a bus architecture and internal slots on its circuit board for expansion. A base model cost under $2,000, and it found a market replacing minicomputers for simple control operations. Around two thousand were sold in the next two years, none of them beyond an industrial market. It is regarded as the first microprocessor-based computer to be sold in the commercial marketplace. Because of the limitations of the 8008, its location in France, and above all, the failure by its creators to see what it "really" was, it never broke out of its niche as a replacement for minicomputers in limited industrial locations.

The perception of the MICRAL as something to replace the mini was echoed at Intel as well. Intel's mental model of its product was this: an *industrial* customer bought an 8080 and wrote specialized software for it, which was then burned into a read-only-memory to give a system with the desired functions. The resulting inexpensive product (no longer programmable) was then put on the market as an embedded controller in an industrial system. A major reason for that mental model was the understanding of how hard it was to program a microprocessor. It seemed absurd to ask untrained consumers to program when Intel's traditional customers, hardware designers, were themselves uncomfortable with programming.

With these embedded uses in mind, microprocessor suppliers developed educational packages intended to ease customers into system design. These kits included the microprocessor, some RAM and ROM chips, and some other chips that handled timing and control, all mounted on a printed circuit board. They also included written material that gave a tutorial on how to program the system. This effort took Intel far from its core business of making chips, but the company hoped to recoup the current losses later on with volume sales of components. These kits were sold for around $200 or given away to engineers who might later generate volume sales.

Intel and the others also built more sophisticated "Development Systems," on which a customer could actually test the software for an application. These were fully assembled products that sold for around $10,000. To use these systems, customers also needed specialized software that would allow them to write programs using a language like FORTRAN, and then "cross-compile" it for the microprocessor—that is, from the FORTRAN program generate machine code, not for the computer on which it was written, but for the microprocessor. The company hired Gary Kildall, an instructor at the Naval Postgraduate School in Monterey, California, to develop a language based on IBM's PL/I. He called it PL/M, and in 1973 Intel offered it to customers. Initially this software was intended to be run on a large mainframe, but it was soon available for minicomputers, and finally to microprocessor-based systems. In 1974 Intel offered a development system, the Intellec 4, which included its own resident PL/M compiler (i.e., one did not need

a mainframe or a mini to compile the code). A similar Intellec-8 introduced the 8-bit microprocessors.

With these development systems, Intel had in fact invented a personal computer. But the company did not realize it. These kits were not marketed as the functional computers they were. Occasionally someone bought one of these systems and used it in place of a minicomputer, but Intel neither supported that effort nor recognized its potential. Intel and the other microprocessor firms made money selling these development systems—for some they were very profitable—but the goal was to use them as a lever to open up volume purchases of chips. The public could not buy one. The chip suppliers were focused on the difficulties in getting embedded systems to do useful work; they did not think that the public would be willing to put up with the difficulties of programming just to own their own computer.

Role of Hobbyists

Here is where the electronics hobbyists and enthusiasts come in. Were it not for them, the two forces in personal computing might have crossed without converging. Hobbyists, at that moment, were willing to do the work needed to make microprocessor-based systems practical.

This community had a long history of technical innovation—it was radio amateurs, for example, who opened up the high-frequency radio spectrum for long-distance radio communications after World War I. After World War II, the hobby expanded beyond amateur radio to include high-fidelity music reproduction, automatic controls, and simple robotics. A cornucopia of war surplus equipment from the U.S. Army Signal Corps found its way into individual hands, further fueling the phenomenon. (A block in lower Manhattan known as "Radio Row," where the World Trade Center was built, was a famous source of surplus electronic gear.) The shift from vacuum tubes to integrated circuits made it harder for an individual to build a circuit on a breadboard at home, but inexpensive TTL chips now contained whole circuits themselves. As the hobby evolved rapidly from analog to digital applications, this group supplied a key component in creating the personal computer: it provided an infrastructure of support that neither the computer companies nor the chip makers could.

This infrastructure included a variety of electronics magazines. Some were aimed at particular segments, for example, *QST* for radio amateurs. Two of them, *Popular Electronics* and *Radio-Electronics*, were of general interest and sold at newsstands; they covered high-fidelity audio, shortwave radio, television, and assorted gadgets for the home and car. Each issue typically had at least one construction project. For these projects the magazine would make arrangements with small electronics companies to supply a printed circuit board, already etched and drilled, as well as specialized components that readers might have difficulty finding locally. By scanning the back issues of these magazines we can trace how hobbyists moved from analog to digital designs.

A machine called the Kenbak-1, made of medium and small-scale integrated circuits, was advertised in the September 1971 issue of *Scientific American*. The advertisement called it suitable for "private individuals," but it was really intended for schools. The Kenbak may be the first personal computer, but it did not use a microprocessor, and its capabilities were quite limited.

The Scelbi-8H was announced in a tiny advertisement in the back of the March 1974 issue of *QST*. It used an Intel 8008, and thus may be the first microprocessor-based computer marketed to the public. According to the advertisement, "Kit prices for the new Scelbi-8H mini-computer start as low as $440!" It is not known how many machines Scelbi sold, but the company went on to play an important part in the early personal computer phenomenon.

In July 1974, *Radio-Electronics* announced a kit based on the Intel 8008, under the headline "Build the Mark-8: Your Personal Minicomputer." The project was much more ambitious than what typically appeared in that magazine. The article gave only a simple description and asked readers to order a separate, $5.00 booklet for detailed instructions. The Mark-8 was designed by Jonathan Titus of Virginia Polytechnic University in Blacksburg. The number of machines actually built may range in the hundreds, although the magazine reportedly sold "thousands" of booklets. At least one Mark-8 users club sprang up, in Denver, whose members designed an ingenious method of storing programs on an audio cassette recorder. Readers were directed to a company in Englewood, New Jersey, that supplied a set of circuit boards for $47.50, and to Intel for the 8008 chip (for $120.00). The Mark-8's appearance in *Radio-Electronics* was a strong factor in the decision by its rival *Popular Electronics* to introduce the Altair kit six months later.

These kits were just a few of many projects described in the hobbyist magazines. They reflected a conscious effort by the community to bring digital electronics, with all its promise and complexity, to amateurs who were familiar only with simpler radio or audio equipment. It was not an easy transition: construction of both the Mark-8 and the TV-typewriter (described next) was too complex to be described in a magazine article; readers had to order a separate booklet to get complete plans. *Radio-Electronics* explained to its readers that "[w]e do not intend to do an article this way as a regular practice." Although digital circuits were more complex than what the magazine had been handling, it recognized that the electronics world was moving in that direction and that its readers wanted such projects.

Other articles described simpler digital devices—timers, games clocks, keyboards, and measuring instruments—that used inexpensive TTL chips. One influential project was the TV-Typewriter, designed by Don Lancaster and published in *Radio-Electronics* in September 1973. This device allowed readers to display alphanumeric characters, encoded in ASCII, on an ordinary television set. It presaged the advent of CRT terminals as the primary input-output device for personal computers—one major distinction between the PC culture and that of the minicomputer, which relied on the Teletype. Lee Felsenstein called the TV-Typewriter "the opening shot of the computer revolution."

Altair

1974 was the *annus mirabilis* of personal computing. In January, Hewlett-Packard introduced its HP-65 programmable calculator. That summer Intel announced the 8080 microprocessor. In July, *Radio-Electronics* described the Mark-8. In late December, subscribers to *Popular Electronics* received their January 1975 issue in the mail, with a prototype of the "Altair" minicomputer on the cover, and an article describing how readers could obtain one for less than $400. This announcement ranks with IBM's announcement of the System/360 a decade earlier as one of the most significant in the history of computing. But what a difference a decade made: the Altair was a genuine personal computer.

H. Edward Roberts, the Altair's designer, deserves credit as the inventor of the personal computer. The Altair was a capable, inexpensive computer designed around the Intel 8080 microprocessor. Although calling Roberts the inventor makes sense only in the context of all that came before him, including the crucial steps described above, he does deserve the credit. Mark Twain said that historians have to rearrange past events so they make more sense. If so, the invention of the personal computer at a small model-rocket hobby shop in Albuquerque cries out for some creative rearrangement. Its utter improbability and unpredictability have led some to credit many other places with the invention, places that are more sensible, such as the Xerox Palo Alto Research Center, or Digital Equipment Corporation, or even IBM. But Albuquerque it was, for it was only at MITS that the technical and social components of personal computing converged.

Consider first the technical. None of the other hobbyist projects had the impact of the Altair's announcement. Why? One reason was that it was designed and promoted as a capable minicomputer, as powerful as those offered by DEC or Data General. The magazine article, written by Ed Roberts and William Yates, makes this point over and over: "a full-blown computer that can hold its own against sophisticated minicomputers"; "not a 'demonstrator' or a souped-up calculator"; "performance competes with current commercial minicomputers." The physical appearance of the Altair computer suggested its minicomputer lineage. It looked like the Data General Nova: it had a rectangular metal case, a front panel of switches that controlled the contents of internal registers, and small lights indicating the presence of a binary one or zero. Inside the Altair's case, there was a machine built mainly of TTL integrated circuits (except for the microprocessor, which was a MOS device), packaged in dual-in-line packages, soldered onto circuit boards. Signals and power traveled from one part of the machine to another on a bus. The Altair used integrated circuits, not magnetic cores, for its primary memory. The *Popular Electronics* cover called the Altair the "world's first minicomputer kit"; except for its use of a microprocessor, that accurately described its physical construction and design.

But the Altair as advertised was ten times cheaper than minicomputers were in 1975. The magazine offered an Altair for under $400 as a kit, and a few hundred more already assembled. The magazine cover said that readers could "save over $1,000." In fact, the cheapest PDP-8 cost several thousand dollars. Of course, a PDP-8 was a fully assembled, operating computer that was considerably more capable than the basic Altair, but that did not really matter in this case. (Just what one got for $400 will be discussed later.) The low cost resulted mainly from its use of the Intel 8080 microprocessor, just introduced. Intel had quoted a price of $360 for small quantities

of 8080s, but Intel's quote was not based on a careful analysis of how to sell the 8080 to this market. MITS bought them for only $75 each.

The 8080 had more instructions and was faster and more capable than the 8008 that the Mark-8 and Scelbi-8 used. It also permitted a simpler design since it required only six instead of twenty supporting chips to make a functional system. Other improvements over the 8008 were its ability to address up to 64 thousand bytes of memory (vs. the 8008's 16 thousand), and its use of main memory for the stack, which permitted essentially unlimited levels of subroutines instead of the 8008's seven levels.

The 8080 processor was only one architectural advantage the Altair had over its predecessors. Just as important was its use of an open bus. According to folklore, the bus architecture almost did not happen. After building the prototype Altair, Roberts photographed it and shipped it via Railway Express to the offices of *Popular Electronics* in New York. Railway Express, a vestige of an earlier American industrial revolution, was about to go bankrupt; it lost the package. The magazine cover issue showed the prototype, with its light-colored front panel and the words "Altair 8800 on the upper left. That machine had a set of four large circuit boards stacked on top of one another, with a wide ribbon cable carrying 100 lines from one board to another. After that machine was lost, Robert redesigned the Altair. He switched to a larger deep blue cabinet and discarded the 100-wire ribbon cable. In the new design, wires connected to a rigid backplane carried the signals from one board to another. That allowed hobbyists to add a set of connectors that could accept other cards besides the initial four.

The $400 kit came with only two cards to plug into the bus: those two, plus a circuit board to control the front panel and the power supply, made up the whole computer. The inside looked quite bare. But laboriously soldering a set of wires to an expansion chassis created a full set of slots into which a lot of cards could be plugged. MITS was already designing cards for more memory, I/O and other functions.

Following the tradition established by Digital Equipment Corporation, Roberts did not hold specifications of the bus as a company secret. That allowed others to design and market cards for the Altair. That decision was as important to the Altair's success as its choice of ar 8080 processor. It also explains one of the great ironies of the Altair, that it inaugurated the PC era although it was neither reliable nor very well-designed. Had it not been possible for other companies to offer plug-in cards that improved on the original MITS design, the Altair might have made no greater impact than the Mark-8 had. The bus architecture also led to the company's demise a few years later, since it allowed other companies to market compatible cards and, later, compatible computers. But by then the floodgates had opened. If MITS was unable lo deliver on its promises of making the Altair a serious machine (though it tried), other companies would step in. MITS continued developing plug-in cards and peripheral equipment, but the flood of orders was too much for the small company.

So while it was true that for $400 hobbyists got very little, they could get the rest—or design and build the rest. Marketing the computer as a bare-bones kit offered a way for thousands of people to bootstrap their way into the computer age, at a pace that they, not a computer company, could control.

Assembling the Altair was much more difficult than assembling other electronics kits, such as those sold by the Heath Company or Dynaco. MITS offered to sell "completely assembled and tested" computers for 5498, but with such a backlog of orders, readers were faced with the choice of ordering the kit and getting something in a couple of months, or ordering the assembled computer and perhaps waiting a year or more. Most ordered the kit and looked to one another for support in finding the inevitable wiring errors and poorly soldered connections that they would make. The audience of electronics hobbyists, at whom the magazine article was aimed, compared the Altair not to the simple Heathkits, but to building a computer from scratch, which was almost impossible: not only was it hard to design a computer, it was impossible to obtain the necessary chips. Chips were inexpensive, but only if they were purchased in large quantities, and anyway, most semiconductor firms had no distribution channels set up for single unit or retail sales. Partly because of this, customers felt, rightly, that they were getting an incredible bargain.

The limited capabilities of the basic Altair, plus the loss of the only existing Altair by the time the *Popular Electronics* article appeared, led to the notion that it was a sham, a "humbug," not a serious product at all. The creators of the Altair fully intended to deliver a serious computer whose capabilities were on a par with minicomputers then on the market. Making those deliveries proved to be a lot harder than they anticipated. Fortunately, hobbyists understood that. But there should be no mistake about it: the Altair was real.

MITS and the editors of *Popular Electronics* had found a way to bring the dramatic advances in integrated circuits to individuals. The first customers were hobbyists, and the first thing they did with these machines, once they got them running, was play games. Roberts was trying to sell it as a machine for serious work, however. In the *Popular Electronics* article he proposed a list of twenty-three applications, none of them games. Because it was several years before anyone could supply peripheral equipment, memory, and software, serious applications were rare at first. That, combined with the primitive capabilities of other machines like the Mark-8, led again to an assumption that the Altair was not a serious computer. Many of the proposed applications hinted at in the 1975 article were eventually implemented. Years later one could still find an occasional Altair (or more frequently, an Altair clone) embedded into a system just like its minicomputer cousins.

The next three years, from January 1975 through the end of 1977, saw a burst of energy and creativity in computing that had almost no equal in its history. The Altair had opened the floodgates, even though its shortcomings were clear to everyone. One could do little more than get it to blink a pattern of lights on the front panel. And even that was not easy: one had to flick the toggle switches for each program step, then deposit that number into a memory location, then repeat that for the next step, and so on—hopefully the power did not go off while this was going on—until the whole program (less than 256 bytes long!) was in memory. Bruised fingers from flipping the small toggle switches were the least of the frustrations. In spite of all that, the bus architecture meant that other companies could design boards to remedy each of these Shortcomings, or even design a copy of the Altair itself, as IMSAI and others did.

But the people at MITS and their hangers-on created more than just a computer. This $400 computer inspired the extensive support of user groups, informal newsletters, commercial magazines, local clubs, conventions, and even retail stores. This social activity went far beyond traditional computer user groups, like SHARE for IBM or DECUS for Digital. Like the calculator users groups, these were open and informal, and offered more to the neophyte. All of this sprang up with the Altair, and many of the publications and groups lived long after the last Altair computer itself was sold.

Other companies, beginning with Processor Technology, soon began offering plug-in boards that gave the machine more memory. Another board provided a way of connecting the machine to a Teletype, which allowed fingers to heal. But Teletypes were not easy to come by—an individual not affiliated with a corporation or university could only buy one secondhand, and even then they were expensive. Before long, hobbyists-led small companies began offering ways of hooking up a television set and a keyboard (although Don Lancaster's TV Typewriter was not the design these followed). The board that connected to the Teletype sent data serially—one bit at a time; another board was designed that sent out data in parallel, for connection to a line printer that minicomputers used, although like the Teletype these were expensive and hard to come by.

The Altair lost its data when the power was shut off, but before long MITS designed an interface that put out data as audio tones, to store programs on cheap audio cassettes. A group of hobbyists met in Kansas City in late 1975 and established a "Kansas City Standard" for the audio tones stored on cassettes, so that programs could be exchanged from one computer to another. Some companies brought out inexpensive paper tape readers that did not require the purchase of a Teletype. Others developed a tape cartridge like the old 8-track audio systems, which looped a piece of tape around and around. Cassette storage was slow and cumbersome—users usually had to record several copies of a program and make several tries before successfully loading it into the computer. Inadequate mass storage limited the spread of PCs until the "floppy" disk was adapted.

The floppy was invented by David L. Noble at IBM for a completely different purpose. When IBM introduced the System/370, which used semiconductor memory, it needed a way to store the computer's initial control program, as well as to hold the machine's microprogram. That had not been a problem for the System/360, which used magnetic cores that held their contents when the power was switched off. From this need came the 8-inch diameter flexible diskette, which IBM announced in 1971. Before long, people recognized that it could be used for other purposes besides the somewhat limited one for which it had been invented. In particular, Alan Shugart, who had once worked for IBM, recognized that the floppy's simplicity and low cost made it the ideal storage medium for low-cost computer systems. Nevertheless, floppy drives were rare in the first few years of personal computing. IBM's hardware innovation was not enough; there had to be an equivalent innovation in system software to make the floppy practical. Before that story is told, we shall look first at the more immediate issue of developing a high-level language for the PC.

Software: BASIC

The lack of a practical mass storage device was one of two barriers that blocked the spread of personal, interactive computing. The other was a way to write applications software. By 1977 two remarkable and influential pieces of software—Microsoft BASIC and the CP/M Operating System—overcame those barriers.

In creating the Altair, Ed Roberts had to make a number of choices: what processor to use, the design of the bus (even whether to use a bus at all), the packaging, and so on. One such decision was the choice of a programming language. Given the wide acceptance of BASIC it is hard to imagine that there ever was a choice, but there was. BASIC was not invented for small computers. The creators of BASIC abhorred the changes others made to shoehorn the language onto systems smaller than a mainframe. Even in its mainframe version, BASIC had severe limitations—on the numbers and types of variables it allowed, for example. In the view of academic computer scientists, the versions of BASIC developed for minicomputers were even worse—full of ad hoc patches and modifications. Many professors disparaged BASIC as a toy language that fostered poor programming habits, and they refused to teach it. Serious programming was done in FORTRAN—an old and venerable but still capable language.

If, in 1974, one asked for a modern, concise, well-designed language to replace FORTRAN, the answer might have been APL, an interactive language invented at IBM by Kenneth Iverson in the early 1960s. A team within IBM designed a personal computer in 1973 that supported APL, the "SCAMP," although a commercial version of that computer sold poorly. Or PL/I: IBM had thrown its resources into this language, which it hoped would replace both FORTRAN and COBOL. Gary Kildall chose a subset of PL/I for the Intel microprocessor development kit.

BASIC'S strength was that it was easy to learn. More significant, it already had a track record of running on computers with limited memory. Roberts stated that he had considered FORTRAN and APL, before he decided the Altair was to have BASIC.

William Gates III was born in 1955, at a time when work on FORTRAN was just underway. He was a student at Harvard when the famous cover of

Popular Electronics appeared describing the Altair. According to one biographer, his friend Paul Allen saw the magazine and showed it to Gates, and the two immediately decided that they would write a BASIC compiler for the machine. Whether it was Gates's or Roberts's decision to go with BASIC for the Altair, BASIC it was.

In a newsletter sent out to Altair customers, Gates and Allen stated that a version of BASIC that required only 4K bytes of memory would be available in June 1975, and that more powerful versions would be available soon after. The cost, for those who also purchased Altair memory boards, was $60 for 4K BASIC, $75 for 8K, and $150 for "extended" BASIC (requiring disk or other mass storage). Those who wanted the language to run on another 8080-based system had to pay $500.

In a burst of energy, Gates and Allen, with the help of Monte Davidoff, wrote not only a BASIC that fit into very little memory; they wrote a BASIC with a lot of features and impressive performance. The language was true to its Dartmouth roots in that it was easy to learn. It broke with those roots by providing a way to move from BASIC commands to instructions written in machine language. That was primarily through a USR command, which was borrowed from software written for DEC minicomputers (where the acronym stood for user service routine). A programmer could even directly put bytes into or pull data out of specific memory locations, through the PEEK and POKE commands—which would have caused havoc on the time-shared Dartmouth system. Like USR, these commands were also derived from prior work done by DEC programmers, who came up with them for a time-sharing system they wrote in BASIC for the PDP-11. Those commands allowed users to pass from BASIC to machine language easily—a crucial feature for getting a small system to do useful work.

These extensions kept their BASIC within its memory constraints while giving it the performance of a more sophisticated language. Yet it remained an interactive, conversational language that novices could learn and use. The BASIC they wrote for the Altair, with its skillful combination of features taken from Dartmouth and from the Digital Equipment Corporation, was the key to Gates's and Allen's success in establishing a personal computer software industry.

The developers of this language were not formally trained in computer science or mathematics as were Kemeny and Kurtz. They were introduced to computing in a somewhat different way. Bill Gates's private

school in Seattle had a General Electric time-sharing system available for its pupils in 1968, a time when few students even in universities had such access. Later on he had access to an even better time-shared system: a PDP-10 owned by the Computer Center Corporation. Later still, he worked with a system of PDP-10s and PDP-11s used to control hydroelectric power for the Bonneville Power Administration. One of his mentors at Bonneville Power was John Norton, a TRW employee who had worked on the Apollo Program and who was a legend among programmers for the quality of his work.

When he was writing BASIC for the Altair, Gates was at Harvard. He did not have access to an 8080-based system, but he did have access to a PDP-10 at Harvard's computing center (named after Howard Aiken). He and fellow student Monte Davidoff used the PDP-10 to write the language, based on the written specifications of the Intel 8080. In early 1975 Paul Allen flew to Albuquerque and demonstrated it to Roberts and Yates. It worked. Soon after, MITS advertised its availability for the Altair. Others were also writing BASIC interpreters for the Altair and for the other small computers now flooding the market, but none was as good as Gates's and Allen's, and it was not long before word of that got around.

It seemed that Roberts and his company had made one brilliant decision after another: the 8080 processor, the bus architecture, and now BASIC. However, by late 1975 Gates and Allen were not seeing it that way. Gates insists that he never became a MITS employee (although Allen was until 1976), and that under the name "Micro Soft," later "Micro-Soft," he and Allen retained the rights to their BASIC. In a now-legendary "Open Letter to Hobbyists," distributed in early 1976, Gates complained about people making illicit copies of his BASIC by duplicating the paper tape. Gates claimed "the value of the computer time we have used [to develop the language] exceeds $40,000." He said that if he and his programmers were not paid, they would have little incentive to develop more software for personal computers, such as an APL language for the 8080 processor. He argued that illicit copying put all personal computing at risk: "Nothing would please me more than to hire ten programmers and deluge the hobby market with good software."

Gates did his initial work on the PDP-10 while still an undergraduate at Harvard. Students were not to use that computer for commercial purposes, although these distinctions were not as clear then as they would be later. The language itself was the invention of Kemeney and Kurtz of Dartmouth; the extensions that were crucial to its success came from programmers at the Digital Equipment Corporation, especially Mark Bramhall, who led the effort to develop a time-sharing system (RSTS-11) for the PDP-11. Digital, the only commercial entity among the above group, did not think of its software as a commodity to sell; it was what the company did to get people to buy hardware.

Bill Gates had recognized what Roberts and all the others had not: that with the advent of cheap, personal computers, software could and should come to the fore as the principal driving agent in computing. And only by charging money for it—even though it had originally been free—could that happen. By 1978 his company, now called "Microsoft," had severed its relationship with MITS and was moving from Albuquerque to the Seattle suburb of Bellevue. (MITS itself had lost its identity, having been bought by Pertec in 1977.) Computers were indeed coming to "the people," as Stewart Brand had predicted in 1972. But the driving force was not the counterculture vision of a Utopia of shared and free information; it was the force of the marketplace. Gates made good on his promise to "hire ten programmers and deluge the . . . market".

System Software: The Final Piece of the Puzzle

Gary Kildall's entree into personal computing software was as a consultant for Intel, where he developed languages for system development. While doing that he recognized that the floppy disk would make a good mass storage device for small systems, if it could be properly adapted. To do that he wrote a small program that managed the flow of information to and from a floppy disk drive. As with the selection of BASIC, it appears in hindsight to be obvious and inevitable that the floppy disk would be the personal computer's mass storage medium. That ignores the fact that it was never intended for that use. As with the adaptation of BASIC, the floppy had to be recast into a new role. As with BASIC, doing that took the work of a number of individuals, but the primary effort came from one man, Gary Kildall.

A disk had several advantages over magnetic or paper tape. For one, it was faster. For another, users could both read and write data on it. Its primary advantage was that a disk had "random" access: Users did not have to run through the entire spool of tape to get at a specific piece of data. To accomplish this, however, required tricky programming—something IBM had called, for one of its mainframe systems, a Disk Operating System, or DOS.

A personal computer DOS had little to do with mainframe operating systems. There was no need to schedule and coordinate the jobs of many users: an Altair had one user. There was no need to "spool" or otherwise direct data to a roomful of chain printers, card punches, and tape drives: a personal computer had only a couple of ports to worry about. What *was* needed was rapid and accurate storage and retrieval of files from a floppy disk. A typical file would, in fact, be stored as a set of fragments, inserted at whatever free spaces were available on the disk. It was the job of the operating system to find those free spaces, put data there, retrieve it later on, and reassemble the fragments. All that gave the user an illusion that the disk was just like a traditional file cabinet filled with folders containing paper files.

Once again, Digital Equipment Corporation was tire pioneer, in part because of its culture; because of the experience many of its employees had had with the TX-0 at MIT, one of the first computers to have a conversational, interactive feel to it. For its early systems DEC introduced DECtape, which although a tape, allowed programmers rapid access to data written in the middle, as well as at the ends, of the reel. The PDP-10s had powerful DECtape as well as disk storage abilities; its operating systems were crucial in creating the illusion of personal computing that had so impressed observers like Stewart Brand.

In the late 1960s DEC produced OS/8 for the PDP-8, which had the feel of the PDP-10 but ran on a machine with very limited memory. OS-8 opened everyone's eyes at DEC; it showed that small computers could have capabilities as sophisticated as mainframes, without the bloat that characterized mainframe system software. Advanced versions of the PDP-11 had an operating system called RT-11 (offered in 1974), which was similar to OS/8, and which further refined the concept of managing data on disks. These were the roots of personal computer operating systems. DEC's role in creating this software ranks with its invention of the minicomputer as major contributions to the creation of personal computing.

Gary Kildall developed PL/M for the Intel 8080. He used an IBM System/360, and PL/M was similar to IBM's PL/I. While working on that project Kildall wrote a small control program for the mainframe's disk drive. "It turned out that the operating system, which was called CP/M for Control Program for

Micros, was useful, too, fortunately." Kildall said that PL/M was "the base for CP/M," even though the commands were clearly derived from Digital's, not IBM's software." For example, specifying the drive in use by a letter; giving file names a period and three-character extension; and using the DIR (Directory) command, PIP, and DDT were DEC features carried over without change. CP/M was announced to hobbyists as "similar to DECSYSTEM 10" in an article by Jim Warren in *Dr. Dobb's Journal of Computer Calisthenics and Orthodontia* [sic] in April 1976. Warren was excited by CP/M, stating that it was "well designed, based on an easy-to-use operating system that has been around for a DECade. [sic]" Suggested prices were well under $100, with a complete floppy system that included a drive and a controller for around $800—not cheap, but clearly superior to the alternatives of cassette, paper tape, or any other form of tape. CP/M was the final piece of the puzzle that, when made available, made personal computers a practical reality.

Gary Kildall and his wife, Dorothy McEwen, eased themselves into the commercial software business while he also worked as an instructor at the Naval Postgraduate School in Monterey, California. As interest in CP/M picked up, he found himself writing variations of it for other customers. The publicity in *Dr. Dobb's Journal* led to enough sales to convince him of the potential market for CP/M. In 1976 he quit his job and with Dorothy founded a company, Digital Research (initially Intergalactic Digital Research), whose main product was CP/M.

The next year, 1977, he designed a version with an important difference. IMSAI, the company that had built a "clone" of the Altair, wanted a license to use CP/M for its products. Working with IMSAI employee Glen Ewing, Kildall rewrote CP/M so that only a small portion of it needed to be customized for the specifics of the IMSAI. The rest would be common code that would not have to be rewritten each time a new computer or disk drive came along. He called the specialized code the BIOS—Basic Input/Output System. This change standardized the system software in the same way that the 100-pin Altair bus had standardized hardware. IMSAI's computer system became a standard, with its rugged power supply, room for expansion with plenty of internal slots, external floppy drive, and CP/M.

End of the Pioneering Phase, 1977

By 1977 the pieces were all in place. The Altair's design shortcomings were corrected, if not by MITS then by other companies. Microsoft BASIC allowed programmers to write interesting and, for the first time, serious software for these machines. The ethic of charging money for this software gave an incentive to such programmers, although software piracy also became established. Computers were also being offered with BASIC supplied on a read-only-memory (ROM), the manufacturer paying Microsoft a simple royalty fee. (With the start-up codes also in ROM, there was no longer a need for the front panel, with its array of lights and switches.) Eight-inch floppy disk drives, controlled by CP/M, provided a way to develop and exchange software that was independent of particular models. Machines came with standardized serial and parallel ports, and connections for printers, keyboards, and video monitors. Finally, by 1977 there was a strong and healthy industry of publications, software companies, and support groups to bring the novice on board. The personal computer had arrived.

Code Is Law

Lawrence Lessig

Like the other authors in this section, Lawrence Lessig believes that technology and society are intertwined. Instead of only examining the past, however, Lessig is especially concerned about the present and future. In tackling a relatively new and still evolving technology—the Internet—Lessig goes as far as to say that computer code will not simply affect our lives, but that it will be as powerful as law in enabling and limiting our actions. He argues that an individual's behavior is influenced by four forms of regulation—law, norms, markets, and what he calls architecture. Architecture is the term he uses for computer code though what he claims applies equally well and more broadly to the built environment. He urges us to act while the technology is still developing so that we can direct it to create a world we want rather than one dictated by those involved in writing code. Lessig's analysis illustrates just one more way in which the social and technical are intertwined and reveals an important implication: In order to get the kind of future we want, we can't focus only on laws, norms and markets. We also must pay attention to technology, technological design, and technical details. Some of Lessig's arguments seem deterministic, but Lessig recognizes that humans have power to control the technology that will ultimately shape us.

Code Is Law

A decade ago, in the spring of 1989, communism in Europe died—collapsed, as a tent would fall if its main post were removed. No war or revolution brought communism to its end. Exhaustion did. Born in its place across Central and Eastern Europe was a new political regime, the beginnings of a new political society.

For constitutionalists (as I am), this was a heady time. I had just graduated from law school in 1989, and in 1991 I began teaching at the University of Chicago. Chicago had a center devoted to the study of the emerging democracies in Central and Eastern Europe. I was a part of that center. Over the next five years I spent more hours on air-planes, and more mornings drinking bad coffee, than I care to remember.

Eastern and Central Europe were filled with Americans telling former Communists how they should govern. The advice was endless and silly. Some of these visitors literally sold constitutions to the emerging constitutional republics; the balance had innumerable half-baked ideas about how the new nations should be governed. These Americans came from a nation where constitutionalism had worked, yet apparently had no clue why.

The center's mission, however, was not to advise. We knew too little to guide. Our aim was to watch and gather data about the transitions and how they progressed. We wanted to understand the change, not direct it.

What we saw was striking, if understandable. Those first moments after communism's collapse were filled with antigovernmental passion—with a surge of anger directed against the state and against state regulation. Leave us alone, the people seemed to say. Let the market and nongovernmental organizations—a new society—take government's place. After generations of communism, this reaction was completely understandable. What compromise could there be with the instrument of your repression?

A certain American rhetoric supported much in this reaction. A rhetoric of libertarianism. Just let the market reign and keep the government out of the way, and freedom and prosperity would inevitably grow. Things would take care of themselves. There was no need, and could be no place, for extensive regulation by the state.

But things didn't take care of themselves. Markets didn't flourish. Governments were crippled, and crippled governments are no elixir of freedom. Power didn't disappear—it simply shifted from the state to mafiosi, themselves often created by the state. The need for traditional state functions—police, courts, schools, health care—didn't magically go away. Private interests didn't emerge to fill the need. Instead, needs were unmet. Security evaporated. A modern if plodding anarchy replaced the bland communism of the previous three generations: neon lights flashed advertisements for Nike; pensioners were swindled out of their life savings by fraudulent stock deals; bankers were murdered in broad daylight on Moscow streets. One system of control had been replaced by another, but neither system was what Western libertarians would call freedom.

At just about the time when this post-communist euphoria was waning—in the mid-1990s—there emerged in the West another "new society," to many just as exciting as the new societies promised in post-communist Europe. This was cyberspace. First in universities and centers of research, and then within society generally, cyberspace became the new target of libertarian utopianism. Here freedom from the state would reign. If not in Moscow or Tblisi, then here in cyberspace would we find the ideal libertarian society.

The catalyst for this change was likewise unplanned. Born in a research project in the Defense Department, cyberspace too arose from the displacement of a certain architecture of control. The tolled, single-purpose network of telephones was displaced by the untolled and multipurpose network of packet-switched data. And thus the old one-to-many architectures of publishing (television, radio, newspapers, books) were supplemented by a world where everyone could be a publisher. People could communicate and associate in ways that they had never done before. The space promised a kind of society that real space could never allow—freedom without anarchy, control without government, consensus without power. In the words of a manifesto that will define our generation: "We reject: kings, presidents and voting. We believe in: rough consensus and running code."[1]

As in post-Communist Europe, first thoughts about cyberspace tied freedom to the disappearance of the state. But here the bond was even stronger than in post-Communist Europe. The claim now was that government *could not* regulate cyberspace, that cyberspace was essentially, and unavoidably, free. Governments could threaten, but behavior could not

be controlled; laws could be passed, but they would be meaningless. There was no choice about which government to install—none could reign. Cyberspace would be a society of a very different sort. There would be definition and direction, but built from the bottom up, and never through the direction of a state. The society of this space would be a fully self-ordering entity, cleansed of governors and free from political hacks.

I taught in Central Europe during the summers of the early 1990s; I witnessed the transformation in attitudes about communism that I described at the start of this chapter. And so I felt a bit of déjà vu when in the spring of 1995, I began to teach the law of cyberspace, and saw in my students these very same post-communist thoughts about freedom and government. Even at Yale—not known for libertarian passions—the students seemed drunk with what James Boyle would later call the "libertarian gotcha":[2] no government could survive without the Internet's riches, yet no government could control what went on there. Real-space governments would become as pathetic as the last Communist regimes. It was the withering of the state that Marx had promised, jolted out of existence by trillions of gigabytes flashing across the ether of cyberspace. Cyberspace, the story went, could *only* be free. Freedom was its nature.

But why was never made clear. That *cyberspace* was a place that governments could not control was an idea that I never quite got. The word itself speaks not of freedom but of control. Its etymology reaches beyond a novel by William Gibson (*Neuromancer,* published in 1984) to the world of "cybernetics," the study of control at a distance.[3] Cybernetics had a vision of perfect regulation. Its very motivation was finding a better way to direct. Thus, it was doubly odd to see this celebration of non-control over architectures born from the very ideal of control.

As I said, I am a constitutionalist. I teach and write about constitutional law. I believe that these first

[1]Paulina Borsook, "How Anarchy Works," *Wired* 110 (October 1995): 3.10, available at http://www.wired.com/wired/archive/3.10/ietf_pr.html (visited May 30, 1999), quoting Net-lander, David Clark.

[2]James Boyle, talk at Telecommunications Policy Research Conference (TPRC), Washington, D.C., September 28, 1997. David Shenk discusses the libertarianism that cyberspace inspires (as well as other, more fundamental problems with the age) in a brilliant cultural how-to book that responsibly covers both the technology and the libertarianism; see *Data Smog: Surviving the Information Glut* (San Francisco, Harper Edge, 1997), esp. 174–77. The book also describes technorealism, a responsive movement that advances a more balanced picture of the relationship between technology and freedom.

[3]See Kevin Kelley, *Out of Control: The New Biology of Machines, Social Systems, and the Economic World* (Reading, Mass.: Addison-Wesley, 1994), 119.

thoughts about government and cyberspace are just as misguided as the first thoughts about government after communism. Liberty in cyberspace will not come from the absence of the state. Liberty there, as anywhere, will come from a state of a certain kind.[4] We build a world where freedom can flourish not by removing from society any self-conscious control; we build a world where freedom can flourish by setting it in a place where a particular kind of self-conscious control survives. We build liberty, that is, as our founders did, by setting society upon a certain *constitution.*

But by "constitution" I don't mean a legal text. Unlike my countrymen in Eastern Europe, I am not trying to sell a document that our framers wrote in 1787. Rather, as the British understand when they speak of their constitution, I mean en *architecture*— not just a legal text but a way of life—that structures and constrains social and legal power, to the end of protecting fundamental *values*—principles and ideals that reach beyond the compromises of ordinary politics.

Constitutions in this sense are built, they are not found. Foundations get laid, they don't magically appear. Just as the founders of our nation learned from the anarchy that followed the revolution (remember: our first constitution, the Articles of Confederation, was a miserable failure of do-nothingness), so too are we beginning to see in cyberspace that this building, or laying, is not the work of an invisible hand. There is no reason to believe that the grounding for liberty in cyberspace will simply emerge. In fact, as I will argue, quite the opposite is the case. As our framers learned, and as the Russians saw, we have every reason to believe that cyberspace, left to itself, will not fulfill the promise of freedom. Left to itself, cyberspace will become a perfect tool of control.[5]

Control. Not necessarily control by government, and not necessarily control to some evil, fascist end. But the argument of this book is that the invisible hand of cyberspace is building an architecture that is quite the opposite of what it was at cyberspace's birth. The invisible hand, through commerce, is constructing an architecture that perfects control—an architecture that makes possible highly efficient regulation. As Vernor Vinge warned in 1996, a distributed architecture of regulatory control; as Tom Maddox added, an axis between commerce and the state.[6]

This book is about that change, and about how we might prevent it. When we see the path that cyberspace is on . . . we see that much of the "liberty" present at cyberspace's founding will vanish in its future. Values that we now consider fundamental will not necessarily remain. Freedoms that were foundational will slowly disappear.

If the original cyberspace is to survive, and if values that we knew in that world are to remain, we must understand how this change happens, and what we can do in response. . . . Cyberspace presents something new for those who think about regulation and freedom. It demands a new understanding of how regulation works and of what regulates life there. It compels us to look beyond the traditional lawyer's scope—beyond laws, regulations, and norms. It requires an account of a newly salient regulator.

That regulator is the obscurity in the book's title—*Code.* In real space we recognize how laws regulate—through constitutions, statutes, and other legal codes. In cyberspace we must understand how code regulates—how the software and hardware that make cyberspace what it is *regulate* cyberspace as it is. As William Mitchell puts it, this code is cyberspace's "law."[7] *Code is law.*

This code presents the greatest threat to liberal or libertarian ideals, as well as their greatest promise.

_____

[4]As Stephen Holmes has put it, "Rights depend upon the competent exercise of . . . legitimate public power. . . . The largest and most reliable human rights organization is the liberal state. . . . Unless society is politically well organized, there will be no individual liberties and no civil society"; "What Russia Teaches Us Now: How Weak States Threaten Freedom," *American Prospect* 33 (1997): 30, 33.

[5]This is a dark picture, I confess, and it contrasts with the picture of control drawn by Andrew Shapiro in *The Control Revolution* (New York: Public Affairs, 1999). As I discuss later, however, the difference between Shapiro's view and my own turns on the extent to which architectures enable top-down regulation. In my view, one highly probable architecture would enable greater regulation than Shapiro believes is likely.

[6]See "We Know Where You Will Live," Computers, Freedom, and Privacy Conference, March 30, 1996, audio link available at http://www-swiss.ai.mit.edu/projects/mac/cfp96/plenary-sf.html.

[7]See William J. Mitchell, *City of Bits: Space, Place, and the Infobahn* (Cambridge, Mass.: MIT Press, 1995), 111. In much of this book, I work out Mitchell's idea, though I drew the metaphor from others as well. Ethan Katsh discusses this notion of software worlds in "Software Worlds and the First Amendment: Virtual Doorkeepers in Cyberspace," *University of Chicago Legal Forum* (1996): 335, 338. Joel Reidenberg discusses the related notion of "lex informatica" in "Lex Informatica: The Formulation of Information Policy Rules Through Technology," *Texas Law Review* 76 (1998): 553. I have been especially influenced by James Boyle's work in the area. I discuss his book in chapter 9, but see also "Foucault in Cyberspace: Surveillance, Sovereignty, and Hardwired Censors," *University of Cincinnati Law Review* 66 (1997): 177. For a recent and powerful use of the idea, see Shapiro, *The Control Revolution.* Mitch Kapor is the father of the meme "architecture is politics" within cyberspace talk. I am indebted to him for this.

We can build, or architect, or code cyberspace to protect values that we believe are fundamental, or we can build, or architect, or code cyberspace to allow those values to disappear. There is no middle ground. There is no choice that does not include some kind of *building*. Code is never found; it is only ever made, and only ever made by us. As Mark Stefik puts it, "Different versions of [cyberspace] support different kinds of dreams. We choose, wisely or not."[8]

My argument is not for some top-down form of control; my claim is not that regulators must occupy Microsoft. A constitution envisions an environment; as Justice Holmes said, it "call[s] into life a being the development of which [can not be] foreseen."[9] Thus, to speak of a constitution is not to describe a one-hundred-day plan. It is instead to identify the values that a space should guarantee. It is not to describe a "government"; it is not even to select (as if a single choice must be made) between bottom-up or top-down control. In speaking of a constitution in cyberspace we are simply asking: What values are protected there? What values will we build into the space to encourage certain forms of life?

The "values" here are of two sorts—substantive and structural. In the American tradition, we worried about the second first. The framers of the Constitution of 1787 (enacted without a Bill of Rights) were focused on structures of government. Their aim was to ensure that a particular government (the federal government) did not become too powerful. And so they built into its design checks on the power of the federal government and limits on its reach over the states.

Opponents of that Constitution insisted that more checks were needed, that the Constitution needed to impose substantive limits on government's power as well as structural limits. And thus the Bill of Rights was born. Ratified in 1791, the Bill of Rights promised that the federal government will not remove certain protections—of speech, privacy, and due process. And it guaranteed that the commitment to these substantive values will remain despite the passing fancy of normal government. These values were to be entrenched, or embedded, in our constitutional design; they can be changed, but only by changing the Constitution's design.

These two kinds of protection go together in our constitutional tradition. One would have been meaningless without the other. An unchecked structure could easily have overturned the substantive protections expressed in the Bill of Rights, and without substantive protections, even a balanced and reflective government could have violated values that our framers thought fundamental.

We face the same questions in constituting cyberspace, but we have approached them from an opposite direction. Already we are struggling with substance: Will cyberspace promise privacy or access? Will it preserve a space for free speech? Will it facilitate free and open trade? These are choices of substantive value.

But structure matters as well. What checks on arbitrary regulatory power can we build into the design of the space? What "checks and balances" are possible? How do we separate powers? How do we ensure that one regulator, or one government, doesn't become too powerful? . . .

Theorists of cyberspace have been talking about these questions since its birth.[10] But as a culture, we are just beginning to get it. We are just beginning to see why the architecture of the space matters—in particular, why the *ownership* of that architecture matters. If the code of cyberspace is owned, . . . it can be controlled; if it is not owned, control is much more difficult. The lack of ownership, the absence of properly, the inability to direct how ideas will be used—in a word, the presence of a commons—is key to limiting, or checking, certain forms of governmental control.

One part of this question of ownership is at the core of the current debate between open and closed source software. In a way that the American founders would have instinctively understood, "free software" or "open source software"—or "open code," to (cowardly) avoid taking sides in a debate I describe later—is itself a check on arbitrary power. A structural guarantee of constitutionalized liberty, it functions as a type of separation of powers in the American constitutional tradition. It stands alongside substantive protections, like freedom of speech or of the press, but its stand is more fundamental. . . . The first intuition of our founders was right: structure builds substance. Guarantee the structural (a space

---

[8]Mark Stefik, "Epilogue: Choices and Dreams," in *Internet Dreams: Archetypes, Myths, and Metaphors,* edited by Mark Stefik (Cambridge, Mass.: MIT Press, 1996), 390.

[9]*Missouri v Holland,* 252 US 416, 433 (1920).

[10]Richard Stallman, for example, organized resistance to the emergence of passwords at MIT. Passwords are an architecture that facilitates control by excluding users not "officially sanctioned." Steven Levy, *Hackers* (Garden City, N.Y.: Anchor Press/Doubleday, 1984), 416–17.

in cyberspace for open code), and (much of) the substance will take care of itself.

* * *

Given our present tradition in constitutional law and our present faith in representative government, are we able to respond collectively to the changes I will have described?

My strong sense is that we are not. We are at a stage in our history when we urgently need to make fundamental choices about values, but we trust no institution of government to make such choices. Courts cannot do it, because as a legal culture we don't want courts choosing among contested matters of values, and Congress should not do it because, as a political culture, we so deeply question the products of ordinary government.

Change is possible. I don't doubt that revolutions remain in our future; the open code movement is just such a revolution. But I fear that it is too easy for the government to dislodge these revolutions, and that too much will be at stake for it to allow the revolutionaries to succeed. Our government has already criminalized the core ethic of this movement, transforming the meaning of *hacker* into something quite alien to its original sense. This, I argue, is only the start.

Things could be different. They are different elsewhere. But I don't see how they could be different for us just now. This no doubt is a simple confession of the limits of my own imagination. I would be grateful to be proven wrong. I would be grateful to watch as we relearn—as the citizens of the former Communist republics are learning—how to escape our disabling ideas about the possibilities for governance.

* * *

What Things Regulate

John Stuart Mill was an Englishman, though one of the most influential political philosophers in America in the nineteenth century. His writings ranged from important work on logic to a still striking text, *The Subjection of Women*. But his continuing influence comes from a relatively short book titled *On Liberty*. Published in 1859, this powerful argument for individual liberty and diversity of thought represents an important view of liberal and libertarian thinking in the second half of the nineteenth century.

"Libertarian," however, has a specific meaning for us. It associates with arguments against government.[11] Government, in the modern libertarian's view, is the threat to liberty; private action is not. Thus, the good libertarian is focused on reducing government's power. Curb the excesses of government, the libertarian says, and you will have ensured freedom for your society.

Mill's view was not so narrow. He was a defender of liberty and an opponent of forces that suppressed it. But those forces were not confined to government. Liberty, in Mill's view, was threatened as much by norms as by government, as much by stigma and intolerance as by the threat of state punishment. His objective was to argue against these private forces of coercion. His work was a defense against liberty-suppressing norms, because in England at the time these were the real threat to liberty.

Mill's method is important, and it should be our own. It asks, What is the threat to liberty, and how can we resist it? It is not limited to asking, What is the threat to liberty *from government?* It understands that more than government can threaten liberty, and that sometimes this something more can be private rather than state action. Mill was not so concerned with the source. His concern was with liberty.

Threats to liberty change. In England norms may have been the problem in the late nineteenth century; in the United States in the first two decades of the twentieth century it was state suppression of speech.[12] The labor movement was founded on the idea that the market is sometimes a threat to liberty—not just because of low wages, but also because the market form of organization itself disables a certain kind of

---

[11]Or more precisely, against a certain form of government regulation. The more powerful libertarian arguments against regulation in cyberspace are advanced, for example, by Peter Huber in *Law and Disorder in Cyberspace: Abolish the FCC and Let Common Law Rule the Telecosm*. Huber argues against agency regulation and in favor of regulation by the common law. See also Thomas Hazlett in "The Rationality of U.S. Regulation of the Broadcast Spectrum," *Journal of Law and Economics* 33 (1990): 133, 133–39. For a lawyer, it is hard to understand precisely what is meant by "the common law." The rules of the common law are many, and the substantive content has changed. There is a common law process, which lawyers like to mythologize, in which judges make policy decisions in small spaces against the background of binding precedent. It might be this that Huber has in mind, and if so, there are, of course, benefits to this system. But as he plainly understands, it is a form of *regulation* even if it is constituted differently.

[12]The primary examples are the convictions under the 1917 Espionage Act; see, for example, *Schenck v United States*, 249 US 47 (1919) (upholding conviction for distributing a leaflet attacking World War I conscription); *Frohwerk v United States*, 249 US 204 (1919) (upholding conviction based on newspaper alleged to cause disloyalty); *Debs v United States*, 249 US 211 (1919) (conviction upheld for political speech said to cause insubordination and disloyalty).

freedom.[13] In other societies, at other times, the market is the key, not the enemy, to liberty.

Thus, rather than think of an enemy in the abstract, we should understand the particular threat to liberty that exists in a particular time and place. And this is especially true when we think about liberty in cyberspace. For my argument is that cyberspace teaches a new threat to liberty. Not new in the sense that no theorist has conceived of it before. Others have.[14] But new in the sense of newly urgent. We are coming to understand a newly powerful regulator in cyberspace, and we don't yet understand how best to control it.

This regulator is code—or more generally, the "built environment" of social life, its architecture.[15] And if in the middle of the nineteenth century it was norms that threatened liberty, and at the start of the twentieth state power that threatened liberty, and during much of the middle twentieth century the market that threatened liberty, my argument is that we understand how in the late twentieth century, and into the twenty-first, it is a different regulator—code—that should be our concern.

But it is not my aim to say that this should be our new single focus. My argument is not that there is a new single enemy different from the old. Instead,

I believe we need a more general understanding of how regulation works. One that focuses on more than the single influence of any one force such as government, norms, or the market, and instead integrates these factors into a single account.

This chapter is a step toward that more general understanding.[16] It is an invitation to think beyond the narrow threat of government. The threats to liberty have never come solely from government, and the threats to liberty in cyberspace certainly will not.

A Dot's Life

There are many ways to think about constitutional law and the limits it may impose on government regulation. I want to think about it from the perspective of someone who is regulated or constrained. That someone regulated is represented by this (pathetic) dot—a creature (you or me) subject to the different constraints that might regulate it. By describing the various constraints that might bear on this individual, I hope to show you something about how these constraints function together.

Here then is the dot.

How is this dot "regulated"?

Let's start with something easy: smoking. If you want to smoke, what constraints do you face? What factors regulate your decision to smoke or not?

One constraint is legal. In some places at least, laws regulate smoking—if you are under eighteen, the law says that cigarettes cannot be sold to you. If you are under twenty-six, cigarettes cannot be sold to you unless the seller checks your ID. Laws also regulate where smoking is permitted—not in O'Hare Airport, on an airplane, or in an elevator, for instance. In these two ways at least, laws aim to direct smoking behavior. They operate as a kind of constraint on an individual who wants to smoke.[17]

But laws are not the most significant constraints on smoking. Smokers in the United States certainly feel their freedom regulated, even if only rarely by

---

[13]See, for example, the work of John R. Commons, *Legal Foundations of Capitalism* (1924), 296–98, discussed in Herbert Hovenkamp, *Enterprise and American Law, 1836–1937* (Cambridge, Mass.: Harvard University Press, 1991), 235; see also John R. Commons, *Institutional Economics: Its Place in Political Economy* (1934).

[14]The general idea is that the tiny corrections of space enforce a discipline, and that this discipline is an important regulation. Such theorizing is a tiny part of the work of Michel Foucault; see *Discipline and Punish: The Birth of the Prison* (1979), 170–77, though his work generally inspires this perspective. It is what Oscar Gaudy speaks about in *The Panoptic Sort: A Political Economy of Personal Information* (Boulder, Colo.: Westview Press, 1993), 23. David Brin makes the more general point that I am arguing—that the threat to liberty is broader than a threat by the state; see David Brin, *The Transparent Society: Will Technology Force Us to Choose between Privacy and Freedom?* (New York: Basic Books, 1990), 110.

[15]See, for example, *The Built Environment: A Creative Inquiry into Design and Planning*, edited by Tom J. Bartuska and Gerald L. Young (Menlo Park, Calif.: Crisp Publications, 1994); *Preserving the Built Heritage: Tools for Implementation*, edited by J. Mark Schuster et al. (Hanover, N.H.: University Press of New England. 1997). In design theory, the notion I am describing accords with the tradition of Andres Duany and Elizabeth Plater-Zyberk; see, for example, William Lennertz. "Town-Making Fundamentals," in *Towns and Town-Making Principles,* edited by Andres Duany and Elizabeth Plater-Zyberk (New York: Rizzoli, 1991): "The work of . . . Duany and . . . Plater-Zyberk begins with the recognition that design affects behavior. [They] see the structure and function of a community as interdependent. Because of this, they believe a designer's decisions will permeate the lives of residents not just visually but in the way residents live. They believe design structures functional relationships, quantitatively and qualitatively, and that it is a sophisticated tool whose power exceeds its cosmetic attributes" (21).

[16]Elsewhere I've called this the "New Chicago School"; see Lawrence Lessig, "The New Chicago School," *Journal of Legal Studies* 27 (1998): 661. It is within the "tools approach" to government action (see John de Monchaux and J. Mark Schuster, "Five Things to Do," in Schuster, *Preserving the Built Heritage,* 3), but it describes four tools whereas Schuster describes five. I develop the understanding of the approach in the appendix to this book.

[17]See generally *Smoking Policy: Law, Politics, and Culture,* edited by Robert L. Rabin and Stephen D. Sugarman (New York: Oxford University Press, 1993); Lawrence Lessig, "The Regulation of Social Meaning." *University of Chicago Law Review* 62 (1995): 943, 1025–34; Cass R. Sunstein, "Social Norms and Social Roles," *Columbia Law Review* 96 (1996): 903.

the law. There are no smoking police, and smoking courts are still quite rare. Rather, smokers in America are regulated by norms. Norms say that one doesn't light a cigarette in a private car without first asking permission of the other passengers. They also say, however, that one needn't ask permission to smoke at a picnic. Norms say that others can ask you to stop smoking at a restaurant, or that you never smoke during a meal.

European norms are savagely different. There the presumption is in the smoker's favor; vis-à-vis the smoker, the norms are laissez-faire. But in the States the norms effect a certain constraint, and this constraint, we can say, *regulates* smoking behavior.

Law and norms are still not the only forces regulating smoking behavior. The market too is a constraint. The price of cigarettes is a constraint on your ability to smoke. Change the price, and you change this constraint. Likewise with quality. If the market supplies a variety of cigarettes of widely varying quality and price, your ability to select the kind of cigarette you want increases; increasing choice here reduces constraint.

Finally, there are the constraints created, we might say, by the technology of cigarettes, or by the technologies affecting their supply.[18] Unfiltered cigarettes present a greater constraint on smoking than filtered cigarettes if you are worried about your health. Nicotine-treated cigarettes are addictive and therefore create a greater constraint on smoking than untreated cigarettes. Smokeless cigarettes present less of a constraint because they can be smoked in more places. Cigarettes with a strong odor present more of a constraint because they can be smoked in fewer places. In all of these ways, how the cigarette *is* affects the constraints faced by a smoker. How it is, how it is designed, how it is built—in a word, its *architecture*.

Thus, four constraints regulate this pathetic dot—the law, social norms, the market, and architecture—and the "regulation" of this dot is the sum of these four constraints. Changes in any one will affect the regulation of the whole. Some constraints will support others; some may undermine others. A complete view, however, should consider them together.

So think of the four together like this:

In this drawing, each oval represents one kind of constraint operating on our pathetic dot in the center. Each constraint imposes a different kind of cost on the dot for engaging in the relevant behavior—in this case, smoking. The cost from norms is different from the market cost, which is different from the cost from law and the cost from the (cancerous) architecture of cigarettes.

The constraints are distinct, yet they are plainly interdependent. Each can support or oppose the others. Technologies can undermine norms and laws; they can also support them. Some constraints make others possible; others make some impossible. Constraints work together, though they function differently and the effect of each is distinct. Norms constrain through the stigma that a community imposes; markets constrain through the price that they exact; architectures constrain through the physical burdens they impose; and law constrains through the punishment it threatens.

We can call each constraint a "regulator," and we can think of each as a distinct modality of regulation. Each modality has a complex nature, and the interaction among these four is hard to describe. I've worked through this complexity more completely in the appendix. But for now, it is enough to see that they are linked and that, in a sense, they combine to produce the regulation to which our pathetic dot is subject in any given area.

The same model describes the regulation of behavior in cyberspace.

Law regulates behavior in cyberspace. Copyright law, defamation law, and obscenity laws all continue to threaten *ex post* sanction for the violation of legal rights. How well law regulates, or how efficiently, is a different question: in some cases it does so more efficiently, in some cases less. But whether better or not, law continues to threaten a certain consequence if it is defied. Legislatures enact;[19] prosecutors threaten;[20] courts convict.[21]

---

[18]These technologies are themselves affected, no doubt, by the market. Obviously, these constraints could not exist independently of each other but affect each other in significant ways.

[19]The ACLU lists twelve states that passed Internet regulations between 1995 and 1997; see http://www.aclu.org/issues/cyber/censor/stbills .html#bills (visited May 31, 1999).

[20]See, for example, the policy of the Minnesota attorney general on the jurisdiction of Minnesota over people transmitting gambling information into the state; available at http://www.ag.state.mn.us/home/consumer /consumernews/OnlineScams/memo.html (visited May 31, 1999).

[21]See, for example, *Playboy Enterprises v Chuckleberry Publishing, Inc.*, 939 FSupp 1032 (SDNY 1996); *United States v Thomas*, 74 F3d 1153 (6th Cir 1996); *United States v Miller*, 166 F3d 1153 (11th Cir 1999); *United States v Lorge*, 166 F3d 516 (2d Cir 1999); *United States v Whiting*, 165 F3d 631 (8th Cir 1999); *United States v Hibbler*, 159 F3d 233 (6th Cir 1998); *United States v Fellows*, 157 F3d 1197 (9th Cir 1998); *United States v Simpson*, 152 F3d 1241 (10th Cir 1998); *United States v Hall*, 142 F3d 988 (7th Cir 1998); *United States v Hockings*, 129 F3d 1069 (9th Cir 1997); *United States v Lacy*, 119 F3d 742 (9th Cir 1997); *United States v Smith*, 47 MJ 588 (CrimApp 1997); *United States v Ownby*, 926 FSupp 558 (WDVa 1996).

Norms also regulate behavior in cyberspace. Talk about democratic politics in the alt.knitting newsgroup, and you open yourself to flaming; "spoof" someone's identity in a MUD, and you may find yourself "toaded";[22] talk too much in a discussion list, and you are likely to be placed on a common bozo filter. In each case, a set of understandings constrain behavior, again through the threat of *ex post* sanctions imposed by a community.

Markets regulate behavior in cyberspace. Pricing structures constrain access, and if they do not, busy signals do. (AOL learned this quite dramatically when it shifted from an hourly to a flat rate pricing plan.)[23] Areas of the Web are beginning to charge for access, as online services have for some time. Advertisers reward popular sites; online services drop low-population forums. These behaviors are all a function of market constraints and market opportunity. They are all, in this sense, regulations of the market.

And finally, an analog for architecture regulates behavior in cyberspace—*code*. The software and hardware that make cyberspace what it is constitute a set of constraints on how you can behave. The substance of these constraints may vary, but they are experienced as conditions on your access to cyberspace. In some places (online services such as AOL, for instance) you must enter a password before you gain access; in other places you can enter whether identified or not.[24] In some places the transactions you engage in produce traces that link the transactions (the "mouse droppings") back to you; in other places this link is achieved only if you want it to be.[25] In

some places you can choose to speak a language that only the recipient can hear (through encryption);[26] in other places encryption is not an option.[27] The code or software or architecture or protocols set these features; they are features selected by code writers; they constrain some behavior by making other behavior possible, or impossible. The code embeds certain values or makes certain values impossible. In this sense, it too is regulation, just as the architectures of real-space codes are regulations.

As in real space, then, these four modalities regulate cyberspace. The same balance exists. As William Mitchell puts it (though he omits the constraint of the market); Architecture, laws, and customs maintain and represent whatever balance has been struck [in real space]. As we construct and inhabit cyberspace communities, we will have to make and maintain similar bargains—though they will be embodied in software structures and electronic access controls rather than in architectural arrangements.[28]

Laws, norms, the market, and architectures interact to build the environment that "Netizens" know. The code writer, as Ethan Katsh puts it, is the "architect."[29]

But how can we "make and maintain" this balance between modalities? What tools do we have to achieve a different construction? How might the mix of real-space values be carried over to the world of cyberspace? How might the mix be changed if change is desired?

From *Code: And Other Laws of Cyberspace* (New York: Basic Books, 1999), pp. 3–8; 85–90, 241–242, 254–255. Reprinted by permission of Basic Books, a member of Perseus Books Group.

---

[22]See Julian Dibbell, "A Rape in Cyberspace," *Village Voice*, December 23, 1993, 36.

[23]See, for example, "AOL Still Suffering but Stock Price Rises," *Network Briefing*, January 31, 1997; David S. Hilzenrath, "'Free' Enterprise, Online Style; AOL, CompuServe, and Prodigy Settle FTC Complaints," *Washington Post*, May 2, 1997, G1; "America Online Plans Better Information About Price Changes," *Wall Street Journal*, May 29, 1998, B2; see also Swisher, *Aol.com*, 206–8.

[24]USENET postings can be anonymous; see Henry Spencer and David Lawrence, *Managing USENET* (Sebastopol, Calif.: O'Reilly & Associates, 1998), 366–67.

[25]Web browsers make this information available, both in real time and archived in a cookie file; see http://www.cookiecentral.com/faq.htm (visited May 31, 1999). They also permit users to turn this tracking feature off.

[26]PGP is a program to encrypt messages that is offered both commercially and free.

[27]Encryption, for example, is illegal in some international contexts; see Stewart A. Baker and Paul R. Hurst, *The Limits of Trust: Cryptography, Governments, and Electronic Commerce* (Boston: Kluwer Law International, 1998) 130–6.

[28]William J. Mitchell, *City of Bits: Space, Place, and the Infobahn* (Cambridge, MA: MIT Press, 1995), 159.

[29]See M. Ethan Katsh, "Software Worlds and the First Amendment: Virtual Doorkeepers in Cyberspace," *University of Chicago legal Forum* (1996): 335, 340. "If a comparison to the physical world is necessary, one might say that the software designer is the architect, the builder, and the contractor, as well as the interior decorator."

Rosalind Franklin and the Double Helix

Lynne Osman Elkin

Although she made essential contributions toward elucidating the structure of DNA, Rosalind Franklin is known to many only as seen through the distorting lens of James Watson's book, The Double Helix.

In 1962, James Watson, then at Harvard University, and Cambridge University's Francis Crick stood next to Maurice Wilkins from King's College, London, to receive the Nobel Prize in Physiology or Medicine for their "discoveries concerning the molecular structure of nucleic acids and its significance for information transfer in living material." Watson and Crick could not have proposed their celebrated structure for DNA as early in 1953 as they did without access to experimental results obtained by King's College scientist Rosalind Franklin. Franklin had died of cancer in 1958 at age 37, and so was ineligible to share the honor. Her conspicuous absence from the awards ceremony—the dramatic culmination of the struggle to determine the structure of DNA—probably contributed to the neglect, for several decades, of Franklin's role in the DNA story. She most likely never knew how significantly her data influenced Watson and Crick's proposal.

Capsule Biography

Franklin, shown in figures 1 and 2, was born 25 July 1920 to Muriel Waley Franklin and merchant banker Ellis Franklin, both members of educated and socially conscious Jewish families. They were a close immediate family, prone to lively discussion and vigorous debates at which the politically liberal, logical, and determined Rosalind excelled: She would even argue with her assertive, conservative father. Early in life, Rosalind manifested the creativity and drive characteristic of the Franklin women, and some of the Waley women, who were expected to focus their education, talents, and skills on political, educational, and charitable forms of community service. It was

thus surprising when young Rosalind expressed an early fascination with physics and chemistry classes at the academically rigorous St. Paul's Girls' School in London, and unusual that she earned a bachelor's degree in natural sciences with a specialty in physical chemistry. The degree was earned at Newnham College, Cambridge in 1941.

From 1942 to 1946, Franklin did war-related graduate work with the British Coal Utilization Research Association. That work earned her a PhD from Cambridge in 1945, and an offer to join the Laboratoire Central des Services Chimiques de l'Etat in Paris. She worked there, from 1947 to 1950, with Jacques Mering and became proficient at applying x-ray diffraction techniques to imperfectly crystalline matter such as coal. In the period 1946–49, she published five landmark coal-related papers, still cited today, on graphitizing and nongraphitizing carbons. By 1957, she had published an additional dozen articles on carbons other than coals. Her papers changed the way physical chemists view the microstructure of coals and related substances.

Franklin made many friends in the Paris laboratory and often hiked with them on weekends. She preferred to live on her own modest salary and frustrated her parents by continually refusing to accept money from them. She excelled at speaking French and at French cooking and soon became more comfortable with intellectual and egalitarian "French ways" than with conventional English middle-class customs. Consequently, she did not fit in well at King's College, where she worked on DNA from 1951 to 1953. Franklin chose to leave King's and, in the spring of 1953, moved to Birkbeck College. Many of the students there were evening students who worked

Lynne Elkin *(lelkin@csuhayward.edu) is a professor of biological sciences at California Stale University, Hayward. She welcomes responses and inquiries about issues pertaining to the story of Rosalind Franklin and DNA structure.*

during the day, and Franklin was impressed with their dedication. After the move to Birkbeck, she began her celebrated work with J. Desmond Bernal on RNA viruses like tobacco mosaic virus (TMV). She was a cautious scientist who began to trust her intuition more as she matured (see box 1 on page 45). She published 14 papers about viruses between 1955 and 1958, and completed the research for three others that colleague Aaron Klug submitted for publication after her death.

In his obituary for Franklin, Bernal described her as a "recognized authority in industrial physico-chemistry." In conclusion, he wrote, "As a scientist, Miss Franklin was distinguished by extreme clarity and perfection in everything she undertook. Her photographs are among the most beautiful of any substances ever taken [1]."

Discovery of Two Forms for DNA

Franklin's most famous and controversial work yielded critical data that Watson and Crick used to determine DNA's structure. DNA is a double-helical molecule roughly in the form of a spiral staircase. The double-helical molecule, consisting of two unbranched polynucleotide chains, is best visualized by imagining it straightened into a ladder. The side rails of the ladder are each made up of alternating sugar and phosphate groups, linked by so-called 3' or 5' phosphodiester bonds. The sequence of the atoms in each rail runs in opposite directions, so the two sides of the molecular backbone are often described as antiparallel to each other. The rungs of the ladder consist of specific hydrogen-bonded horizontal pairs of nitrogenous bases that are attached to the deoxy-ribose sugars in the backbone's side rails. Birkbeck's Sven Furberg, who studied both the nucleoside and nucleotide of one of the bases, cytosine, in 1949, discovered that the base would be perpendicular to the sugar: The result can be extrapolated to hydrated DNA [2].

The pairs of nitrogenous bases that make up the rungs are in the keto, as opposed to enol, tautomeric configurations. (The two are distinguished by the locations of hydrogen atoms available for hydrogen bonding.) The smaller single-ringed pyrimidines, cytosine (C) and thymine (T), are always paired with larger double-ringed purines, guanine (G) and adenine (A). Indeed, the consistent pairing of G with C and of A with T, as first proposed by Watson, explains the identical size of the ladder rungs and also Erwin Chargaff's 1952 observation that G and

C (and likewise A and T) are always present in DNA in approximately equal amounts. The consistent pairings, along with the irregular linear vertical sequence of the bases, underlie DNA's genetic capacity.

In an experiment carried out shortly after she arrived at King's, Franklin identified two distinct configurations, called by her the A and B forms, in which DNA could exist. Her work, first presented in an internal King's seminar in November 1951 and published in *Nature* in 1953, was essential for determining the structure of DNA. Researchers working prior to Franklin's discovery invariably had to deal with confusing x-ray diffraction patterns that resulted from a mixture of the A and B forms.

The drier crystalline A form contains about 20% water by weight and is optimally produced at about 75% relative humidity. Cation (for example, Na^+) bridges between ionized phosphates are probably responsible for intermolecular linking in the crystalline structure. The less ordered, fully hydrated, paracrystalline B form—typically the configuration that occurs in vivo—is obtained from the crystalline A form when DNA fibers absorb water in excess of 40% of their weight. Optimal production of the B form occurs at approximately 90% relative humidity. Extra hydration makes it easier for the molecule to assume the lowest-energy, helical configuration. It also keeps the two helical backbone chains farther apart than in the A form and elongates the molecule by about 30% until the B form appears with its bases oriented perpendicular to the fiber axis.

Franklin slowly and precisely hydrated then dehydrated her DNA sample to obtain her best pictures of the A form. To get her samples, though, she had to extract DNA fibers from a gel-like undenatured DNA sample that Wilkins had acquired from Rudolf Signer of the University of Berne in Switzerland. Franklin pulled exceptionally thin single fibers and controlled the humidity in her specimen chamber by bubbling hydrogen gas through salt solutions and then flooding the chamber with the humid gas that resulted. Franklin's PhD student, Raymond Gosling, told me that the chamber leaked so much hydrogen gas that he was afraid they would blow themselves up accidentally and take half of King's College with them.

Franklin's careful treatment during the transformation from crystalline A-form to hydrated B-form DNA resulted in such a drastic size change that, according to Gosling, the elongating specimen practically "leaped off the stage." After designing a tilting microfocus camera and developing a technique for improving the orientation of her DNA fibers in the camera's collimated beam, Franklin took x-ray diffraction photographs of the B form.

Franklin's B-form data, in conjunction with cylindrical Patterson map calculations that she had applied to her A-form data, allowed her to determine DNA's density, unit-cell size, and water content. With those data, Franklin proposed a double-helix structure with precise measurements for the diameter, the separation between each of the coaxial fibers along the fiber axis direction, and the pitch of the helix [3].

The resolution of the B-form photograph #51 shown in figure 3 allowed Franklin to determine that each turn of the helix in the B form is 34 Å long and contains 10 base pairs separated by 3.4 Å each,[3] in accordance with less precise data obtained by William Astbury and Florence Bell in 1938 [4]. Wilkins, photographing living sperm cells in 1952, obtained an X-shaped B-form diffraction pattern similar to Franklin's. Her photograph, though, showed much more detail.

Additional Contributions

The cylindrical Patterson map calculations that Franklin applied to the A-form of DNA were the first such calculations applied to any molecule. They confirmed her suggestions that the hydrophilic sugar phosphates form the external backbone of the DNA molecule and that the hydrophobic base pairs are protected inside that backbone from the cell's aqueous environment. The calculations also allowed her to deduce that the A-form helix has two antiparallel chains. With Gosling, Franklin provided details of the physical distortion accompanying the dehydration transformation from B-form to A-form DNA [5]. She also showed that the bases of the A form are tilted and curved slightly, and that 11 pairs of bases are compacted within a repeat distance of 28.1 Å.

In her section of the 1952 King's Medical Research Council (MRC) report, Franklin gave quantitative measurements for the interphosphate distances and discussed the external placement of the phosphates. Her presentation was instrumental in getting Watson and Crick to abandon their earlier attempts at placing the bases on the outside of their model. Initially, they (and Linus Pauling, too) mistakenly thought that the bases would have to be externally accessible in order to pass genetic information.

In May 1952, Franklin presented her clearest evidence of the helical backbone, with her diffraction photograph #51. Although she did not yet realize how the nitrogenous bases are paired or that the helical backbone rails of B-form DNA are antiparallel, her notebook entries starting in January 1951 clearly show that she was making significant progress toward solving those two final aspects of DNA structure. After reading an article by June Broomhead [6], and studying other related papers, she had used the keto configuration for at least three of the four bases. She was aware both of Jerry Donohue's work concerning tautomeric forms of bases and of Chargaff's work.

Astbury and Bell's earlier, less clear diffraction photographs and later data of Wilkins suggested some of the data that Franklin derived from her photograph #51. But Franklin's results were much more precise than the Astbury and Bell data, which showed neither an X pattern nor layer lines. Astbury and Bell themselves described their results as "still rather obscure." After Oxford crystallographer Dorothy Hodgkin helped her to eliminate two of three possibilities she had calculated, Franklin described the correct crystallographic space group for DNA in the 1952 MRC report.

Only after Crick obtained Franklin's data—his thesis adviser, Max Perutz, agreed to give him a copy of the 1952 report and Watson had seen photograph #51—was he sufficiently convinced to start constructing the backbone of the successful DNA model. He recognized the similarity of the space group Franklin had calculated to that of his thesis molecule, hemoglobin, and immediately deduced that there would be an antiparallel orientation between the two DNA coaxial fibers. Within one week, he started modeling the correct backbone in a manner compatible with Franklin's data. On several occasions, Crick has acknowledged that the data and conclusions in the 1952 report were essential.

Franklin's 17 March 1953 Draft

On 18 March 1953, Wilkins penned a letter acknowledging receipt of the Watson and Crick manuscript that described the structure of DNA. A day earlier, Franklin, who was preparing to leave for Birkbeck, polished an already written draft manuscript outlining her conclusions about the double-helix backbone chain of B-form DNA [7]. Franklin only slightly modified her draft to prepare her April 1953 *Nature* paper, which appeared as the third in a series that led off with the famous Watson and Crick proposal. Partly as a consequence of its placement, Franklin's paper seemed merely to support Watson and Crick's work. But Franklin's data played far more than just a supporting role—as early as 1968, Watson's *The Double Helix* tells us so.

Ironically, despite its negative portrayal of Franklin, *The Double Helix* was what first brought widespread attention to Franklin's key contributions to the Watson and Crick proposal. The book describes how Watson and Crick built their first, and incorrect, model right after Watson inaccurately reported Franklin's November 1951 seminar data to Crick. It also details how, after 13 months of inactivity, they built their correct model once Wilkins showed Franklin's photograph #51 to Watson, and Perutz showed Crick the 1952 MRC report.

The importance of Franklin's work to the discovery of DNA structure has not been well documented until recently for a variety of reasons too long to discuss here. Relevant issues include women's being underrepresented in historical accounts, although several authors have striven to correct that imbalance [8]; Watson and Crick's routinely citing the more senior Wilkins before Franklin; and Wilkins's repeating much of Franklin's work. In addition, Wilkins, not Franklin, was nominated for membership in the Royal Society even though, at the time of his nomination, Franklin was famous for her TMV accomplishments.

Conflict within the King's MRC

Franklin was an outstanding and accomplished scientist—a fascinating individual with a strong personality who made a lasting impression on almost everyone she met. Throughout her career, she routinely ate lunch amicably with both male and female colleagues and most of her acquaintances liked her. Her numerous lifelong friends thought her bright, fascinating, witty, and fun. Most of her lunchtime colleagues at King's would agree with that description, but only as it pertained to her lunchtime persona. When Rosalind headed for the laboratory, she shed almost every vestige of lightheartedness as she focused exclusively on her work.

Furthermore, in what amounted almost to social heresy in England, en route to her laboratory she typically bypassed the morning coffeepot and afternoon tea in favor of a direct assault on her work. Franklin was considered by King's colleagues to be "too French" in her dress, in her intellectual interests, and in her temperament. She was exceedingly direct, intent, and serious, with the tendency to leap into passionate debate. She could be assertive, uniquely stubborn, argumentative, and abrasive to the point that colleagues, especially Wilkins, sometimes found her unpleasant.

On the rare occasions when Franklin departed from her typical behavior, King's colleagues usually did not even notice—they took her seriously at all times. That became especially important when she created a "death of the helix" funeral invitation as a joke after obtaining some data indicating that A-form DNA is nonhelical (see box 2 on page 46).

Whereas Franklin was quick, intense, assertive, and directly confrontational, Wilkins was exceedingly shy, indirect, and slowly calculating to the point of appearing plodding. Almost all testimony from King's staff indicates that any blame for their perpetual conflict needs to be shared. John Randall, director of the MRC, also deserves some share of the blame. Without informing Wilkins, he wrote a letter to Franklin assigning DNA structural studies to her. He also did not warn Franklin about Wilkins's continuing interest in DNA.

Randall and Wilkins did make some important accommodations for Franklin. They hired her into a senior position based on her expertise. They gave her an excellent laboratory, the highest quality DNA, and a decent budget. However, numerous MRC women, although very well treated for that era, did not receive precisely equal treatment with men. Taken as a whole, the King's MRC women did not rank quite as highly as the men. Also, Gosling told me that women were not allowed upstairs for after-lunch coffee in the smoking room, "a wonderful long room," he explained, "with window seats." That exclusion undoubtedly cut off a natural route for easy scientific conversation.

Franklin's tenure at the MRC ended on an unfortunate note. As a condition to agreeing to transfer her fellowship to Birkbeck, Randall made Franklin promise not to perform additional experiments on DNA, or even to think about DNA. Moreover, Franklin was forced to leave her diffraction photograph behind at King's and to leave the work of confirming DNA's structure to Wilkins. Personality conflicts were the major source of Franklin's difficulty at King's, and her status as a woman may have made her problems worse. Even today, I and many of my women colleagues find that forcefully aggressive behavior, for example, is usually considered merely irritating when exhibited by a man but is often deemed unacceptable when demonstrated by a woman.

Box 21.1 The Evolution of Franklin's Intuition

Every scientist I interviewed, except James Watson and Max Perutz, agreed that Rosalind Franklin was a superb scientist with the sharp mind and vision needed to plan, execute, and analyze a good experiment. Questions about her capabilities have centered on her ability to make intuitive leaps when interpreting results, mainly because she seemed hesitant to do so in her DNA work. Franklin often expressed the opinion that the facts should speak for themselves. Her desire to have solid proof for her ideas before publishing them helps explain her highly successful publication record. She argued that a scientist need not be highly speculative, which gave the impression that she was incapable of speculation.

Franklin's manner and scientific approach while at King's College, London were uncannily close to those described by Frederick Dainton, Franklin's physical chemistry don when she was an undergraduate at Cambridge University. Writing to biographer Anne Sayre, Dainton says that "he was attracted by [Franklin's] directness . . . including the way she defended her point of view, something she was never loath to do." Dainton found Franklin to be a "very private person with very high personal and scientific standards, and uncompromisingly honest. . . . She once told me the facts will speak for themselves, but never fully accepted my urging that she must try to help the receptor of her messages. . . . As you point out, the logical sequential arguments meant everything to Rosalind [13]. . . ."

Francis Crick has suggested that Franklin's aversion to speculative thinking about DNA excludes her from the rank of the great scientists. He has noted, though, that the problem might not have been that Franklin lacked intuition, but rather that she might not have trusted it. Moreover, Crick has suggested that Franklin's intuition seemed to be rapidly improving as she matured as a scientist while working with tobacco mosaic virus. And if Franklin did not trust her intuition while working on DNA, it may have been at least partially because she did not have anyone at King's with whom she could properly discuss her ideas.

Box 21.2 The Helix Funeral Invitation

Maurice Wilkins insisted that DNA was helical before there was clear evidence. That greatly annoyed Rosalind Franklin, as did so many things about him. Franklin was well aware that the x-ray diffraction photographs of the A form, unlike those of the B form, lacked the clear X-shaped pattern indicative of a helix. Instead, they displayed a detailed and confusing assortment of reflections that she could only interpret through the complicated and daunting procedure of cylindrical Patterson analysis. Therefore, when, over a period of about five months starting on 18 April 1952, Franklin recorded some misleading, apparently asymmetrical double orientation data in the A form, she got perverse pleasure out of possibly annoying Wilkins with her result. As a joke, she penned the "death of the helix" funeral invitation reproduced here. That she wrote the invitation is clearly substantiated in interviews with Raymond Gosling and Wilkins and in many other professional accounts.

The extent and significance of the distribution of the invitation is another matter. During interviews, only diffraction expert Alec Stokes said that he had received one. A few people said that they saw an invitation posted on a bulletin board, and most said that they had never even seen one. That testimony is incompatible with the often repeated claim that a multitude of these cards were sent out.

Historians Horace Judson and Robert Olby argued that the invitation indicated Franklin was antihelical, although Olby notes it was a joke as well [10, 11]. To the contrary: Franklin's student Raymond Gosling is adamant that Franklin considered the B form, with its striking x-ray pattern, to be helical. Examinations of Franklin's notebooks have led to the same conclusion, although it is also clear that, for a while, Franklin definitely had her doubts about the helicity of the A form.

Watson's view: *The Double Helix*

In *The Double Helix*, Watson bases his account of Franklin on recollections of their three brief meetings between 1951 and 1953, and on repeated complaints about her from Wilkins. The "Rosy" that Watson describes is a caricature based on the more difficult aspects of Franklin's personality. His portrayal—a far cry from the competent scientist described by her colleagues or the fascinating person described by her friends—is an effective device for promoting the idea that Watson and Crick had to rescue DNA data from—as Watson's book puts it—this "belligerent" woman who could not "keep her emotions under control" and who did not know how to interpret her own data. Watson falsely depicts Franklin as Wilkins's assistant, incapable and unworthy of Nobel Prize-caliber work. His book was published against the vehement protest of key DNA participants, who were upset about its numerous inaccuracies [9].

Unfortunately, Watson's admittedly fascinating, irreverent, and concise book has sold millions of copies and is for many, especially in the US, the primary source of information about DNA history. That such a one-sided account both is presented as historical fact and has had tremendous influence is worrisome. Watson's treatment of Franklin, who was then deceased and not protected by libel laws, is especially troublesome.

Watson's depiction of Franklin went largely unchallenged, at least in print, until Anne Sayre published her 1975 biography *Rosalind Franklin and DNA* (W. W. Norton, 1975). Sayre asked awkward, science-related questions: Why would Franklin give her data to Watson, Crick, and Wilkins, three scientists who seemed to have nothing to offer in return at that time? Where would Watson and Crick be without Franklin's results? Why did they not acknowledge Franklin's contributions clearly and appropriately? Regrettably, Sayre's influence was diminished because she was misled about the conditions for women at the MRC. She mistakenly assumed those conditions to be as problematic as they were at the rest of King's, and therefore incorrectly interpreted Franklin's problems at King's in terms of gender issues. Subsequent accounts, though in many ways excellent, typically are dismissive of gender concerns [10] and personality issues [11]. They don't address the awkward questions Sayre raises nor do they sufficiently emphasize that the Watson and Crick model was made possible by Franklin's data. Fortunately, Brenda Maddox's outstanding and comprehensive 2002 biography of Franklin [12] does consider questions of acknowledgment and corrects misconceptions about gender and personality issues. Perhaps it will educate the public more fully than have previous texts. (See PHYSICS TODAY, February 2003, page 61, for a review of Maddox's book.)

Inadequate Acknowledgment

In their 1953 paper, Watson and Crick state that they had been "stimulated by a knowledge of the general nature of the unpublished experimental results and ideas of Dr. M. H. F. Wilkins, Dr. R. E. Franklin and their co-workers at King's College, London." That oblique acknowledgment misrepresented Franklin's role and, whatever its intention, left most people with the impression that her work mainly served to confirm that of Watson and Crick. It has to be one of the greatest understatements in the history of scientific writing.

Given her temper, it is likely that Franklin would have been very angry if she had known the extent to which Watson and Crick used her data. In a 1951 incident, Franklin was furious that a conference acquaintance planned on publishing an idea of hers without giving proper acknowledgment. She shocked him by the tone of her letter in which she demanded coauthorship.

In 1954, Crick and Watson published a detailed methods paper in the *Proceedings of the Royal Society*. In that paper, their acknowledgment of Franklin is often ambiguous. Three of the four times they mention the importance of her data, they link it with mention of Wilkins's data first in a way that suggests the two scientists' contributions were of equal importance. Twice they follow what appears to be a clear acknowledgment of Franklin's contributions with a recanting qualifying statement, for example, "we should at the same time mention that the details of [the King's College's] X-ray photographs were not shown to us and that the formulation of the structure was largely the result of extensive model building in which the main effort was to find any structure which was stereochemically feasible." They might not have known all of the details, but they had access to a significant number of them (see also box 3 below).

Another lost opportunity for acknowledging Franklin occurred during the 1962 Nobel Prize ceremony. Neither Watson nor Crick thanked Franklin for making their discovery possible. Indeed, neither mentioned her name, although, according to Wilkins, Crick did ask him to mention Franklin. That request was a dubious shifting of responsibility given Wilkins's antipathy toward Franklin, and Wilkins in fact made only minor mention of her. Crick spoke on the genetic code, which had nothing to do with Franklin's work. Watson spoke on RNA, including RNA viruses. Franklin was an expert on those viruses, yet, in his 59 citations, Watson managed to omit any reference to her work. To this day, Watson emphasizes the opinion that Franklin, although a gifted experimentalist, could not properly interpret all of her own DNA data.

Franklin is prominent in virtually every telling of DNA history, but she is painted differently in various accounts. Watson and Crick made one of the most important and impressive scientific discoveries of the 20th century, but their golden helix is tarnished by the way they have treated Franklin and Wilkins. A meaningful gesture, given that it was Franklin's data that Watson and Crick most directly used, would be for scientists to refer to the "Watson, Crick, and Franklin structure for DNA."

It is important to stop demeaning Franklin's reputation, but equally important to avoid obscuring her more difficult personality traits. She should not be put on a pedestal as a symbol of the unfair treatment accorded to many women in science. Her complicated relationship with Wilkins has been treated in overly simplistic ways. Distorted accounts, which inaccurately portray the three Nobel Prize winners as well as Franklin, are unfortunate and unnecessary: There was enough glory in the work of the four to be shared by them all.

Box 21.3 Frustrated Contributors to DNA Structure

Jerry Donohue, the chemist who had insisted that James Watson try modeling DNA with the keto form of nitrogenous bases, expressed to historian Horace Judson his frustration at not being sufficiently acknowledged. "Let's face it," Donohue wrote, "if the fates hadn't ordained that I share an office with Watson and [Francis] Crick in the Cavendish [Laboratory] in 1952–3, they'd *still* be puttering around trying to pair 'like-with-like' enol forms of the bases [14]." Erwin Chargaff similarly felt insufficiently acknowledged for discovering that the DNA bases guanine and cytosine (and likewise adenine and thymine) are present in approximately equal amounts.

Maurice Wilkins was frustrated because he, understandably, had expected to be collaborating with Watson and Crick. In 1951, Rosalind Franklin joined the Medical Research Council unit at King's College, London. At that time, Wilkins lacked her years of experimental diffraction experience; like most of his MRC colleagues at the King's biophysics unit, he was an accomplished scientist struggling to apply his prior experience in physics to biological problems. As a result, other MRC units jokingly called the biophysics unit "Randall's Circus," after director John Randall. Wilkins acknowledged the nature of Randall's unit when he said of Franklin, "We were amateurs, but she was a professional."

Wilkins, though, learned quickly and acquired some valuable collaborators. By early 1953, approximately eight months after Franklin recorded her famous photograph #51, he and Herbert Wilson were able to use a variety of DNA sources to obtain x-ray photographs sufficiently detailed for structural studies [15]. Later, Wilkins designed a higher-resolution x-ray camera that enabled his team to take diffraction photographs that helped to confirm and refine the Watson and Crick structure. Wilkins had initiated DNA structural studies at King's and, in 1950 with Raymond Gosling, had taken the best DNA photograph to date—one that showed a high degree of crystallinity and thus implied that DNA structure could be solved by x-ray diffraction. Wilkins had recognized DNA's genetic importance long before Crick, obtained high-quality DNA from Rudolf Signer, and supported the hiring of experienced x-ray diffraction expert Rosalind Franklin. Wilkins is well respected by the King's staff that I interviewed and his contributions were deemed worthy of a Nobel Prize.

I thank Barbara Low, Mary Singleton, Martha Breed, Marilyn Goldfeather, and Paulina Miner for their many helpful editorial suggestions, and Richard Hasbrouck for his assistance in preparing figures.

References

1. J. D. Bernal, *Nature* **182**, 154 (1958).
2. S. Furberg, *Acta Crystallogr.* **3**, 325 (1950); *Acta Chem. Scand.* **6**, 634 (1952).
3. These figures are quoted in Franklin's 1952 Medical Research Council report and her unpublished 17 March 1953 draft paper. The draft was discovered many years later and written about by A. Klug in *Nature* **219**, 808 (1968); **219**, 843 (1968); **248**, 787 (1974).
4. W. T. Astbury, F. O. Bell, *Nature* **141**, 747 (1938); *Cold Spring Harb. Symp. Quant. Biol.* **6**, 112 (1938).
5. R. E. Franklin, R. G. Gosling, *Acta Crystallogr.* **6**, 673 (1953); **6**, 678 (1953).
6. J. Broomhead, *Acta Crystallogr.* **4**, 92 (1951).
7. A. Klug, *Nature* **219**, 808 (1968); **219**, 843 (1968); **248**, 787 (1974).
8. M. Bailey, *American Women in Science: A Biographical Dictionary*, ABC-CLIO, Santa Barbara, Calif. (1994); G. Kass-Simon, R Farnes, eds., *Women of Science: Righting the Record* Indiana U. Press, Bloomington, Ind. (1990); M. B. Ogilvie, *Women in Science: Antiquity Through the Nineteenth Century. A Biographical Dictionary With Annotated Bibliography*, MIT Press, Cambridge, Mass. (1986); M. W. Rossiter, *Women Scientists in America: Struggles and Strategies to 1940*, Johns Hopkins U. Press, Baltimore, Md. (1982); M. W. Rossiter, *Women Scientists in America: Before Affirmative Action, 1940–1972*, Johns Hopkins U. Press, Baltimore, Md. (1995).
9. See for example, W. Sullivan in J. D. Watson's *The Double Helix: A Personal Account of the Discovery of the Structure of DNA*, Norton Critical Edition, G. S. Stent, ed., W. W. Norton, New York (1980). Copies of original letters are in the Norman Collection on the History of Molecular Biology in Novato, Calif.
10. F. H. Portugal, J. S. Cohen, *A Century of DNA: A History of the Discovery of the Structure and Function of the Genetic Substance*, MIT Press, Cambridge, Mass. (1977); H. F. Judson, *The Eighth Day of Creation: Makers of the Revolution in Biology*, CSHL Press, Plainview, N.Y. (1996).
11. R. C. Olby, *The Path to the Double Helix: The Discovery of DNA*, Dover, New York (1994).
12. B. Maddox, *Rosalind Franklin: The Dark Lady of DNA*, HarperCollins, New York (2002).
13. The original letter is in the Anne Sayre Collection of the Microbiological Society Archives at the University of Maryland, Baltimore County. See also ref. 12.
14. The original 5 March 1976 letter is in the Anne Sayre Collection of the Microbiological Society Archives at the University of Maryland, Baltimore County. See also ref. 12.
15. H. R. Wilson, *Trends Biochem. Sci.* **13**, 275 (1988).

Reading 22

Genomics and Its Impact on Science and Society

US Department of Energy Genome Research Program

The Human Genome Project and Beyond

A Primer

Cells are the fundamental working units of every living system. All the instructions needed to direct their activities are contained within the chemical DNA (deoxyribonucleic acid).

DNA from all organisms is made up of the same chemical and physical components. The **DNA sequence** is the particular side-by-side arrangement of bases along the DNA strand (e.g., ATTCCGGA). This order spells out the exact instructions required to create a particular organism with its own unique traits.

The **genome** is an organism's complete set of DNA. Genomes vary widely in size: The smallest known genome for a free-living organism (a bacterium) contains about 600,000 DNA base pairs, while human and mouse genomes have some 3 billion (see p. 3). Except for mature red blood cells, all human cells contain a complete genome.

DNA in each human cell is packaged into 46 **chromosomes** arranged into 23 pairs. Each chromosome is a physically separate molecule of DNA that ranges in length from about 50 million to 250 million base pairs. A few types of major chromosomal abnormalities, including missing or extra copies or gross breaks and rejoinings (translocations), can be detected by microscopic examination. Most changes in DNA, however, are more subtle and require a closer analysis of the DNA molecule to find perhaps single-base differences.

Each chromosome contains many **genes**, the basic physical and functional units of heredity. Genes are specific sequences of bases that encode instructions on how to make proteins. Genes comprise only about 2% of the human genome; the remainder consists of noncoding regions, whose functions may include providing chromosomal structural integrity and regulating where, when, and in what quantity proteins are made. The human genome is estimated to contain some 25,000 genes.

Although genes get a lot of attention, the **proteins** perform most life functions and even comprise the majority of cellular structures. Proteins are large, complex molecules made up of chains of small chemical compounds called amino acids. Chemical properties that distinguish the 20 different amino acids cause the protein chains to fold up into specific three-dimensional structures that define their particular functions in the cell.

The constellation of all proteins in a cell is called its **proteome**. Unlike the relatively unchanging genome, the dynamic proteome changes from minute to minute in response to tens of thousands of intra- and extracellular environmental signals. A protein's chemistry and behavior are determined by the gene sequence and by the number and identities of other proteins made in the same cell at the same time and with which it associates and reacts. Studies to explore protein structure and activities, known as proteomics, will be the focus of much research for decades to come and will help elucidate the molecular basis of health and disease.

The Human Genome Project, 1990–2003

A Brief Overview

Though surprising to many, the Human Genome Project (HGP) traces its roots to an initiative in the U.S. Department of Energy (DOE). Since 1947, DOE and its predecessor agencies have been charged by Congress with developing new energy resources and technologies and pursuing a deeper understanding of potential health and environmental risks posed by their production and use. Such studies, for example, have provided the scientific basis for individual risk assessments of nuclear medicine technologies.

In 1986, DOE took a bold step in announcing the Human Genome Initiative, convinced that its missions would be well served by a reference human genome sequence. Shortly thereafter, DOE joined with the National Institutes of Health to develop a plan for a

Genomics and Its Impact on Science and Society. U.S. DOE Human Genome Project, HYPERLINK "http://www.ornl.gov/sci/techresources/ Human_Genome/publicat/primer2001/1.shtml" www.ornl.gov/sci/techresources/Human_Genome/publicat/primer2001/1.shtml.

joint HGP that officially began in 1990. During the early years of the HGP, the Wellcome Trust, a private charitable institution in the United Kingdom, joined the effort as a major partner. Important contributions also came from other collaborators around the world, including Japan, France, Germany, and China.

Ambitious Goals

The HGP's ultimate goal was to generate a high-quality reference DNA sequence for the human genome's 3 billion base pairs and to identify all human genes. Other important goals included sequencing the genomes of model organisms to interpret human DNA, enhancing computational resources to support future research and commercial applications, exploring gene function through mouse-human comparisons, studying human variation, and training future scientists in genomics.

The powerful analytical technology and data arising from the HGP present complex ethical and policy issues for individuals and society. These challenges include privacy, fairness in use and access of genomic information, reproductive and clinical issues, and commercialization (see p. 8). Programs that identify and address these implications have been an integral part of the HGP and have become a model for bioethics programs worldwide.

A Lasting Legacy

In June 2000, to much excitement and fanfare, scientists announced the completion of the first working draft of the entire human genome. First analyses of the details appeared in the February 2001 issues of the journals *Nature* and *Science*. The high-quality reference sequence was completed in April 2003, marking the end of the Human Genome Project— 2 years ahead of the original schedule. Coincidentally, it also was the 50th anniversary of Watson and Crick's publication of DNA structure that launched the era of molecular biology.

Available to researchers worldwide, the human genome reference sequence provides a magnificent and unprecedented biological resource that will serve throughout the century as a basis for research and discovery and, ultimately, myriad practical applications. The sequence already is having an impact on finding genes associated with human disease (see p. 3). Hundreds of other genome sequence projects— on microbes, plants, and animals—have been completed since the inception of the HGP, and these data now enable detailed comparisons among organisms, including humans.

Many more sequencing projects are under way or planned because of the research value of DNA sequence, the tremendous sequencing capacity now available, and continued improvements in technologies. Sequencing projects on the genomes of many microbes, as well as the chimpanzee, pig, sheep, and domestic cat, are in progress.

Beyond sequencing, growing areas of research focus on identifying important elements in the DNA sequence responsible for regulating cellular functions and providing the basis of human variation. Perhaps the most daunting challenge is to begin to understand how all the "parts" of cells—genes, proteins, and many other molecules—work together to create complex living organisms. Future analyses of this treasury of data will provide a deeper and more comprehensive understanding of the molecular processes underlying life and will have an enduring and profound impact on how we view our own place in it.

Insights from the Human DNA Sequence

The first panoramic views of the human genetic landscape have revealed a wealth of information and some early surprises. Much remains to be deciphered in this vast trove of information; as the consortium of HGP scientists concluded in their seminal paper, ". . . the more we learn about the human genome, the more there is to explore." A few highlights follow from the first publications analyzing the sequence.

- The human genome contains 3.2 billion chemical nucleotide base pairs (A, C, T, and G).
- The average gene consists of 3,000 base pairs, but sizes vary greatly, with the largest known human gene being dystrophin at 2.4 million base pairs.
- Functions are unknown for more than 50% of discovered genes.
- The human genome sequence is almost exactly the same (99.9%) in all people.
- About 2% of the genome encodes instructions for the synthesis of proteins.
- Repeat sequences that do not code for proteins make up at least 50% of the human genome.
- Repeat sequences are thought to have no direct functions, but they shed light on chromosome structure and dynamics. Over time, these repeats reshape the genome by rearranging it, thereby creating entirely new genes or modifying and reshuffling existing genes.

- The human genome has a much greater portion (50%) of repeat sequences than the mustard weed (11%), the worm (7%), and the fly (3%).
- Over 40% of predicted human proteins share similarity with fruit-fly or worm proteins.
- Genes appear to be concentrated in random areas along the genome, with vast expanses of noncoding DNA between.
- Chromosome 1 (the largest human chromosome) has the most genes (3,168), and Y chromosome has the fewest (344).
- Particular gene sequences have been associated with numerous diseases and disorders, including breast cancer, muscle disease, deafness, and blindness.
- Scientists have identified millions of locations where single-base DNA differences (see p. 9) occur in humans. This information promises to revolutionize the processes of finding DNA sequences associated with such common diseases as cardiovascular disease, diabetes, arthritis, and cancers.

How Does the Human Genome Stack Up?

Organism	Genome Size (Base Pairs)	Estimated Genes
Human (*Homo sapiens*)	3.2 billion	25,000
Laboratory mouse (*M. musculus*)	2.6 billion	25,000
Mustard weed (*A. thaliana*)	100 million	25,000
Roundworm (*C. elegans*)	97 million	19,000
Fruit fly (*D. melanogaster*)	137 million	13,000
Yeast (*S. cerevisiae*)	12.1 million	6,000
Bacterium (*E. coli*)	4.6 million	3,200
Human immunodeficiency virus (HIV)	9,700	9

The estimated number of human genes is only one-third as great as previously thought, although the numbers may be revised as more computational and experimental analyses are performed.

Scientists suggest that the genetic key to human complexity lies not in gene number but in how gene parts are used to build different products in a process called alternative splicing. Other underlying reasons for greater complexity are the thousands of chemical modifications made to proteins and the repertoire of regulatory mechanisms controlling these processes.

Managing and Using the Data

Bioinformatics Boom

Massive quantities of genomic data and high-throughput technologies are now enabling studies on a vastly larger scale than ever before. Examples include simultaneously monitoring and comparing the activity of tens of thousands of genes in cancerous and noncancerous tissue. Advanced computational tools and interdisciplinary experts are needed to capture, represent, store, integrate, distribute, and analyze the data.

Bioinformatics is the term coined for the new field that merges biology, computer science, and information technology to manage and analyze the data, with the ultimate goal of understanding and modeling living systems. Computing and information demands will continue to rise with the explosive torrent of data from large-scale studies at the molecular, cellular, and whole-organism levels.

Gene Gateway: A User-Friendly Guide to Genome, Gene, and Protein Databases

All Human Genome Project data and much related information are freely available on the web, but how do you find and use these rich resources? The Gene Gateway website provides introductory guides and step-by-step tutorials that show how to access and explore genome, gene, and protein databases used by scientists. Gene Gateway demonstrates how to gather information from different databases to gain a better understanding of the molecular biology behind life processes. The site offers a free workbook downloadable in PDF format.

Tutorials

- Identify genes associated with various genetic conditions and biological processes
- Learn about mutations that cause genetic disorders
- Browse a genome and find a gene's location on a chromosome map
- View the DNA sequence of a gene or amino acid sequence of a gene's protein product
- Visualize and modify three-dimensional representations of protein structures

Medicine and the New Genetics

Gene Testing, Pharmacogenomics, and Gene Therapy

DNA underlies almost every aspect of human health, both in function and dysfunction. Obtaining a detailed picture of how genes and other DNA sequences work

http://genomics.energy.gov/genegateway/

Gene Gateway was created as a companion to the Human Genome Landmarks wall poster (see back page).

together and interact with environmental factors ultimately will lead to the discovery of pathways involved in normal processes and in disease pathogenesis. Such knowledge will have a profound impact on the way disorders are diagnosed, treated, and prevented and will bring about revolutionary changes in clinical and public health practice. Some of these transformative developments are described below.

Gene Testing

DNA-based tests are among the first commercial medical applications of the new genetic discoveries. Gene tests can be used to diagnose and confirm disease, even in asymptomatic individuals; provide prognostic information about the course of disease; and, with varying degrees of accuracy, predict the risk of future disease in healthy individuals or their progeny.

Currently, several hundred genetic tests are in clinical use, with many more under development, and their numbers and varieties are expected to increase rapidly over the next decade. Most current tests detect mutations associated with rare genetic disorders that follow Mendelian inheritance patterns. These include myotonic and Duchenne muscular dystrophies, cystic fibrosis, neurofibromatosis type 1, sickle cell anemia, and Huntington's disease.

Recently, tests have been developed to detect mutations for a handful of more complex conditions such as breast, ovarian, and colon cancers. Although they have limitations, these tests sometimes are used to make risk estimates in presymptomatic individuals with a family history of the disorder. One potential benefit to these gene tests is that they could provide information to help physicians and patients manage the disease or condition more effectively. Regular colonoscopies for those having mutations associated with colon cancer, for instance, could prevent thousands of deaths each year.

Some scientific limitations are that the tests may not detect every mutation associated with a particular condition (many are as yet undiscovered), and the ones they do detect may present different risks to various people and populations. Another important consideration in gene testing is the lack of effective treatments or preventive measures for many diseases and conditions now being diagnosed or predicted.

Knowledge about the risk of potential future disease can produce significant emotional and psychological impacts. Because genetic tests reveal information about individuals and their families, test results can affect family dynamics. Results also can pose risks for population groups if they lead to group stigmatization.

Other issues related to gene tests include their effective introduction into clinical practice, the regulation of laboratory quality assurance, the availability of testing for rare diseases, and the education of healthcare providers and patients about correct interpretation and attendant risks.

Families and individuals who have genetic disorders or are at risk for them often seek help from medical geneticists (an M.D. specialty) and genetic counselors (graduate-degree training). These professionals can diagnose and explain disorders, review available options for testing and treatment, and provide emotional support. (For more information, see the URL for Medicine and the New Genetics, p. 12.)

Pharmacogenomics: Moving Away from "One-Size-Fits-All" Therapeutics

Within the next decade, researchers will begin to correlate DNA variants with individual responses to medical treatments, identify particular subgroups of patients, and develop drugs customized for those populations. The discipline that blends pharmacology with genomic capabilities is called pharmacogenomics.

More than 100,000 people die each year from adverse responses to medications that may be beneficial to others. Another 2.2 million experience serious reactions, while others fail to respond at all. DNA variants in genes involved in drug metabolism, particularly the cytochrome P450 multigene family, are the focus of much current research in this area. Enzymes encoded by these genes are responsible for metabolizing most drugs used today, including many for treating psychiatric, neurological, and cardiovascular diseases. Enzyme function affects patient responses to both the drug and the dose. Future advances will enable rapid testing to determine the patient's genotype and guide treatment with the most effective drugs, in addition to drastically reducing adverse reactions.

Genomic data and technologies also are expected to make drug development faster, cheaper, and more effective. Most drugs today are based on about 500 molecular targets, but genomic knowledge of genes involved in diseases, disease pathways, and drug-response sites will lead to the discovery of thousands of additional targets. New drugs, aimed at specific sites in the body and at particular biochemical events leading to disease, probably will cause fewer side effects than many current medicines. Ideally, genomic drugs could be given earlier in the disease process. As knowledge becomes available to select patients most likely to benefit from a potential drug, pharmacogenomics will speed the design of clinical trials to market the drugs sooner.

Gene Therapy, Enhancement

The potential for using genes themselves to treat disease or enhance particular traits has captured the imagination of the public and the biomedical community. This largely experimental field—gene transfer or gene therapy—holds potential for treating or even curing such genetic and acquired diseases as cancers and AIDS by using normal genes to supplement or replace defective genes or to bolster a normal function such as immunity.

Almost 1,200 clinical gene-therapy trials were identified worldwide in 2006.* The majority (67%) take place in the United States, followed by Europe (29%). Although most trials focus on various types of cancer, studies also involve other multigenic and monogenic, infectious, and vascular diseases. Most current protocols are aimed at establishing the safety of gene-delivery procedures rather than effectiveness.

Gene transfer still faces many scientific obstacles before it can become a practical approach for treating disease. According to the American Society of Human Genetics' Statement on Gene Therapy, effective progress will be achieved only through continued rigorous research on the most fundamental mechanisms underlying gene delivery and gene expression in animals.

Other Anticipated Benefits of Genetic Research

Expanding Impacts of New Technologies, Resources

Rapid progress in genome science and a glimpse into its potential applications have spurred observers to predict that biology will be the foremost science of the 21st Century. Technology and resources generated by the Human Genome Project and other genomic research already are having major impacts across the life sciences. The biotechnology industry employed more than 250,000 people in 2006, and revenues for 2005 totaled more than $50.7 billion.* Future revenues are expected to reach trillions of dollars.

A list of some current and potential applications of genome research follows. More studies and public discussion are required for eventual validation and implementation of some of these uses (see p. 8).

---

*Source: *Journal of Gene Medicine* website (www.wiley.co.uk/gene-therapy/clinical/), August 2006.
*Source: Biotechnology Industry Organization website (www.bio.org), June 2008.

Molecular Medicine

- Improve diagnosis of disease
- Detect genetic predispositions to disease
- Create drugs based on molecular information
- Use gene therapy and control systems as drugs
- Design "custom drugs" based on individual genetic profiles

Microbial Genomics

- Rapidly detect and treat pathogens (disease-causing microbes) in clinical practice
- Develop new energy sources (biofuels)
- Monitor environments to detect pollutants
- Protect citizenry from biological and chemical warfare
- Clean up toxic waste safely and efficiently

Risk Assessment

- Evaluate the health risks faced by individuals who may be exposed to radiation (including low levels in industrial areas) and to cancer-causing chemicals and toxins

Bioarchaeology, Anthropology, Evolution, and Human Migration

- Study evolution through germline mutations in lineages
- Study migration of different population groups based on maternal genetic inheritance
- Study mutations on the Y chromosome to trace lineage and migration of males
- Compare breakpoints in the evolution of mutations with population ages and historical events

DNA Identification

- Identify potential suspects whose DNA may match evidence left at crime scenes
- Exonerate people wrongly accused of crimes
- Identify crime, catastrophe, and other victims
- Establish paternity and other family relationships
- Identify endangered and protected species as an aid to wildlife officials (e.g., to prosecute poachers)
- Detect bacteria and other organisms that could pollute air, water, soil, and food
- Match organ donors with recipients in transplant programs
- Determine pedigree for seed or livestock breeds
- Authenticate consumables such as caviar and wine

Agriculture, Livestock Breeding, and Bioprocessing

- Grow disease-, insect-, and drought-resistant crops
- Optimize crops for bioenergy production
- Breed healthier, more productive, disease-resistant farm animals
- Grow more nutritious produce

- Develop biopesticides
- Incorporate edible vaccines into food products
- Develop new environmental cleanup uses for plants such as tobacco

Societal Concerns Arising from the New Genetics

Critical Policy and Ethical Issues

From its inception, the Human Genome Project dedicated funds to identify and address the ethical, legal, and social issues surrounding the availability of new genetic data and capabilities. Examples of such issues follow.*

- **Privacy and confidentiality of genetic information.** Who owns and controls genetic information? Is genetic privacy different from medical privacy?
- **Fairness in the use of genetic information by insurers, employers, courts, schools, adoption agencies, and the military, among others.** Who should have access to personal genetic information, and how will it be used?
- **Psychological impact, stigmatization, and discrimination due to an individual's genetic makeup.** How does personal genetic information affect self-identity and society's perceptions?
- **Reproductive issues including adequate and informed consent and the use of genetic information in reproductive decision making.** Do healthcare personnel properly counsel parents about risks and limitations? What larger societal issues are raised by new reproductive technologies?
- **Clinical issues including the education of doctors and other health-service providers, people identified with genetic conditions, and the general public; and implementation of standards and quality-control measures.** How should health professionals be prepared for the new genetics? How can the public be educated to make informed choices? How will genetic tests be evaluated and regulated for accuracy, reliability, and usefulness? (Currently, there is little regulation.) How does society balance current scientific limitations and social risk with long-term benefits?
- **Fairness in access to advanced genomic technologies.** Who will benefit? Will there be major worldwide inequities?

New Genetics Privacy Act Becomes Law

The Genetic Information Nondiscrimination Act (GINA) became law on May 21, 2008. GINA prohibits U.S. health insurance companies and employers from discrimination on the basis of information derived from genetic tests. In addition, insurers and employers are not allowed under the law to request or demand a genetic test.

- **Uncertainties associated with gene tests for susceptibilities and complex conditions (e.g., heart disease, diabetes, and Alzheimer's disease).** Should testing be performed when no treatment is available or when interpretation is unsure? Should children be tested for susceptibility to adult-onset diseases?
- **Conceptual and philosophical implications regarding human responsibility, free will vs genetic determinism, and understanding of health and disease.** Do our genes influence our behavior, and can we control it? What is considered acceptable diversity? Where is the line drawn between medical treatment and enhancement?
- **Health and environmental issues concerning genetically modified (GM) foods and microbes.** Are GM foods and other products safe for humans and the environment? How will these technologies affect developing nations' dependence on industrialized nations?
- **Commercialization of products including property rights (patents, copyrights, and trade secrets) and accessibility of data and materials.** Will patenting DNA sequences limit their accessibility and development into useful products?

Beyond the Human Genome Project—What's Next?

Genome Sequences: Paving the Way for a More Comprehensive Understanding

Building a "Systems Level" View of Life

DNA sequences generated in hundreds of genome projects now provide scientists with the "parts lists" containing instructions for how an organism builds, operates, maintains, and reproduces itself while responding to various environmental conditions. We still have very little knowledge of how cells use this information to "come alive," however, and the functions of most genes remain unknown. Nor do we understand how genes and the proteins they encode interact with each other and with the environment. If we are to realize the potential of the genome projects, with far-ranging applications to such diverse fields as medicine, energy, and the environment, we must obtain this new level of knowledge.

One of the greatest impacts of having whole-genome sequences and powerful new genomic

---

*For more information, see the Ethical, Legal, and Social Issues URL, p. 12.

technologies may be an entirely new approach to conducting biological research. In the past, researchers studied one or a few genes or proteins at a time. Because biological processes are intertwined, these strategies provided incomplete—and often inaccurate—views. Researchers now can approach questions systematically and on a much grander scale. They can study all the genes expressed in a particular environment or all the gene products in a specific tissue, organ, or tumor. Other analyses will focus on how tens of thousands of genes and proteins work together in interconnected networks to orchestrate the chemistry of life. These holistic studies are the focus of a new field called "systems biology" (see DOE Genomics:GTL Program, p. 10).

Charting Human Variation

Slight variations in our DNA sequences can have a major impact on whether or not we develop a disease and on our responses to such environmental factors as infectious microbes, toxins, and drugs. One of the most common types of sequence variation is the single nucleotide polymorphism (SNP). SNPs are sites in a genome where individuals differ in their DNA sequence, often by a single base. For example, one person might have the DNA base A where another might have C, and so on. Scientists believe the human genome has at least 10 million SNPs, and they are generating different types of maps of these sites, which can occur in both genes and noncoding regions.

Sets of SNPs on the same chromosome are inherited in blocks (haplotypes). In 2005 a consortium of researchers from six countries completed the first phase of a map of SNP patterns that occur across populations in Africa, Asia, and the United States. Researchers hope that dramatically decreasing the number of individual SNPs to be scanned will provide a shortcut for tracking down the DNA regions associated with such common complex diseases as cancer, heart disease, diabetes, and some forms of mental illness. The new map also may be useful in understanding how genetic variation contributes to responses to environmental factors.

DOE Genomics:GTL Program

Exploring Genomes for Energy and Environmental Applications

The Genomics:GTL (formerly Genomes to Life) program of the U.S. Department of Energy (DOE) is using the Human Genome Project's technological achievements to help solve our growing energy and environmental challenges.

Today, genomics is the starting point for a new level of exploration across the life sciences. The GTL research program uses genomic (DNA) sequences of microbes and plants to launch large-scale investigations into their wide-ranging biochemical capabilities having potential applications in bioenergy and the environment (see sidebar below). Before these biological processes can be safely and economically harnessed for such uses, however, they must be understood in far greater detail and in the context of their operations within a dynamic, living organism.

To obtain this whole-systems knowledge, GTL investigates relevant plant and microbial properties on multiple levels. Starting with the DNA sequence, studies follow its expression (e.g., protein production, interactions, and regulation) in individual cells and populations of cells or organisms in ecosystems. Integrating genomic and many other data types into a computerized knowledgebase will stimulate new research strategies and insights needed for specialized applications.

GTL Investigations of Microbial and Plant Genomes

Microbes and plants have evolved unique biochemistries, offering a rich resource that can be applied to diverse national needs. Some recent projects funded by the DOE Genomics:GTL program highlight the potential wealth of natural capabilities available.

Plants for Biomass, Carbon Storage

Understanding the genes and regulatory mechanisms controlling growth and other traits in the recently sequenced poplar tree may lead to its use for bioethanol production and for sequestration (storage) of carbon.

Microbes Living in Termites: A Potential Source of Enzymes for Bioenergy Production

GTL researchers are investigating bacteria that live in termite hindguts and churn out wood-digesting enzymes. These proteins may be usable for breaking down plant cellulose into sugars needed for ethanol production. Termites also produce hydrogen as a by-product, a process that potentially could be reproduced on a larger scale.

Synthetic Nanostructures: Harnessing Microbial Enzyme Functions

Enzymes incorporated into synthetic membranes can carry out some of the functions of living cells and may be useful for generating energy, inactivating contaminants, and sequestering atmospheric carbon.

Genomes for Bioenergy

Cellulosic Biomass: An Abundant, Secure Energy Source to Reduce U.S. Dependence on Gasoline

Bioethanol made from cellulosic biomass—the inedible, fibrous portions of plants—offers a renewable, sustainable, and expandable domestic resource to meet the growing demand for transportation fuels and reduce our dependence on oil.

The United States now produces 7 billion gallons of corn-grain ethanol per year, a fraction of the 142 billion gallons of transportation fuel used annually. Cellulosic ethanol has the potential to dramatically increase the availability of ethanol and help meet the national goal of displacing 30% of gasoline by 2030.

Cellulose is the most abundant biological material on earth. The crops used to make cellulosic ethanol (e.g., postharvest corn plants—not corn grain—and switchgrass) can be grown in most states and often on marginal lands. As with ethanol from corn grain, cellulose-based ethanol can be used as a fuel additive to improve gasoline combustion in today's vehicles. Modest engine modifications are required to use higher blends (85% ethanol). Additionally, the amount of carbon dioxide emitted to the atmosphere from producing and burning ethanol is far less than that released from gasoline.

To accelerate technological breakthroughs, the DOE Genomics:GTL program will establish research centers to target specific DOE mission challenges. Three DOE Bioenergy Research Centers are focused on overcoming biological challenges to cellulosic ethanol production. In addition to ethanol, these centers are exploring ways to produce a new generation of petroleum-like biofuels and other advanced energy products from cellulosic biomass.

For More Information

Related Websites

- **Human Genome Project Information**
  www.ornl.gov/hgmis/home.shtml
- **Medicine and the New Genetics**
  www.ornl.gov/hgmis/medicine/
- **Ethical, Legal, and Social Issues**
  www.ornl.gov/hgmis/elsi/
- **Genetics Privacy and Legislation**
  www.ornl.gov/hgmis/elsi/legislat.shtml
- **Gene Gateway**
  genomics.energy.gov/genegateway/

- **Image Gallery (downloadable)**
  genomics.energy.gov/gallery/
- **Resources for Teachers**
  www.ornl.gov/hgmis/education/
- **Resources for Students**
  www.ornl.gov/hgmis/education/students.shtml
- **Careers in Genetics and the Biosciences**
  www.ornl.gov/hgmis/education/careers.shtml
- **DOE Joint Genome Institute**
  www.jgi.doe.gov
- **NIH National Human Genome Research Institute**
  www.genome.gov
- **Genomes OnLine Database (GOLD)**
  www.genomesonline.org
- **National Center for Biotechnology Information**
  www.ncbi.nlm.nih.gov

Genomics and Its Impact on Science and Society: The Human Genome Project and Beyond

This document was revised in June 2008 by the Genome Management Information System (GMIS) at Oak Ridge National Laboratory, Oak Ridge, Tennessee, for the Office of Biological and Environmental Research within the U.S. Department of Energy Office of Science.

Free Copies and Presentation Materials

- **Individual print copies or class sets**
  Contact: genome@ornl.gov, 865/574-0597
- **Downloadable files**
  A PDF version of this document and accompanying PowerPoint file are accessible via the web (www.ornl.gov/hgmis/publicat/primer).
- **DOE Genomics:GTL Program**
  genomicsgtl.energy.gov
- **Focus on Biofuels**
  genomicsgtl.energy.gov/biofuels/
- **National Geographic: The Genographic Project**
  www.nationalgeographic.com/genographic/
- **International HapMap Project**
  www.hapmap.org

Free Wall Poster of Human Chromosomes and Genes

The poster also features sidebars explaining genetic terms, with URLs for finding more detailed information (see web companion. Gene Gateway, p. 4).

Order free copies via the web:
- genomics.energy.gov/posters/
  Or see contact information below left.

Ecology and Environmentalism

Peter J. Bowler & Iwan Rhys Morus

At first sight it might seem obvious that the two topics listed in the title above should be linked together. The environmentalist movement has sought to warn of the dangers posed by humanity's evermore powerful efforts to exploit the world and its inhabitants through industry and intensive agriculture. It points to the increasingly common catastrophes that can be attributed to the uncontrolled exploitation of the world's resources and notes that we are now witnessing a mass extinction of geological proportions caused by the destruction of species' natural habitats. If we are not careful, the environmentalists warn, we shall wipe ourselves out by rendering the whole world uninhabitable. To make this point they sometimes call on the science of ecology, which seeks to describe and understand the relationships between organisms and their environment. Indeed the term "ecological" is often taken to mean "environmentally beneficial," as though the science went hand in hand with the social philosophy that seeks to defend the natural world (see the title of Bramwell's 1989 book, which is actually about environmentalism). Many assume that ecology is a science created by environmentalists to provide them with the information they need about the balance of nature and the ways in which disturbing influences such as human exploitation upset and ultimately destroy that balance. Such an interpretation of the origins of ecology would take it for granted that the science is based on a holistic worldview that seeks to understand how everything in nature interacts to produce a harmonious and self-sustaining whole. Ecology is the science behind James Lovelock's image of the earth as "Gaia"—a sustaining mother to all living things who will not hesitate to discipline one of her children if it gets out of line and threatens the whole.

One pioneering study by Donald Worster (1985) sought to present such a unified picture of the origins of both environmentalist thought and scientific ecology. But subsequent work has uncovered a more complex and far less coherent pattern of relationships. To a large extent, the environmentalist movement has opposed modern science as the handmiden of industrialization, seeking its image of nature in a romantic impressionism rather than in scientific analysis. To the extent that it has had an impact on science, it has done so by encouraging a holistic methodology that openly challenges the materialistic and reductionist approach favored by the majority of scientists. There are thus some forms of scientific ecology that do draw inspiration from environmentalist concerns—but there are others that owe their origins to the reductionist viewpoint that is anathema to the romantic vision of natural harmony. Many of the first professional ecologists used physiology as a model, arguing that just as the physiologists saw the body as a machine, so they should apply a purely naturalistic methodology to studying how the body interacted with its environment. Some schools of ecology have remained resolutely materialistic, depicting natural relationships in terms more of a Darwinian struggle for existence than of harmony. Ecologists from these backgrounds are among the leading critics of Lovelock's efforts to depict nature as a purposeful whole that seeks to maintain the earth as an abode for life.

Modern historical studies force us to see ecology as a complex science with many historical roots. Indeed, it is not really a unified branch of science at all, since its various schools of thought have such different origins that they still find it hard to communicate with one another. Providing hard evidence for the environmentalist campaign is certainly not on most scientific ecologists' agenda. As in so many other areas, a historical study forces us to contextualize the rise of science, breaking down the more obvious links, such as those assumed to exist between ecology, holism, and environmentalism. Instead, we see the science emerging from a number of different research programs instituted in different places and times and for different purposes, some of them designed more to encourage the exploitation of the environment than to promote its protection. Far from originating as a unified response to a single philosophical message, ecology is a composite of many rival approaches that even today have not coalesced into a single discipline with a coherent methodology.

We begin with an overview of how science became associated with the drive to exploit the world's resources, then move on to an account of how the environmentalist movement emerged to counter this program. The second half of this chapter then outlines the emergence of scientific ecology from the late nineteenth century onward, showing how different research problems and different philosophical an ideological agendas promoted theoretical disagreement almost from the beginning.

Science and the Exploitation of Resources

From the Scientific Revolution of the seventeenth century onward, the rise of science has been linked to the hope that better knowledge of the world would allow a more effective use of natural resources. The ideology promoted by Francis Bacon stressed the use of observation and experimentation to build up practical knowledge that could be applied through improvements to industry and agriculture. The world was depicted as a passive source of raw materials to be exploited by humanity for its own benefit. Even the methodology of science stressed the dominance of humanity and the passivity of the natural world: the experimenter sought to isolate particular phenomena so they could be manipulated at will. There was no expectation that everything might interact in a way that would negate the insights gained from the study of the particular. If the whole universe was just a machine, there was no reason why humanity should not tinker with individual parts for its own benefit. Carolyn Merchant (1980) sees this attitude as characteristic of an increasingly "masculine" attitude toward nature (see chap. 21, "Science and Gender"). By the end of the eighteenth century, this attitude was already bearing fruit as the Industrial Revolution got underway, and in the course of the following century the role that science could play in promoting technological development became obvious to all (see chap. 17, "Science and Technology").

At the same time, science was increasingly involved in the effort to locate and exploit natural resources around the world. The voyages of discovery undertaken by navigators such as Captain James Cook were intended to bring back information on the plants and animals of remote regions for Europeans to study and classify, but they were also intended to locate new territories that might be colonized. Sir Joseph Banks accompanied Cook on his first voyage to the South Seas (1768–71) as a naturalist. In his later capacity as president of the Royal Society, he helped to coordinate the British navy's efforts to explore and map the world, often with a view to discovering useful natural resources (MacKay 1985). The voyage of H.M.S. *Beagle,* which provided Darwin with crucial insights, was undertaken to map the coast of South America, a region vital to British trade. In the 1870s, the British navy provided a vessel, H.M.S. *Challenger,* for the first deep-sea oceanographic expedition. Although much information of scientific interest was generated, funding for marine science was increasingly provided in the expectation that there would be benefits for navigation, fisheries, and other practical concerns.

On land, too, there were many expeditions designed to explore remote regions to satisfy curiosity about the world (see below), but there were also explicit signs of science's growing involvement with imperialism. Many European nations established botanical gardens at home and in their colonies with the deliberate intention of identifying commercially useful plant species and studying how foreign species could be imported as new cash crops. Kew Gardens in London was the center of the British effort, under the direction of botanists such as Joseph Dalton Hooker, a leading supporter of Darwin (Brockway 1979). The cinchona plant, source of the anti-malarial drug quinine and hence vital to European efforts to colonize the tropics, was transported via Kew from its home in South America to found commercial plantations in India. The rubber plant was smuggled out of Brazil despite a government prohibition to create the worldwide rubber-production industry. North America was transformed as European farming methods were adapted to its wide range of different environments. By the early twentieth century, the Bureau of Biological Survey under C. Hart Merriam was coordinating deliberate attempts to eradicate native "pests" such as the prairie dog that destroyed the farmers' crops. Europeans and Americans were now interfering on an unprecedented scale with natural ecosystems, destroying native habitats and importing alien species as cash crops (for a survey of these developments, see Bowler [1992]).

The Rise of Environmentalism

These developments were not without their critics, and gradually an articulate movement evolved to criticize the unrestricted exploitation—and often the consequent destruction—of the natural environment (McCormick 1989). The Romantic thinkers of the early nineteenth century celebrated the wilderness as a source of spiritual renewal and hated the industrialists who destroyed it for profit. Significantly, writers such as William Blake saw mechanistic science as a key component of the unrestrained exploitation of the natural world. A later generation of writers such as Henry Thoreau also celebrated the recuperative value of wilderness for a humanity increasingly alienated by an urban and industrialized lifestyle. In 1864, the American diplomat George Perkins Marsh wrote his *Man and Nature* to protest against the destruction of the natural environment. He warned that, contrary to early optimistic expectations, there was a degree of human destructiveness that nature might never be able to repair: "The earth is fast becoming an unfit home for its noblest inhabitant, and another era of equal human crime and human improvidence . . . would reduce it to such a condition of impoverished productiveness, of shattered surface, of climatic excess, as to threaten the depravation, barbarism and perhaps even extinction of the species" (Marsh 1965, 43). Marsh was not calling for a halt to all human interference but for better management that would allow the earth to retain its self-sustaining capacities. Partly as a result of his efforts, the U.S. government set up the Forestry Commission to manage the nation's resources, and eventually areas of woodland were set aside to be protected from logging. Public concern also led to the designation of areas of outstanding natural beauty as national parks, Yosemite Valley in California in 1864 and Yellowstone in Wyoming in 1872. The Sierra Club, founded in 1892 by John Muir, was dedicated to the protection of wilderness areas. In Europe, where there was little true wilderness left to protect, efforts were nevertheless made to create nature reserves where stable environments that had existed for centuries could be conserved (on nature reserves in Britain, see Sheal [1976]).

There was considerable tension between those who called for a more careful management of nature in order to allow resources to be renewed and an increasingly vocal movement that depicted all human interference as evil and potentially damaging to the earth as a whole. The former group was willing to call in science, in the form of the newly developed ecology, to help better understand the ways in which natural ecosystems would respond to human interference. But a more extreme form of environmentalism developed out of an alternative, more romantic vision of nature that, if it had any use for science at all, insisted that it must be a science based on holistic rather than mechanistic principles. This movement cut across all traditional political divisions and was by no means always sympathetic to a democratic approach to government. After all, the common people may well vote for more industrialization out of a short-sighted desire for more material goods. In Germany, a "religion of nature" often linked to the philosophy of the evolutionist Ernst Haeckel, became part of Nazi ideology—and the Nazis created nature reserves on ground cleared of Jews and Poles sent to the death camps. Soviet Russia had a strong environmentalist policy until Stalin's drive for industrialization led to unrestricted exploitation of the country's resources (on European environmentalism, see Bramwell [1989]).

In America, there were debates between those who saw the "dust bowl" on the Great Plains in the 1930s as part of a natural climatic cycle and those who insisted that it was a consequence of the unsuitability of the prairies for farming. The latter position was increasingly typical of the more active environmentalist movement, which allied itself with those who saw the preservation of wilderness as essential for human psychological health, to say nothing of the health of the planet as a whole. In America, Aldo Leopold's *Sand County Almanac,* published posthumously in 1949, recorded the transition of a Wisconsin game manager into an environmentalist with an emotional and aesthetic attachment to wilderness. For Leopold, scientific ecology was not enough because it needed to be supplemented by an ethical commitment that recognized that all species have a right to exist, a right that should not be compromised by human expediency: "Conservation is getting nowhere because it is incompatible with our Abrahamic concept of land. We abuse land because we regard it as a commodity belonging to us. When we see land as a community to which we belong, we may begin to use it with love and respect. There is no other way for land to survive the impact of mechanized man, nor for us to reap from it the esthetic harvest it is capable, under science, of contributing to culture" (Leopold 1966, x). Leopold's environmentalism did not rule out a role for the scientific study of nature, but that had to take place within a framework in which humanity was part of nature, not dominant over it.

Such an attitude has grown in influence, as more people have become aware of the dangers of the unrestricted exploitation of the environment. Rachel Carson's *Silent Spring* of 1962 highlighted the damage done to many species by the use of insecticides. Numerous environmental catastrophes have driven home the same message, although there are still significant differences between the ways in which different communities have responded. In America, despite the activities of those who cherish the wilderness, the public seems content to let corporate agriculture manipulate nature in the interests of producing cheaper food. In Europe, by contrast, the use of chemical fertilizers and insecticides has become unpopular, while genetic manipulation of food crops is restricted. In the Third World, however, genetic engineering is seen as perhaps the lesser of the two evils, since it might increase yields without leaving farmers dependent on expensive and potentially dangerous chemicals.

The Origins of Ecology

A distinct science of ecology only began to emerge at the end of the nineteenth century, although concepts we associate with the discipline had long been recognized. The Swedish naturalist Linnaeus wrote of the "balance of nature" in the mid-eighteenth century, noting that it one species increased its numbers due to favorable conditions, its predators would also increase and tend to restore the equilibrium. For Linnaeus, this was all part of God's plan of creation, and the natural theologians routinely described the adaptation of species to their physical and biological environment as an illustration of divine benevolence.

Systematic study of such relationships was also part of Alexander von Humboldt's project for a coordinated science of the natural world, which focused especially on the geographical factors that shaped different environments. Humboldt was impressed by the Romantic movement popular in the arts around 1800, with its emphasis on the ability of wilderness to inspire human emotions, but he insisted that a serious study of the natural world must use the scientific techniques of measurement and rational coordination. His aim was a science that focused on material interactions but interpreted them as parts of a coordinated whole in which each natural phenomenon was interlinked with all the others. He spent the years 1799–1804 exploring South and Central America, taking numerous scientific measurements in a variety of environments that were used to throw light on the interactions between their geological structure, physical conditions, and biological inhabitants. Humboldt made important contributions to geology—he was a follower of A. G. Werner and named the Jurassic system of rocks after the Jura Mountains of Switzerland (see chap. 5, "The Age of the Earth"). He also produced maps showing the variations of temperature and other climatic factors on a worldwide scale and others showing cross sections of mountainous regions illustrating how the characteristic vegetation changed with altitude. Humboldt's accounts of his South American voyage inspired many European scientists, including Darwin, and his emphasis on the earth as an integrated whole encouraged a whole generation to undertake systematic surveys of a variety of physical and biological phenomena. Under the influence of "Humboldtian science" biologists were taught to think in what we would now call ecological terms, looking for the ways in which the distribution of animals and plants was determined by the character of the soil and underlying rocks, the local climate, and the other native inhabitants of the region.

In the next generation, Darwinism, too, stressed the adaptation of the species to its environment but encouraged a more materialistic view of each population in competition not just with its predators but also with rivals seeking to exploit the same resources (see chap. 6, "The Darwinian Revolution"). Darwin also focused attention on biogeography, which illustrated how species adapted to new environments. It was the German Darwinist Ernst Haeckel who coined the term "oecology" in 1866 from the Greek *oikos,* referring to the operations of the family household—the ecology of a region showed how the species there interacted to exploit its natural resources. But unlike Darwin, Haeckel adopted a nonmaterialistic view of nature in which living things were active agents within a unified and progressive world. The tension between the materialistic and holistic worldviews ensured that the science of ecology would he driven by theoretical disagreements from its inception. There were a number of different research programs, each trying to tackle the complex relationships between species and their environment in a different way. Because they began different origins, they often adopted different theoretical outlooks.

The stimulus for the creation of the new biological discipline that would adopt the name ecology came from the breakdown of the descriptive or morphological approach to nature at the end of the nineteenth century. At that juncture, the emphasis was on experimentation, with physiology as the model, and a number of new biological disciplines arose in response to this challenge, including genetics. It was much harder to apply the experimental method to the study of how species relate to their environment, but there were several avenues that pointed the way to a more scientific approach to this topic. One was the increasing refinement of Humboldt's biogeographical techniques. In America, C. Hart Merriam of the Bureau of Biological Survey developed detailed maps showing the various "life zones" or habitats stretching from east to west across the continent. In 1896, Oscar Drude of the Dresden botanical garden published a fine-grained plant geography of Germany that showed how local factors such as rivers and hills shaped the vegetation of each region.

Plant physiology provided the model for other pioneers of plant ecology. Experimental studies had produced a much better understanding of how the internal functions of a plant operate, but by the end of the century a number of botanists began to realize that it would also be necessary to look at how the plant's physical environment affected these functions. This insight was especially obvious to those who worked in botanical gardens established in the tropics and other extreme environments, where the role of adaptation was crucial (Cittadino 1991). The founder of plant ecology, botanist Eugenius Warming, was trained in plant physiology in Denmark and had worked for a time in Brazil. He developed his approach as an alternative both to pure physiology and to the traditional focus of most botanists on classification (Coleman 1986). His *Plantesamfund,* published in 1895, was translated into German the following year and into English as *Oecology of Plants* in 1909. Warming could see how the physical conditions of an area determined which plants could live there, but he also realized that there was a network of interactions between the plants that were characteristic of a particular environment. These typical plants formed a natural community, each dependent in various ways on the others. The concept of a natural community had already been described by naturalists such as Stephen A. Forbes of Illinois, whose 1887 address to the Peoria Scientific Association, "The Lake as a Microcosm," had stressed that all the species inhabiting a lake were dependent on one another. It was a concept that was all too easily taken up by the opponents of materialism to argue that the community formed a kind of superorganism with a life and purpose of its own. But Warming resolutely opposed this almost mystical view of the community; for him the relationships were just a natural consequence of evolution adapting species to the biological as well as the physical environment. He acknowledged that all the species were competing with each other in a constant struggle for existence and that when the original community was disturbed (as by human interference) there was no guarantee that the original collection of species would reestablish itself. If we cut down a forest, the trees may never get a chance to grow again because the soil has been modified in a way that prevents them from reseeding themselves. This view was also characteristic of one of the first American schools of ecology founded at the University of Chicago by Henry C. Cowles.

There was another American research tradition, however, that developed around a very different viewpoint. At the state university in Nebraska, Frederic E. Clements sought to put the study of grassland ecology on a more scientific footing (Tobey 1981). The European techniques were not suited to the vast uniform areas of the prairies, and Clements realized that in these conditions the only way to get really accurate information about the plant population was iterally to count every single plant growing in a series of sample areas. He marked out measured squares or quadrats spread over a wide region and compounded the information to give a much more precise assessment of the overall population. By clearing quadrats of all vegetation, he was able to see how the natural plant community reestablished itself and became convinced that in these circumstances there was a definite sequence by which the natural or "climax" population was built up. Clements's *Research Methods in Ecology* (1905) publicized the new techniques, and the school of grassland ecology established itself, especially in institutions dealing with the practical problems of the farmers whose activities inevitably destroyed the natural climax grassland of the prairies. Clements was an influential writer and he promoted a philosophy of ecology that was very different from the materialistic approach of Warming and Cowles. He saw the natural climax population of a region in almost mystical terms: nature was predestined to move toward this community whenever it was disturbed, and the community had a reality of its own that required it to be seen as something more than a collection of competing species. Here was an ecology that seemed to derive from the romantic image of nature as a purposeful whole that resisted human interference, yet it was being used to give advice to the farmers whose activities had destroyed the natural environment of the plains.

Consolidation and Conflict

In the early decades of the twentieth century, the rival approaches to ecology pioneered by Warming and Clements gained enough attention for the area as a whole to become recognized as an important branch of science. But new developments continued the original tensions, and there was competition among the different research schools for control of its journals and societies and for access to government and university departments where it might flourish. In fact, despite a promising start, expansion was slow until after World War II. The British Ecological Society was the first ecological society to be founded, in 1913 (Sheal 1987), followed two years later by the Ecological Society of America (whose journal, *Ecology,* first appeared in 1920). But the new discipline's bid to establish itself in academic departments was slow, except in America, and even here the membership of the Ecological Society remained static through the interwar years. In Britain, pioneer ecologists such as Arthur G. Tansley had to struggle for academic recognition; Tansley spent some time as a Freudian psychologist and blamed the slow growth of ecology in part on the loss of promising young scientists in World War I.

In America, Clements's school of grassland ecology continued to flourish into the 1930s, when it provided support for the claim that the prairies should be returned to their natural climax of grassland to recover from the erosion of the Dust Bowl. The idealist notion of the climax community as a superorganism with a life of its own was linked by his student John Phillips to the holistic philosophy being popularized by the South African statesman Jan Christiaan Smuts, whose *Holism and Evolution* appeared in 1926. Smuts made an emotional appeal to a vision of nature as a creative process with inbuilt spiritual values and depicted evolution as a process designed to bring about complex entities whose properties were of a higher level than anything visible in their individual parts. In Britain, Tansley had to compete with South African ecologists wedded to Smuts's philosophy who were threatening to dominate ecology throughout the British Empire (Anker 2001).

Although Clements and his supporters tried to explain the Dust Bowl, the fact that the soil had disappeared effectively undermined their claim that the natural climax vegetation could reestablish itself. Other schools of ecology developed, especially in university departments that did not have to deal with the problems of the prairie farmers. Henry Allan Gleason and James C. Malin both challenged Clements's ideas by arguing that changes could take place in the vegetation of a region due to fluctuations in the climate and the natural invasion or species from other regions. In Britain, Tansley—who eventually gained a chair at Oxford—argued strenuously against Phillips's use of the superorganism concept, openly dismissing it as little more than mysticism. Yet Tansley used research methods very similar to those of the Clements's school, and it was he who coined the term "ecosystem" in 1935 to denote the system of interactions holding the species of a particular area together. For any European biologist, it seemed obvious that most apparently "natural" communities were to some extent the product of human activity, perhaps extended over centuries, so there was little point in trying to claim that a particular ecosystem had some sort of prior claim to be recognized as the only one appropriate for a certain area. Tansley and other critics also worried that promoting the idea of a superorganism would play into the hands of mystics who wanted to block any scientific study of the natural world. In continental Europe, an entirely different form of ecology based on the precise classification of all the plants in an area was developed, and in this the notion of a superorganism was simply irrelevant.

A clear indication of the fragmentary origins of ecology can be seen in the fact that it was not until the 1920s that systematic study of animal ecology began. But here, too, the tensions between the materialistic and holistic viewpoints immediately asserted themselves. At the University of Chicago, Victor E. Shelford applied Clements's approach to the study of animal communities and their dependence on the local vegetation. Also at Chicago, Warder Clyde Allee began to study animal communities on the assumption that cooperation between the members of the population is an integral part of how a species deals with its environment. Allee dismissed the Darwinian view of individual competition as the driving force of behavior and of evolution—he explicitly rejected the notion of a "pecking order" determining individuals' rank within the group. For him, evolution promoted cooperation, not competition, a view closely allied with the holistic philosophy characteristic of Clements's group. Allee and his followers also developed the political implications of their vision of natural relationships as an alternative to the "social Darwinism" that presented individual competition as natural and inevitable (Mitman 1992).

A very different approach was developed in Britain by Charles Elton, who worked at the Bureau of Animal Populations at Oxford from 1932 (Crowcroft 1991). His book *Animal Ecology* (1927) established itself as a textbook for the field and popularized the term "niche" to denote the particular way in which a species interacted with its environment. Elton had worked with the records of the Hudson's Bay Company that gave details of fluctuations in the numbers of fur-bearing animals trapped over many years. These revealed occasional massive increases in numbers (plagues of lemmings are the classic example) caused when rapidly reproducing species outstrip their natural predators in a time of plentiful resources. The occurrence of such episodes made nonsense out of the old idea of a "balance of nature" and confirmed Darwin's Malthusian image of populations constantly tending to expand to the limit of the available resources.

Elton made common cause with Tansley and with the young Julian Huxley to promote their vision of ecology, which Huxley was also concerned to link with the new Darwinism emerging in evolution theory. By denying the existence of a natural ecosystem characteristic of any environment, their approach made it easier to see the natural world as something that could be adjusted to human activity through scientific planning. Such a vision had clear social implications and was popularized in the science fiction novels being written by H. G. Wells (who also collaborated with Huxley on a major popular work, *The Science of Life,* in 1931). At this point, however, they did not envisage ecology as a subject that could be analyzed using mathematical models, partly because the rapid fluctuations in population density observed by Elton seemed unpredictable. But others were becoming more interested in the possibility of using mathematics, perhaps by seeing an analogy between the behavior of individual molecules in a gas and of individual animals interacting with their environment. The American physical chemist Alfred J. Lotka published a book on this topic in 1925, and this approach was subsequently taken up by the Italian mathematical physicist Vico Volterra, who had become interested in predicting the fluctuations in commercial fish populations. In the late 1930s, the Russian biologist G. F. Gause performed experiments on protozoa to test the "Lotka-Volterra equations," and his efforts to substantiate the mathematical techniques would play a vital role in stimulating the expansion of ecology after World War II (Kingsland 1985). For the time being, however, there were many who shared Elton's suspicions, feeling that the unpredictable dynamics of natural population changes were an unsuitable field for the application of abstract mathematical models.

Modern Ecology

Ecology expanded rapidly in the 1950s and 1960s as the world became more aware of the pressing environmental problems created by human activity. But the pressure was not necessarily coming from environmentalist groups. Those who sought to control and exploit nature also wanted information that would help them manage the ever more complex problems that they were confronting (Bocking 1997). The ecologists exploited the new image of a more "scientific" approach made possible by the mathematical techniques developed by Lotka and Volterra before the war. They were also able to make common cause with the Darwinian synthesis now beginning to dominate evolutionary biology following the emergence of the genetical theory of natural selection (itself based on mathematical modeling of populations). A school of population ecology emerged based on the exploitation of the Darwinian idea that competition was the driving force of natural relationships. There was no overall theoretical consensus, however, because at the same time a rival school of systems ecology emerged, exploiting analogies between ecological relationships and the stable economic structures existing in human society. Here there was a renewed focus on the harmonious nature of communities, drawing not on the old vitalistic philosophy but on the models of purposeful natural systems created in cybernetics. When James Lovelock's Gaia theory extended this approach into something that looked like the old mysticism, he was violently criticized by most biologists for abandoning the materialist ethos of science and pandering to the romanticized image of nature favored by the extreme environmentalists.

The Lotka-Volterra equations reinforced the lessons of Darwinism by implying that in a world dominated by competition, the best-adapted species in any environment would drive all rivals to extinction. This became known as the "principle of competitive exclusion," which states that there can be only one species occupying a particular niche in a particular location. This principle was tested by David Lack, a student of Julian Huxley, in the case of "Darwin's finches" on the Galapagos Islands. Although Darwin had used these birds as a classic example of specialization, later studies had shown that there were often several different species feeding in apparently the same way on the same island. Lack showed that this was not the case because each species was actually exploiting a different way of feeding—just because they were all mingling together did not mean they were taking the same food in the same way. His book *Darwin's Finches* (1947) helped to establish the new Darwinian synthesis in evolutionism and the principle of competitive exclusion in ecology, while at the same time renewing interest in Darwin's role as the founder of the selection theory.

The British-trained ecologist G. Evelyn Hutchinson, who had moved to America in 1928, launched an attack on Elton's refusal to use mathematical models in animal ecology. He argued that where there were difficulties in applying the Lotka-Volterra equations, the best approach was to modify the mathematical models, not reject the technique altogether. Hutchinson wanted to use the mathematical models to unify ecology and evolution theory, as proclaimed in the title of his 1965 book *The Ecological Theatre and the Evolutionary Play*. His student Robert MacArthur went on to found a new science of community ecology based on Darwinian principles of struggle and competitive exclusion (Collins 1986; Palladino 1991). MacArthur used mathematical models to address questions such as how close the niches could be in a particular environment and whether the niches evolved along with the species. Like Lack, MacArthur became interested in the problems posed by the structure of populations on isolated islands. He teamed up with Edward O. Wilson to develop a theory that predicted that the diversity of species on an oceanic island was directly proportional to its area. The number of species was maintained by a balance between immigration and extinction, the latter always a threat to small isolated populations. Wilson became interested in the way in which different reproductive strategies would help or hinder a species trying to establish itself on a new island and subsequently went on to develop the science of sociobiology.

Hutchinson had other interests, however, and these helped to create a rival school of systems ecology based on very different theoretical principles. He wanted to study communities using not an organismic analogy but an economic one, which traced the flow of energy and resources through the system and sought to identify feedback loops that maintained the stability of the whole. This was an approach pioneered by the Russian earth scientist V. Vernadskii, who had coined the term "biosphere" earlier in the century. The concept of feedback loops was central to the new science of cybernetics founded by Norbert Weiner to explain the activity of self-regulating machines. Hutchinson imagined such feedback loops working on a global scale to maintain the various ecosystems in a stable state. He also saw an analogy between this model of nature and the economists' attempts to depict human society as a stable system based on the cooperative use of resources. Hutchinson's student Raymond Lindemann wrote an influential paper in 1942 analyzing the flow of energy derived from the sun through the ecosystem of Cedar Bog Lake in Minnesota. This model of energy flow was then built on by the brothers Howard and Eugene Odum, the founders of systems ecology. The Odums studied the energy and resource circulations in a wide variety of environments, basing their work on the assumption that large-scale ecosystems would have a substantial robustness in the face of external threats. Some of their studies were funded by the U.S. Atomic Energy Commission, anxious about the potential damage that might be caused by nuclear war or accident. Systems ecology saw the human economy as just one aspect of a global network of energy and resource consumption and presented models suggesting that all levels of the process could be managed successfully if the flow patterns could be understood. Howard Odum's *Environment, Power and Society* (1971) presented a technocrat's dream of a society carefully structured and managed so that it could maintain itself even in the face of the more restricted levels of resources that will be available to humanity in the future (Taylor 1988).

Community ecology and systems ecology thus represented rival visions of how to construct a model of the ecosystem, the one based on the Darwinian principle of competition, the other on a more holistic vision of apparently purposeful feedback loops. Philosophically and politically, they invoked very different implications about nature and human society. The result was a deep level of conflict in which each side dismissed the other as philosophically naive and scientifically incompetent. The later twentieth century thus did not witness a unification of ecology around a coherent paradigm. There were still different schools with different research programs, methodologies, and philosophies. The one thing they all seemed to agree on was that scientific ecology had to present itself as essentially materialistic, offering no opening for communication with the kind of nature mysticism favored by the extreme environmentalist movement. Although systems ecology retained a holistic approach reminiscent of Clements's vision of the ecosystem as an organism in its own right, the advent of cybernetics and the link to economics allowed even this school to distance itself from the old idealism.

It is in this context that we can judge the reaction to James Lovelock's Gaia hypothesis of 1979, in which the whole earth is seen as a self-regulating system designed to maintain life. Gaia is the name of the ancient Greek earth goddess and was chosen to imply that the earth is mother to all living things, humans included. Lovelock made no secret of his support for environmentalism, criticizing those who advocate unrestricted exploitation of nature by implying that Gaia will, if necessary, take steps to eliminate humanity if it becomes a threat to the whole biosphere. Lovelock had impeccable scientific credentials, having worked in the space program developing systems to monitor the earth's surface from satellites, but the rhetoric with which he presented his theory clearly touched a raw nerve with many scientists. Although apparently similar to the systems approach, Gaia seemed to go beyond the cybernetic analogy and return to the older organicism in which ecosystems (in this case the biosphere as a whole) have a real existence and can act on their own behalf to achieve their own purposes. Critics were not slow to point out these implications, dismissing the whole theory as a perversion of science that pandered to the romanticism of the environmentalist movement. For Lovelock, it was as though a dogmatic scientific establishment had closed its ranks in defense of materialism: "I had a faint hope that Gaia might be denounced from the pulpit; instead I was asked to deliver a sermon on Gaia at the Cathedral of St. John the Divine in New York. By contrast Gaia was condemned by my peers and the journals, *Nature* and *Science,* would not publish papers on the subject. No satisfactory reasons for rejection were given; it was as if the establishment, like the theological establishment of Galileo's time, would no longer tolerate radical or eccentric notions" (Lovelock 1987, vii–viii). Nothing could more clearly indicate the gulf that still existed between scientific ecology (in all its forms) and radical environmentalism.

Conclusions

Although many people associate the term "ecology" with the environmentalist movement, we have seen that scientific ecology has a variety of origins, most of which were not linked to the defense of the natural environment. Science has more often been associated with efforts to exploit natural resources, and historical studies show that ecology emerged more from a desire to manage that process than to block it. At best, the majority of biologists have been concerned to ensure that humanity's engagement with the natural world does not do too much damage: sustainable yields are preferable to the wholesale destruction of a resource. Even those ecologists who imagined the ecosystem as a purposeful entity with a life of its own were willing to offer advice to farmers and others whose activities necessarily interfered with the untouched state of nature. In Europe, the whole idea of a purely natural landscape seemed meaningless, so ancient and so pervasive was the human role in shaping the environment. Although the more radical environmentalists can draw comfort from theories such as Lovelock's Gaia, they cannot lay claim to ecology as a science that inevitably lends support to their view that nature should be left untouched.

Equally interesting for the historian of science is the diversity of origins and theoretical perspectives from which the various branches of ecology emerged. Here was no single discipline shaped by a common research program and methodology. On the contrary, the movement toward what became known as ecology occurred in different places and at different times. The various locations of the scientists who became involved shaped the problems they sought to answer and hence the methodologies they thought appropriate. A technique that made sense on the open prairie of the American Midwest would have been inappropriate for the much-tilled landscape of Europe or the tundra of Hudson's Bay. Into these diverse environments came scientists with different backgrounds and interests; some were plant physiologists seeking to extend the experimental method to the interaction between plant and environment, some were biogeographers or taxonomists. All were driven by a determination to make the study of the inter-actions between organisms and their environment more scientific, but what they defined as "scientific" depended on their background and the problems they confronted. There was much suspicion to begin with over the application of mathematical techniques to modeling ecosystems. The majority of ecologists wanted to portray their science as materialistic, and this eventually led to link with the revived Darwinism of the evolutionary synthesis. But there has been a persistent current of philosophical opposition to this movement, paralleling similar doubts in other areas of biology. Smuts's holism was by no means unchar-acteristic of a nonmaterialist current of thought in early twentieth-century science. It certainly appealed to some of the early ecologists, and although that way of thought became less fashionable in the late twentieth century, its revival in the form of the Gaia hypothesis ignited a new level of debate. This debate reminds us of the gulf that still exists between the majority of scientists and the almost mystical vision of nature that has sustained the more radical environ-mentalist movement.

References and Further Reading

Anker, Peder. 2001. *Imperial Ecology: Environmental Order in the British Empire, 1895–1945.* Cambridge, MA: Harvard University Press.

Bocking, Stephen. 1997. *Ecologists and Environ-mental Politics: A History of Contemporary Ecology.* New Haven, CT: Yale University Press.

Bowler, Peter J. 1992. *The Fontana/Norton History of the Environmental Sciences.* London: Fontana; New York: Norton. Norton ed. subsequently retitled *The Earth Encompassed.*

Bramwell, Anna. 1989. *Ecology in the Twentieth Century: A History.* New Haven, CT: Yale University Press.

Brockway, Lucille. 1979. *Science and Colonial Expansion: The Role of the British Royal Botanical Gardens.* New York: Academic Press.

Cittadino, Eugene. 1993. *Nature as the Laboratory: Darwinian Plant Ecology in the German Empire, 1880–1900.* Cambridge: Cambridge University Press.

Coleman, William. 1986. "'Evolution into Ecology?' The Strategy of Warming's Ecological Plant Geography." *Journal of the History of Biology* 19:181–96.

Collins, James P. 1986. "Evolutionary Ecology and the Use of Natural Selection in Ecological Theory." *Journal of the History of Biology* 19:257–88.

Crowcroft, Peter. 1991. *Elton's Ecologists: A History of the Bureau of Animal Population.* Chicago: University of Chicago Press.

Kingsland, Sharon E. 1985. *Modeling Nature: Episodes in the History of Population Ecology.* Chicago: University of Chicago Press.

Lovelock, James. 1987. *Gaia: A New Look at Life on Earth.* New ed. Oxford: Oxford University Press.

Leopold, Aldo. 1966. *A Sand County Almanac: With Other Essays on Conservation from Round River.* Reprint, New York: Oxford University Press.

Marsh, George Perkins. 1965. *Man and Nature.* Edited by David Lowenthal. Reprint, Cambridge, MA: Harvard University Press.

Mackay, David. 1985. *In the Wake of Cook: Exploration, Science and Empire, 1780–1801.* London: Croom Helm.

McCormick, John. 1989. *The Global Environment Movement: Reclaiming Paradise.* Bloomington: Indiana University Press; London: Belhaven.

Merchant, Carolyn. 1980. *The Death of Nature: Women, Ecology and the Scientific Revolution.* London: Wildwood House.

Mitman, Greg. 1992. *The State of Nature: Ecology, Community, and American Social Thought, 1900–1950.* Chicago: University of Chicago Press.

Palladino, Paolo. 1991. "Defining Ecology: Ecological Theories, Mathematical Models, and Applied Biology in the 1960s and 1970s." *Journal of the History of Biology* 24:223–43.

Sheal, John. 1976. *Nature in Trust: The History of Nature Conservancy in Britain.* Glasgow: Blackie.

———. 1987. *Seventy-five Years in Ecology: The British Ecological Society.* Oxford: Blackwell.

Taylor, Peter J. 1988. "Technocratic Optimism, H. T. Odum, and the Partial Transformation of Ecological Metaphor after World War II." *Journal of the History of Biology* 21:213–44.

Tobey, Ronald C. 1981. *Saving the Prairies: The Life Cycle of the Founding School of American Plant Ecology.* Berkeley: University of California Press.

Worster, Donald. 1985. *Nature's Economy: A History of Ecological Ideas.* Reprint, Cambridge: Cambridge University Press.